T0245198

VISCOELASTIC WAVES IN LAYERED MEDIA

This book is a rigorous, self-contained exposition of the mathematical theory for wave propagation in layered media with arbitrary amounts of intrinsic absorption. The theory, previously not published in a book, provides solutions for fundamental wave-propagation problems in the general context of any media with a linear response (elastic or anelastic). It reveals physical characteristics for two- and three-dimensional anelastic body and surface waves, not predicted by commonly used models based on elasticity or one-dimensional anelasticity. It explains observed wave characteristics not explained by previous theories.

This book may be used as a textbook for graduate-level courses and as a research reference in a variety of fields such as solid mechanics, seismology, civil and mechanical engineering, exploration geophysics, and acoustics. The theory and numerical results allow the classic subject of fundamental elastic wave propagation to be taught in the broader context of waves in any media with a linear response, without undue complications in the mathematics. They provide the basis to improve a variety of anelastic wave-propagation models, including those for the Earth's interior, metal impurities, petroleum reserves, polymers, soils, and ocean acoustics. The numerical examples and problems facilitate understanding by emphasizing important aspects of the theory for each chapter.

Roger D. Borcherdt is a Research Scientist at the U.S. Geological Survey and Consulting Professor, Department of Civil and Environmental Engineering at Stanford University, where he also served as visiting Shimizu Professor. Dr. Borcherdt is the author of more than 180 scientific publications including several on the theoretical and empirical aspects of seismic wave propagation pertaining to problems in seismology, geophysics, and earthquake engineering. He is the recipient of the Presidential Meritorious Service and Distinguished Service awards of the Department of Interior for Scientific Leadership in Engineering Seismology, and the 1994 and 2002 Outstanding Paper Awards of Earthquake Spectra. He is an honorary member of the Earthquake Engineering Research Institute, a past journal and volume editor, and an active member of several professional societies.

VISCOELASTIC WAVES IN LAYERED MEDIA

ROGER D. BORCHERDT, PH.D.

United States Geological Survey
Stanford University, USA

CAMBRIDGE
UNIVERSITY PRESS

University Printing House, Cambridge CB2 8BS, United Kingdom

One Liberty Plaza, 20th Floor, New York, NY 10006, USA

477 Williamstown Road, Port Melbourne, VIC 3207, Australia

314-321, 3rd Floor, Plot 3, Splendor Forum, Jasola District Centre, New Delhi - 110025, India

79 Anson Road, #06-04/06, Singapore 079906

Cambridge University Press is part of the University of Cambridge.

It furthers the University's mission by disseminating knowledge in the pursuit of education, learning and research at the highest international levels of excellence.

www.cambridge.org
Information on this title: www.cambridge.org/9781108462112

First published 2009
First paperback edition 2018

A catalogue record for this publication is available from the British Library

Library of Congress Cataloging in Publication data
Borcherdt, Roger D.
Viscoelastic waves in layered media / Roger D. Borcherdt.
p. cm.
Includes bibliographical references and index.
ISBN 978-0-521-89853-9
1. Waves – Mathematics. 2. Viscoelasticity. 3. Viscoelastic materials. I. Title.
QA935.B647 2008
532′.0533–dc22
2008037113

ISBN 978-0-521-89853-9 Hardback
ISBN 978-1-108-46211-2 Paperback

In Memory of My Mother and Father
Dedicated to My Family

Contents

Preface

This book provides a self-contained mathematical exposition of the theory of monochromatic wave propagation in layered viscoelastic media. It provides analytic solutions and numerical results for fundamental wave-propagation problems in arbitrary linear viscoelastic media not published previously in a book. As a text book with numerical examples and problem sets, it provides the opportunity to teach the theory of monochromatic wave propagation as usually taught for elastic media in the broader context of wave propagation in any media with a linear response without undue complications in the mathematics. Formulations of the expressions for the waves and the constitutive relation for the media afford considerable generality and simplification in the mathematics required to derive analytic solutions valid for any viscoelastic solid including an elastic medium. The book is intended for the beginning student of wave propagation with prerequisites being knowledge of differential equations and complex variables.

As a reference text, this book provides the theory of monochromatic wave propagation in more than one dimension developed in the last three to four decades. As such, it provides a compendium of recent advances that show that physical characteristics of two- and three-dimensional anelastic body and surface waves are not predictable from the theory for one-dimensional waves. It provides the basis for the derivation of results beyond the scope of the present text book. The theory is of interest in the broad field of solid mechanics and of special interest in seismology, engineering, exploration geophysics, and acoustics for consideration of wave propagation in layered media with arbitrary amounts of intrinsic absorption, ranging from low-loss models of the deep Earth to moderate-loss models for soils and weathered rock.

The phenomenological constitutive theory of linear viscoelasticity dates to the nineteenth century with the application of the superposition principle in 1874 by Boltzmann. He proposed a general mathematical formulation that characterizes linear anelastic as well as elastic material behavior. Subsequent developments in the theory did not occur until interest increased in the response of polymers and other synthetic materials during the middle of the twentieth century. During this time considerable simplification of the mathematical aspects of the theory occurred with developments in the theory of linear functionals (Volterra 1860–1940) and the introduction of integral transform techniques. Definitive accounts and contributions to the mathematical theory of viscoelasticity include the works of Volterra (2005), Gross (1953), Bland (1960), Fung (1965), and Flugge (1967), and the rigorous account by Gurtin and Sternberg (1962). Gross (1953) suggested that "The theory of viscoelasticity is approaching completion. Further progress is likely to be made in applications rather than on fundamental principles." Hunter (1960) indicated that the application of the general theory of linear viscoelasticity to other than one-dimensional wave-propagation problems was incomplete. Subsequent to Hunter's review, significant advances occurred in the 1970s and 1980s concerning application of viscoelasticity to wave propagation in layered two- and three-dimensional media. This exposition provides a self-contained mathematical account of these developments as well as recent solutions derived herein. It provides closed-form analytic solutions and numerical results for fundamental monochromatic wave-propagation problems in arbitrary layered viscoelastic media.

Recent theoretical developments show that the physical characteristics of the predominant types of body waves that propagate in layered anelastic media are distinctly different from the predominant type in elastic media or one-dimensional anelastic media. For example, two types of shear waves propagate, each with different amounts of attenuation. Physical characteristics of anelastic waves refracted across anelastic boundaries, such as particle motion, phase and energy velocities, energy loss, and direction of energy flux, vary with angle of incidence. Hence, the physical characteristics of the waves propagating through a stack of layers are no longer unique in each layer as they are in elastic media, but instead depend on the angle at which the wave entered the stack. These fundamental differences explain laboratory observations not explained by elasticity or one-dimensional viscoelastic wave-propagation theory. They have important implications for some forward-modeling and inverse problems. These differences lend justification to the need for a book with a rigorous mathematical treatment of the fundamental aspects of viscoelastic wave propagation in layered media.

Viscoelastic material behavior is characterized herein using Boltzmann's formulation of the constitutive law. Analytic solutions derived for various problems are valid for both elastic and linear anelastic media as might be modeled by an infinite number of possible configurations of elastic springs and viscous dashpots. The general analytic solutions are valid for viscoelastic media with an arbitrary amount of intrinsic absorption. The theory, based on the assumption of linear strain, is valid for media with a nonlinear response to the extent that linear approximations are valid for sufficiently small increments in time. Wave-propagation results in this book are based on those of the author as previously published and explicitly derived herein including recent results for multilayered media and Love-Type surface waves.

The book is intended for use in a graduate or upper-division course and as a reference text for those interested in wave propagation in layered media. The book provides a self-contained treatment of energy propagation and dissipation and other physical characteristics of general P, Type-I S, and Type-II S waves in viscoelastic media with arbitrary amounts of absorption. It provides analytic and numerical results for fundamental reflection–refraction and surface-wave problems. The solutions and resultant expressions are derived from first principles. The book offers students the opportunity to understand classic elastic results in the broader context of wave propagation in any material with a linear response.

Chapter 1 provides an introduction for new students to basic concepts of a linear stress–strain law, energy dissipation, and wave propagation for one-dimensional linear viscoelastic media. It provides examples of specific models derivable from various configurations of springs and dashpots as special cases of the general formulation.

Chapter 2 extends the basic concepts for viscoelastic media to three dimensions. It provides the general linear stress–strain law, notation for components of stress and strain, the equation of motion, and a rigorous account of energy balance as the basis needed for a self-contained treatment of viscoelastic wave propagation in more than one dimension.

Chapter 3 provides a thorough account of the physical characteristics of harmonic waves in three-dimensional viscoelastic media. It provides closed-form expressions and corresponding quantitative estimates for characteristics of general (homogeneous or inhomogeneous) P, Type-I S, and Type-II S waves in viscoelastic media with both arbitrary and small amounts of intrinsic absorption. It includes expressions for the physical displacement and volumetric strain associated with various wave types. This chapter is a prerequisite for analytic solutions derived for reflection–refraction and surface-wave problems in subsequent chapters.

Chapter 4 specifies the expressions for monochromatic P, Type-I S, and Type-II S wave solutions needed in subsequent chapters to solve reflection–refraction and surface-wave problems in layered viscoelastic media. The solutions are expressed in terms of the propagation and attenuation vectors and in terms of the components of the complex wave numbers.

Chapter 5 provides analytic closed-form solutions for the problems of general (homogeneous or inhomogeneous) P, Type-I S, and Type-II S waves incident on a plane boundary between viscoelastic media. Conditions for homogeneity and inhomogeneity of the reflected and refracted waves and the characteristics of energy flow at the boundary are derived. Careful study of results for the problems of an incident Type-I S and Type-II S wave is useful for understanding reflection–refraction phenomena and energy flow due to an inhomogeneous wave incident on a viscoelastic boundary.

Chapter 6 provides numerical results for various single-boundary reflection–refraction problems using the analytic solutions derived in the previous chapter. Examples are chosen to provide quantitative estimates of the physical characteristics for materials with moderate and small amounts of intrinsic material absorption as well as a comparison with laboratory measurements in support of theoretical predictions. Study of these examples, especially the first three, is recommended for developing an improved understanding of the physical characteristics of waves reflected and refracted at viscoelastic boundaries.

Chapter 7 provides theoretical solutions and quantitative results for problems of a general Type-I S, P, or Type-II S wave incident on the free surface of a viscoelastic half space. Results are included to facilitate understanding and interpretation of measurements as might be detected on seismometers and volumetric strain meters at or near the free surface of a viscoelastic half space.

Chapter 8 presents the analytic solution and corresponding numerical results for a Rayleigh-Type surface wave on a viscoelastic half space. Analytic and numerical results illustrate fundamental differences in the physical characteristics of viscoelastic surface waves versus those for elastic surface waves as originally derived by Lord Rayleigh.

Chapter 9 provides the analytic solution for the response of multilayered viscoelastic media to an incident general Type-II S wave. It provides numerical results for elastic, low-loss, and moderate-loss viscoelastic media useful in understanding variations in response of a single layer due to inhomogeneity and angle of incidence of the incident wave.

Chapter 10 provides the analytic solution and corresponding numerical results for a Love-Type surface wave on multilayered viscoelastic media. It derives roots of the resultant complex period equation for a single layer needed to provide curves showing the dependencies of wave speed, absorption coefficient, and amplitude distribution on frequency, intrinsic absorption, and other material parameters for the fundamental and first higher mode.

New students desiring a basic understanding of Rayleigh-Type surface waves, the response of a stack of viscoelastic layers to incident inhomogeneous waves, and Love-Type surface waves will benefit from a thorough reading of Chapters 8, 9, and 10.

Chapter 11 provides appendices that augment material presented in preceding chapters. They include various integral and vector identities and lengthy derivations relegated to the appendices to facilitate readability of the main text.

A special note of respect is due those who have developed the elegant constitutive theory of linear viscoelasticity, such as Boltzmann, Volterra, Gurtin, and Sternberg. Their important contributions make applications such as that presented here straightforward. I would like to express my appreciation to Professor Jerome L. Sackman, whose guidance and expertise in viscoelasticity initiated a theoretical journey I could not have imagined. Review comments received on advance copies from colleagues are appreciated, especially those of Professors J. Sackman, W. H. K. Lee, J. Bielak, and C. Langston. Discussions with colleagues, contributions of former coauthors, and assistance with formatting some figures in Chapters 6 and 7 by G. Glassmoyer are appreciated.

Advice and encouragement are appreciated, as received during preparation from friends and colleagues, including Professors H. Shah, A. Kiremidjian, G. Deierlein, H. Krawinkler, and K. Law, of Stanford University, Professors J. Sackman, A. Chopra, J. Moehle, J. Penzien, A. Der Kiureghian, the late B. A. Bolt, T. V. McEvilly, and P.W. Rodgers, of the University of California, Berkeley, and Dr. F. Naeim of John A. Martin and Associates.

A special note of appreciation is due my family (Judy, Darren, Ryan, and Debbie) for their patience and support for the many late hours needed to finish the book. Without their understanding the opportunity to experience the personal excitement associated with discovering characteristics of waves basic to seismology and engineering through the elegance and rigors of mathematics would not have been possible.

1

One-Dimensional Viscoelasticity

The behavior of many materials under an applied load may be approximated by specifying a relationship between the applied load or stress and the resultant deformation or strain. In the case of elastic materials this relationship, identified as Hooke's Law, states that the strain is proportional to the applied stress, with the resultant strain occurring instantaneously. In the case of viscous materials, the relationship states that the stress is proportional to the strain rate, with the resultant displacement dependent on the entire past history of loading. Boltzmann (1874) proposed a general relationship between stress and strain that could be used to characterize elastic as well as viscous material behavior. He proposed a general constitutive law that could be used to describe an infinite number of elastic and linear anelastic material behaviors derivable from various configurations of elastic and viscous elements. His formulation, as later rigorously formulated in terms of an integral equation between stress and strain, characterizes all linear material behavior. The formulation, termed linear viscoelasticity, is used herein as a general framework for the derivation of solutions for various wave-propagation problems valid for elastic as well as for an infinite number of linear anelastic media.

Consideration of material behavior in one dimension in this chapter, as might occur when a tensile force is applied at one end of a rod, will provide an introduction to some of the well-known concepts associated with linear viscoelastic behavior. It will provide a general stress–strain relation from which stored and dissipated energies associated with harmonic behavior can be inferred as well as the response of an infinite number of viscoelastic models. It will permit the derivation of solutions for one-dimensional viscoelastic waves as a basis for comparison with those for two- and three-dimensional waves to be derived in subsequent chapters.

1.1 Constitutive Law

A general linear viscoelastic response in one spatial dimension is defined mathematically as one for which a function $r(t)$ of time exists such that the constitutive equation relating strain to stress is given by

$$p(t) = \int_{-\infty}^{t} r(t-\tau)de(\tau), \tag{1.1.1}$$

where $p(t)$ denotes stress or force per unit area as a function of time, $e(t)$ denotes strain or displacement per unit displacement as a function of time, and $r(t)$, termed a relaxation function, is causal and does not depend on the spatial coordinate.

The physical principle of causality imposed on the relaxation function $r(t)$ implies the function is zero for negative time, hence the constitutive relation may be written using a Riemann–Stieltjes integral as

$$p(t) = \int_{-\infty}^{\infty} r(t-\tau)de(\tau) \tag{1.1.2}$$

or more compactly in terms of a convolution operator as

$$p = r * de. \tag{1.1.3}$$

Properties of the convolution operator are summarized in Appendix 1.

A corresponding constitutive equation relating stress to strain is one defined mathematically for which a causal spatially independent function $c(t)$, termed a creep function, exists such that the corresponding strain time history may be inferred from the following convolution integral

$$e(t) = \int_{-\infty}^{\infty} c(t-\tau)dp(\tau), \tag{1.1.4}$$

which may be written compactly in terms of the convolution operator as

$$e = c * dp. \tag{1.1.5}$$

Linear material behavior is behavior in which a linear superposition of stresses leads to a corresponding linear superposition of strains and vice versa. Such a material response is often referred to as one which obeys Boltzmann's superposition principle. Boltzmann's formulation of the constitutive relation between stress and strain as expressed by the convolution integrals (1.1.2) and (1.1.4) is general in the sense that all linear behavior may be characterized by such a relation. Conversely, if the material response is characterized by one of the convolution integrals then Boltzmann's superposition principle is valid. To show this result explicitly, consider the following arbitrary linear superposition of strains

$$e(t) = \sum_{i=1}^{n} b_i \, e_i(t), \tag{1.1.6}$$

where b_i corresponds to an arbitrary but fixed constant independent of time. Substitution of this expression into (1.1.3) and using the distributive property of the convolution operator, which immediately follows from the corresponding property for the Riemann–Stieltjes integral (see Appendix 1), readily implies the desired result that the resultant stress is a linear superposition of the stresses corresponding to the given linear superposition of strains, namely

$$p = \sum_{i=1}^{n} b_i \, p_i = \sum_{i=1}^{n} b_i (r * de_i) = r * d\left(\sum_{i=1}^{n} b_i \, e_i\right). \tag{1.1.7}$$

Similarly, (1.1.4) implies that a linear superposition of stresses leads to a linear superposition of strains.

The term relaxation function used for the function $r(t)$ derives from physical observations of the stress response of a linear system to a constant applied strain. To show that this physical definition of a relaxation function is consistent with that defined mathematically, consider the stress response to a unit strain applied at some time, say $t=0$, to a material characterized by (1.1.3). Specifically, replace $e(t)$ in (1.1.3) with the Heaviside function

$$h(t) \equiv \left\{ \begin{array}{l} 0 \text{ for } t<0 \\ 1 \text{ for } t \geq 0 \end{array} \right\}. \tag{1.1.8}$$

The fifth property of the Riemann–Stieltjes convolution operator stated in Appendix 1 implies that (1.1.3) simplifies to

$$p(t) = e(0+)r(t) + \int_{0+}^{t} r(t-\tau) \frac{\partial h(\tau)}{\partial \tau} \, d\tau, \tag{1.1.9}$$

hence,

$$p(t) = r(t). \tag{1.1.10}$$

Similarly, the creep function $c(t)$ defined mathematically may be shown to represent the strain response of a linear system to a unit stress applied at $t=0$.

To consider harmonic behavior of a linear viscoelastic material, assume sufficient time has elapsed for the effect of initial conditions to be negligible. Using the complex representation for harmonic functions let

$$p(t) = Pe^{i\omega t} \tag{1.1.11}$$

and

$$e(t) = Ee^{i\omega t}, \tag{1.1.12}$$

where P and E are complex constants independent of time with the physical stress and strain functions determined by the real parts of the corresponding complex numbers. Substitution of (1.1.11) and (1.1.12) into (1.1.2) and (1.1.4), respectively, shows that the corresponding constitutive relations may be written as

$$P = i\omega R(\omega)E \tag{1.1.13}$$

and

$$E = i\omega C(\omega)P, \tag{1.1.14}$$

where $R(\omega)$ and $C(\omega)$ are given by the Fourier transforms

$$R(\omega) = \int_{-\infty}^{\infty} r(\tau)e^{-i\omega\tau}\, d\tau \tag{1.1.15}$$

and

$$C(\omega) = \int_{-\infty}^{\infty} c(\tau)e^{-i\omega\tau}\, d\tau. \tag{1.1.16}$$

In analogy with the definitions given for elastic media the complex modulus M is defined as

$$M(\omega) \equiv \frac{P}{E} = i\omega R(\omega). \tag{1.1.17}$$

The complex compliance is defined as

$$J(\omega) \equiv \frac{E}{P} = i\omega C(\omega), \tag{1.1.18}$$

from which it follows that the complex modulus is the reciprocal of the complex compliance, that is

$$M(\omega) = \frac{1}{J(\omega)} \tag{1.1.19}$$

and the product of the Fourier transforms of the relaxation function and the creep function is given by the negative reciprocal of the circular frequency squared, that is

$$R(\omega)C(\omega) = (i\omega)^{-2}. \tag{1.1.20}$$

A parameter useful for quantifying the anelasticity of a viscoelastic material is the phase angle δ by which the strain lags the stress. This phase angle is given from (1.1.17) by

$$\tan\delta = \frac{M_I}{M_R}, \tag{1.1.21}$$

where the subscripts "I" and "R" denote imaginary and real parts of the complex modulus.

1.2 Stored and Dissipated Energy

Energy in a linear viscoelastic system under a cycle of forced harmonic oscillation is partially dissipated and partially alternately twice stored and returned. To account for the energy in a linear viscoelastic system under a harmonic stress excitation as characterized by a general constitutive relation of the form (1.1.13), consider the complex strain given by

$$e = Jp. \tag{1.2.1}$$

The time rate of change of energy in the system is given by the product of the physical stress and the physical strain rate, namely,

$$p_R \dot{e}_R, \tag{1.2.2}$$

where the dot on $\dot{e}_R$ denotes the derivative with respect to time and the subscript R denotes the real part of the strain rate. Solving (1.2.1) for p, then taking real parts of the resulting equation implies that the physical stress can be expressed as

$$p_R = \frac{J_R e_R + J_I e_I}{|J|^2}. \tag{1.2.3}$$

For harmonic excitation

$$e_I = -\frac{\dot{e}_R}{\omega}. \tag{1.2.4}$$

Substitution of (1.2.3) and (1.2.4) into (1.2.2) shows that the desired expression for the time rate of change of energy in the one-dimensional system is given by

$$p_R \dot{e}_R = \frac{\partial}{\partial t}\left(\frac{1}{2}\frac{J_R}{|J|^2}e_R^2\right) - \left(\frac{1}{\omega}\frac{J_I}{|J|^2}\dot{e}_R^2\right). \tag{1.2.5}$$

Integrating (1.2.5) over one cycle shows that the total rate of change of energy over one cycle equals the integral of the second term on the right-hand side of the equation. Hence, the second term of (1.2.5) represents the rate at which energy is dissipated and the first term represents the time rate of change of the potential energy in the system, that is, the rate at which energy is alternately stored and returned. The second law of thermodynamics requires that the total amount of energy dissipated increase with time, hence the second term in (1.2.5) implies that

$$J_I \leq 0. \tag{1.2.6}$$

A dimensionless parameter, which is useful for describing the amount of energy dissipated, is the fractional energy loss per cycle of forced oscillation or the ratio of the energy dissipated to the peak energy stored during the cycle. Integrating (1.2.5) over one cycle shows that the energy dissipated per cycle as denoted by $\Delta\mathcal{E}/cycle$ is given by

$$\frac{\Delta\mathcal{E}}{cycle} = -\pi|P|^2 J_I. \tag{1.2.7}$$

The first term on the right-hand side of (1.2.5) shows that the peak energy stored during a cycle or the maximum potential energy during a cycle as denoted by $\max[\mathscr{P}]$ is given by

$$\max[\mathscr{P}] = \frac{1}{2}|P|^2 J_R, \tag{1.2.8}$$

where $J_R \geq 0$. Hence, the fractional energy loss for a general linear system may be expressed in terms of the ratio of the imaginary and real parts of the complex compliance or the complex modulus. as

$$\frac{\Delta\mathcal{E}}{cycle}\Big/\max[\mathscr{P}] = 2\pi\frac{-J_I}{J_R} = 2\pi\frac{M_I}{M_R}. \tag{1.2.9}$$

Normalization of the fractional energy loss by 2π yields another parameter often used to characterize anelastic behavior, namely the reciprocal of the quality factor, which may be formally defined as

$$Q^{-1} \equiv \frac{1}{2\pi}\frac{\Delta\mathcal{E}/cycle}{\max[\mathscr{P}]}. \tag{1.2.10}$$

Q^{-1} for a one-dimensional linear system under forced oscillation is, from (1.2.9), given by

$$Q^{-1} = \frac{-J_I}{J_R} = \frac{M_I}{M_R}. \tag{1.2.11}$$

Examination of (1.1.21) shows that Q^{-1} also represents the tangent of the angle by which the strain lags the stress, that is

$$Q^{-1} = \tan \delta. \tag{1.2.12}$$

Another parameter often used to characterize anelastic response is damping ratio ζ, which may be specified in terms of Q^{-1} as

$$\zeta = \frac{Q^{-1}}{2}. \tag{1.2.13}$$

For the special case of an elastic system $J_I = M_I = 0$, hence $Q^{-1} = \tan \delta = \zeta = 0$.

1.3 Physical Models

The characterization of one-dimensional linear material behavior as defined mathematically and presented in the previous sections is general. The considerations apply to any linear behavior for which a relaxation function exists such that the material behavior may be characterized by a convolution integral of the form (1.1.1). Alternatively, the considerations apply to any linear material behavior for which a complex modulus exists such that (1.1.17) is a valid for characterization of harmonic behavior. Specification of the complex modulus for a particular physical model of viscoelastic behavior allows each model to be treated as a special case of the general linear formulation.

The basic physical elements used to represent viscoelastic behavior are an elastic spring and a viscous dashpot. Schematics illustrating springs and dashpots in various series and parallel configurations are shown in Figures (1.3.3)a through (1.3.3)h. In order to derive the viscoelastic response of each configuration one end is assumed anchored with a force applied as a function of time at the other end. Forces are assumed to be applied to a unit cross-sectional area with the resultant elongation represented per unit length, so that force and extension may be used interchangeably with stress and strain.

The elongation of an elastic spring element is assumed to be instantaneous and proportional to the applied load. Upon elimination of the load the spring is assumed to return to its initial state. The constitutive equation for an elastic spring as first proposed by Hooke in 1660 is specified by

$$p = \mu e, \tag{1.3.1}$$

where μ is a constant independent of time. The assumption that the response of an elastic spring is instantaneous implies that for an initial load applied at time $t=0$ the strain at time $t=0$ is $e(0)=p(0)/\mu$. Hence, the creep and relaxation functions for the special case of an elastic model are $h(t)/\mu$ and $\mu h(t)$. For harmonic behavior substitution of (1.1.11) and (1.1.12) into (1.3.1) implies the complex compliance and modulus as specified by (1.1.18) and (1.1.17) are given by $1/\mu$ and μ, where the imaginary parts of each are zero. Hence, Q^{-1} as specified by (1.2.11) for an elastic model is zero.

The rate at which a viscous dashpot element is assumed to elongate is assumed to be proportional to the applied force, with the resultant elongation dependent on the entire past history of loading. The constitutive equation for a viscous element is given by

$$p = \eta \dot{e}, \tag{1.3.2}$$

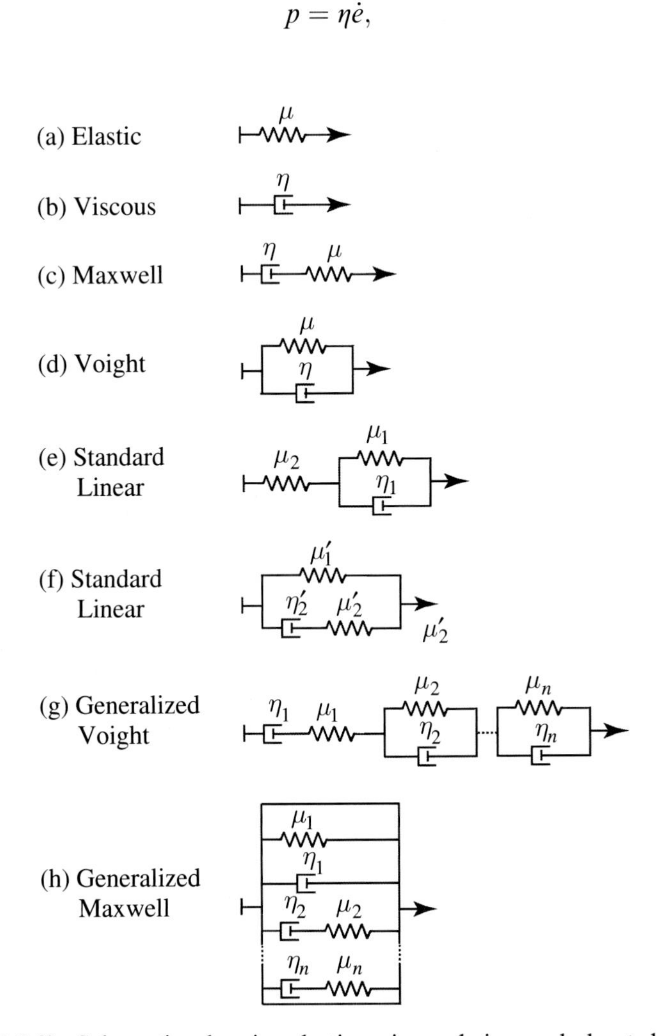

Figure (1.3.3). Schematics showing elastic spring and viscous dashpot elements in series and parallel configurations for various models of linear viscoelasticity.

where $\dot{e}$ denotes the derivative of strain with respect to time or the velocity of the elongation with respect to unit length. The viscous element is assumed not to respond instantaneously, hence its elongation due to an instantaneous load applied at time $t=0$ is $e(0)=0$. Integration of (1.3.2) implies the creep and relaxation functions for the special case of a viscous element are given by

$$c(t) = t\,h(t)/\eta \tag{1.3.4}$$

and

$$r(t) = \eta\,\delta(t), \tag{1.3.5}$$

where $\delta(t)$ denotes the Dirac-delta function, whose integral is unity and whose non-zero values are zero. Substitution of (1.1.11) and (1.1.12) into (1.3.2) implies the complex compliance and modulus for a viscous element are $J(\omega) = 1/(i\omega\eta)$ and $M(\omega)=i\omega\eta$. Equation (1.2.11) implies Q^{-1} is infinite, because no energy is alternately stored and returned in a viscous element.

An infinite number of viscoelastic models may be derived from various serial and parallel configurations of elastic springs and viscous dashpots. Schematics for common models are shown in Figure (1.3.3). Three fundamental viscoelastic models are the Maxwell model, which assumes the basic elements are in series, the Voight model, which assumes the basic elements are in parallel, and a Standard Linear model, which assumes a spring in series with a Voight element or a spring in parallel with a Maxwell element. Generalizations of these models are the Generalized Voight model and the Generalized Maxwell model. The Generalized Voight model includes a Maxwell model in series with a sequence of Voight elements in series. The Generalized Maxwell model includes a Voight element in parallel with a sequence of Maxwell elements in parallel. A Standard Linear model may be considered as a special case of a Generalized Voight model with $n=2$ and $\eta_1^{-1}=0$.

The two configurations shown for a Standard Linear model (Figures (1.3.3)e and f) are equivalent in that the parameters of the elements may be adjusted to give the same response. Similarly, the configuration of springs and dashpots for any model involving more than two of these elements is not unique, in that other configurations of springs and dashpots in series and parallel will yield the same response.

For the Maxwell model the strain resulting from an applied load is the sum of the strains associated with the individual elements in series. Hence, the resultant strain rate is given by

$$\dot{e} = \dot{e}_1 + \dot{e}_2 = \frac{\dot{p}}{\mu} + \frac{p}{\eta}, \tag{1.3.6}$$

where the initial strain, as implied by the assumed instantaneous response of the spring, is $e(0) = p(0)/\mu$. Integration of (1.3.6) and substitution of a unit stress implies that the creep function for a Maxwell model is

$$c(t) = \left(\frac{1}{\mu} + \frac{1}{\eta} t\right) h(t). \tag{1.3.7}$$

Similarly, integration and substitution of a unit strain implies the relaxation function for a Maxwell model is

$$r(t) = \mu \exp[-(\mu/\eta)t]\, h(t). \tag{1.3.8}$$

Substitution of (1.1.11) and (1.1.12) into (1.3.6) together with (1.1.17) and (1.1.18) implies the complex compliance and complex modulus for a Maxwell model are given by $J(\omega) = 1/\mu - i/(\omega\eta)$ and $M(\omega) = (1/\mu - i/(\omega\eta))^{-1}$, from which (1.2.11) implies $Q^{-1} = \mu/(\omega\eta)$.

For a Voight model the applied stress is the sum of the stress associated with each of the elements in parallel. Hence the applied stress is given by

$$p = \mu e + \eta \dot{e}, \tag{1.3.9}$$

where the initial strain is $e(0) = 0$. The creep and relaxation functions inferred from (1.3.9) for the Voight model are

$$c(t) = \frac{1}{\mu}(1 - \exp[-(\mu/\eta)t])h(t) \tag{1.3.10}$$

and

$$r(t) = \eta\delta(t) + \mu h(t). \tag{1.3.11}$$

Substitution of (1.1.11) and (1.1.12) into (1.3.9) implies the complex compliance and modulus for a Voight model are $J(\omega) = 1/(\mu + i\omega\eta)$ and $M(\omega) = \mu + i\omega\eta$. Hence, (1.2.11) implies $Q^{-1} = \omega\eta/\mu$.

For a Standard Linear model with an applied load, the resultant strain is the sum of the strains associated with the spring in series with the Voight element, while the applied stress is the same for the spring and Voight elements in series. The resulting equations for configuration "e" shown in Figure (1.3.3) are

$$p = \mu_1 e_1 + \eta_1 \dot{e}_1 = \mu_2 e_2 \tag{1.3.12}$$

and

$$e = e_1 + e_2, \tag{1.3.13}$$

which upon simplification may be written as

$$p + \tau_p \dot{p} = M_r(e + \tau_e \dot{e}), \tag{1.3.14}$$

where τ_p is the stress relaxation time under constant strain defined by

$$\tau_p \equiv \frac{\eta_1}{\mu_1 + \mu_2}, \tag{1.3.15}$$

τ_e is the strain relaxation time under constant stress defined by

$$\tau_e \equiv \frac{\eta_1}{\mu_1}, \tag{1.3.16}$$

and M_r is known as the relaxed elastic modulus defined by

$$M_r \equiv \frac{\mu_1 \mu_2}{\mu_1 + \mu_2} \tag{1.3.17}$$

with the initial instantaneous response being $e(0) = p(0)/\mu_2$. The creep and relaxation functions for a Standard Linear model inferable from (1.3.14) are

$$c(t) = \frac{1}{M_r}\left(1 - \left(1 - \frac{\tau_p}{\tau_e}\right)\exp[-t/\tau_e]\right)h(t) \tag{1.3.18}$$

and

$$r(t) = M_r\left(1 - \left(1 - \frac{\tau_e}{\tau_p}\right)\exp[-t/\tau_p]\right)h(t). \tag{1.3.19}$$

Substitution of (1.1.11) and (1.1.12) into (1.3.14) implies the corresponding complex compliance is

$$J(\omega) = \frac{1}{M_r}\frac{1 + i\omega\tau_p}{1 + i\omega\tau_e} \tag{1.3.20}$$

and the complex modulus is

$$M(\omega) = M_r \frac{1 + i\omega\tau_e}{1 + i\omega\tau_p}. \tag{1.3.21}$$

Hence, (1.2.11) implies Q^{-1} for a Standard Linear model is

$$Q^{-1} = \frac{\omega(\tau_e - \tau_p)}{1 + \omega^2 \tau_p \tau_e}. \tag{1.3.22}$$

For a Generalized Voight model the strain resulting from an applied stress is the sum of the strains associated with the Maxwell and Voight elements in series. Hence, the creep function for the model can be readily inferred as the sum of those associated with the elements in series as

$$c(t) = \left(\frac{1}{\mu_1} + \frac{1}{\eta_1}t\right)h(t) + \sum_{k=2}^{n}\frac{1}{\mu_k}(1 - \exp[-(\mu_k/\eta_k)t])h(t). \tag{1.3.23}$$

Similarly, the compliance for the elements in series for a Generalized Voight model is the sum of the compliances of the individual elements, namely

$$J(\omega) = \frac{1}{\mu_1} + \frac{1}{i\omega\eta_1} + \sum_{k=2}^{n} \frac{1}{\mu_k + i\omega\eta_k}. \tag{1.3.24}$$

Hence, Q^{-1} for a Generalized Voight model may be readily inferred from (1.2.11) as

$$Q^{-1} = -\frac{J_I}{J_R} = \left(\frac{1}{\omega\eta_1} + \sum_{k=2}^{n} \frac{\omega\eta_k}{|\mu_k + i\omega\eta_k|^2}\right) \Bigg/ \left(\frac{1}{\mu_1} + \sum_{k=2}^{n} \frac{\mu_k}{|\mu_k + i\omega\eta_k|^2}\right). \tag{1.3.25}$$

For a Generalized Maxwell model the applied stress is the sum of the stresses associated with each of the Voight and Maxwell elements in parallel. Hence, the relaxation function for the model can be readily inferred as the sum of those for each of the elements in parallel as

$$r(t) = \eta_1 \delta(t) + \mu_1 h(t) + \sum_{k=2}^{n} \mu_k \exp[-(\mu_k/\eta_k)t] h(t). \tag{1.3.26}$$

Similarly, the modulus for the Generalized Maxwell model is given as the sum of the moduli for each of the elements, namely

$$M(\omega) = \mu_1 + i\omega\eta_1 + \sum_{k=2}^{n} \frac{1}{\mu_k} + \frac{1}{i\omega\eta_k}. \tag{1.3.27}$$

Hence, Q^{-1} for a Generalized Maxwell model may be readily inferred from (1.2.11) as

$$Q^{-1} = \frac{M_I}{M_R} = \left(\omega\eta_1 - \sum_{k=2}^{n} \frac{1}{\omega\eta_k}\right) \Bigg/ \left(\mu_1 + \sum_{k=2}^{n} \frac{1}{\mu_k}\right). \tag{1.3.28}$$

Differential equations, creep functions, relaxation functions, complex compliances, complex moduli, and Q^{-1} for the physical models are tabulated in Tables (1.3.29) and (1.3.30). An infinite number of additional physical models can be derived from various configurations of springs and viscous elements in series and parallel. As a converse of this result, Bland (1960) has shown that for any model of material behavior that can be represented by a general linear convolution integral of the form (1.1.2) the mechanical properties and the stored and dissipated energies of the model can be represented by a set of Voight elements in series or a set of Maxwell elements in parallel.

Table (1.3.29). *Common linear viscoelastic models and their corresponding differential equation, creep function, and relaxation function (see the text for notation and Table (1.3.30) for the corresponding compliance, modulus, and* Q^{-1}*).*

Model name	Differential equation	Creep function $c(t)$: $c(t) = e(t) = H(\tau) * dp(\tau)$	Relaxation function $r(t)$: $r(t) = p(t) = H(\tau) * de(\tau)$
Elastic	$p = \mu e$	$h(t)/\mu$	$\mu h(t)$
Viscous	$p = \eta \dot{e}$	$t\, h(t)/\eta$	$\eta\, \delta(t)$
Maxwell	$\dot{e} = \frac{\dot{p}}{\mu} + \frac{p}{\eta}$	$\left(\frac{1}{\mu} + \frac{1}{\eta} t\right) h(t)$	$\mu \exp[-(\mu/\eta)t] h(t)$
Voight	$p = \mu e + \eta \dot{e}$	$\frac{1}{\mu}(1 - \exp[-(\mu/\eta)t])\, h(t)$	$\eta\, \delta(t) + \mu\, h(t)$
Standard Linear	$p + \tau_p \dot{p} = M_r(e + \tau_e \dot{e})$	$\frac{1}{M_r}\left(1 - \left(1 - \frac{\tau_p}{\tau_e}\right) \exp[-t/\tau_e]\right) h(t)$	$M_r\left(1 - \left(1 - \frac{\tau_p}{\tau_e}\right) \exp[-t/\tau_p]\right) h(t)$
Generalized Voight	–	$\left(\frac{1}{\mu_1} + \frac{1}{\eta_1} t + \sum_{k=2}^{n} \frac{1}{\mu_k}(1 - \exp[-(\mu_k/\eta_k)t])\right) h(t)$	–
Generalized Maxwell	–	–	$\eta_1 \delta(t) + \mu_1 h(t) + \sum_{k=2}^{n} \mu_k \exp[-(\mu_k/\eta_k)t] h(t)$

Table (1.3.30). *Common linear viscoelastic models and their corresponding complex compliance, complex modulus, and reciprocal quality factor (see the text for notation and Table (1.3.29) for the corresponding differential equation, creep, and relaxation functions).*

	Compliance $J(\omega)$	Modulus $M\ (\omega)$	Q^{-1}
Model name	$J(\omega)=\frac{E}{P}=i\omega C(\omega)$	$M(\omega)=\frac{P}{E}=i\omega R(\omega)$	$Q^{-1}=\frac{M_I}{M_R}=\tan\delta$
Elastic	$1/\mu$	μ	0
Viscous	$1/i\omega\eta$	$i\omega\eta$	∞
Maxwell	$\frac{1}{\mu}-\frac{i}{\omega\eta}$	$\left(\frac{1}{\mu}-\frac{i}{\omega\eta}\right)^{-1}$	$\mu/(\omega\eta)$
Voight	$1/(\mu+i\omega\eta)$	$\mu+i\omega\eta$	$\omega\eta/\mu$
Standard Linear	$\frac{1}{M_r}\frac{1+i\omega\tau_p}{1+i\omega\tau_e}$	$M_r\frac{1+i\omega\tau_e}{1+i\omega\tau_p}$	$\frac{\omega(\tau_e-\tau_p)}{1+\omega^2\tau_p\tau_e}$
Generalized Voight	$\frac{1}{\mu_1}+\frac{1}{i\omega\eta_1}+\sum_{k=2}^{n}\frac{1}{\mu_k+i\omega\eta_k}$	$\left(\frac{1}{\mu_1}+\frac{1}{i\omega\eta_1}+\sum_{k=2}^{n}\frac{1}{\mu_k+i\omega\eta_k}\right)^{-1}$	$\left(\frac{1}{\omega\eta_1}+\sum_{k=2}^{n}\frac{\omega\eta_k}{\lvert\mu_k+i\omega\eta_k\rvert^2}\right)\Big/\left(\frac{1}{\mu_1}+\sum_{k=2}^{n}\frac{\mu_k}{\lvert\mu_k+i\omega\eta_k\rvert^2}\right)$
Generalized Maxwell	$\left(\mu_1+i\omega\eta_1+\sum_{k=2}^{n}\frac{1}{\mu_k}+\frac{1}{i\omega\eta_k}\right)^{-1}$	$\mu_1+i\omega\eta_1+\sum_{k=2}^{n}\frac{1}{\mu_k}+\frac{1}{i\omega\eta_k}$	$\left(\omega\eta_1-\sum_{k=2}^{n}\frac{1}{\omega\eta_k}\right)\Big/\left(\mu_1+\sum_{k=2}^{n}\frac{1}{\mu_k}\right)$

1.4 Equation of Motion

To derive an equation of motion that governs the propagation of one-dimensional waves in a general linear viscoelastic medium consider an infinitely long rod constrained to deform only in one direction along the length of the rod (see Figure (1.4.3)). Position along the rod shall be denoted by the coordinate, x, with the displacement of a cross-sectional element as a function of time and position with respect to an equilibrium position denoted by $u(x,t)$.

The stress $p(t)$ and strain $e(t)$ are assumed to vary sinusoidally with time, so in terms of complex notation $p(t) = Pe^{i\omega t}$, $e(t) = Ee^{i\omega t}$, and $u(x,t) = U(x)e^{i\omega t}$ with angular frequency $\omega = 2\pi f$ and the corresponding physical components given by the real parts of the corresponding quantities. The problem is linearized by assuming the displacement of cross-sectional elements is infinitesimal, so that the strain at any time is given by

$$e = \frac{\partial u}{\partial x}. \tag{1.4.1}$$

In terms of the complex notation the general relation between stress and strain is from (1.1.17), given by

$$P = i\omega R(\omega) = M(\omega)E, \tag{1.4.2}$$

where M is the complex modulus specified as a function of $\omega = 2\pi f$ appropriate for any chosen viscoelastic model.

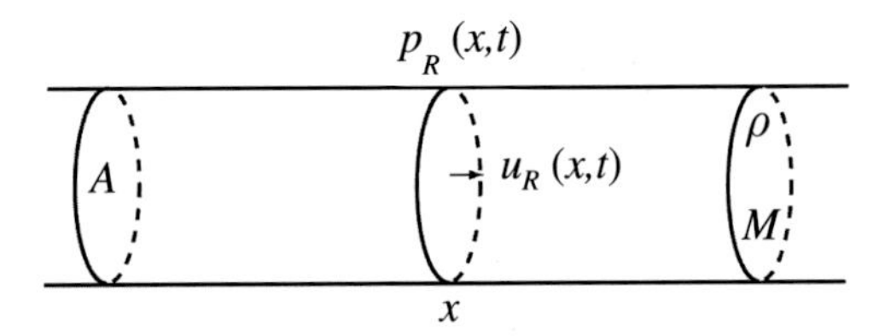

Figure (1.4.3). Cross-sectional element for an infinitely long rod constrained to deform longitudinally, $u_R\,(x,t)$ with respect to an equilibrium position, to a sinusoidal physical stress $p_R\,(x,t)$ acting uniformly on a cross-sectional element of area A.

The law of conservation of linear momentum implies the equation of motion for a cross-sectional element of the rod of density ρ is

$$\frac{\partial p}{\partial x} = \rho \frac{\partial^2 u}{\partial t^2}. \tag{1.4.4}$$

For harmonic motions, this equation of motion may also be written as

$$\frac{\partial P}{\partial x} = \rho(i\omega)^2 U. \tag{1.4.5}$$

The constitutive relation (1.4.2) implies

$$\frac{\partial P}{\partial x} = M \frac{\partial E}{\partial x}. \tag{1.4.6}$$

The assumption of infinitesimal strains (1.4.1) implies

$$\frac{\partial E}{\partial x} = \frac{\partial^2 U}{\partial x^2}. \tag{1.4.7}$$

Hence, (1.4.6) may be written as

$$\frac{\partial P}{\partial x} = M \frac{\partial^2 U}{\partial x^2}, \tag{1.4.8}$$

which upon substitution into (1.4.5) yields the desired equation of motion, namely

$$\frac{\partial^2 U(x)}{\partial x^2} + k^2 U(x) = 0, \tag{1.4.9}$$

where k is termed the complex wave number and defined by

$$k^2 \equiv \frac{\rho \omega^2}{M}. \tag{1.4.10}$$

For steady-state harmonic motion, an equivalent form of (1.4.9) in terms of the complex displacement $u(x,t)$ is

$$\frac{\partial^2 u(x,t)}{\partial x^2} - \frac{1}{v^2} \frac{\partial^2 u(x,t)}{\partial t^2} = 0, \tag{1.4.11}$$

where v is termed the complex velocity and defined by

$$v^2 \equiv \frac{\omega^2}{k^2} = \frac{M}{\rho}. \tag{1.4.12}$$

A steady state solution of (1.4.11) in complex notation is

$$u(x,t) = U(x) e^{i\omega t} = C e^{-ikx} e^{i\omega t} = C e^{i(\omega t - kx)}, \tag{1.4.13}$$

which may be written in terms of the real and imaginary parts of the complex wave number as

$$u(x,t) = C e^{k_I x} e^{i(\omega t - k_R x)}. \tag{1.4.14}$$

Hence, the physical displacement of a cross section in the rod corresponding to this steady state solution is given by the real part of (1.4.14) as

$$u_R(x,t) = C e^{k_I x} \cos(\omega t - k_R x), \tag{1.4.15}$$

where C is without loss of generality assumed to be real.

The steady-state physical displacement described by this solution describes a harmonic wave propagating in the positive x direction with phase speed

$$v_{phase} = \frac{\omega}{k_R}. \tag{1.4.16}$$

The wave decreases in amplitude in the direction of propagation due to the attenuation factor $e^{k_I x}$, where $-k_I$ is often termed the absorption coefficient and sometimes denoted by $\alpha \equiv -k_I$.

For elastic media the modulus M is a real number, hence (1.4.10) implies $k_I = 0$, so there is no attenuation of the wave in the direction of propagation. For a linear anelastic model, M is a complex number and $k_I \neq 0$, so (1.4.15) describes a wave attenuating in amplitude in the direction of propagation. The dependence of the attenuation on the frequency of vibration is determined by the frequency dependence of the modulus M implied by the anelastic viscoelastic model of interest.

The only permissible form of a steady-state solution of the one-dimensional wave equation given by (1.4.11) for anelastic viscoelastic media is one for which the maximum attenuation of the wave must necessarily be in the direction of propagation. This constraint does not apply to wave fields propagating in two or three dimensions. It will be shown in subsequent chapters that solutions of the corresponding wave equation in two and three dimensions describe waves for which the direction of phase propagation is not necessarily the same as that of maximum attenuation. This important result will be shown to be the basis for fundamental differences in the nature of one-dimensional anelastic wave-propagation problems and two- and three-dimensional problems.

The physical characteristics of waves in one dimension in which the direction of propagation coincides with the direction of maximum attenuation are analogous to those for waves in two and three dimensions in which the two directions coincide. Hence, for brevity further derivation of the characteristics of one-dimensional waves will be presented in later chapters.

1.5 Problems

(1) For the Standard Linear model shown in Figure (1.3.3)f derive the corresponding
   - (a) differential equation,
   - (b) creep function,
   - (c) relaxation function,
   - (d) compliance,
   - (e) modulus, and
   - (f) Q^{-1}.

(2) Explain why the configurations of elastic and viscous elements shown in Figures (1.3.3)g and (1.3.3)f for a Standard Linear solid yield equivalent responses to an applied force.

(3) Consider the infinitely long rod shown in Figure (1.4.3) with arbitrary, but fixed, viscoelastic material parameters of density ρ and modulus M constrained to deform longitudinally with harmonic displacements of frequency f and maximum amplitude C. For the one-dimensional waves propagating along the rod find
   (a) the rate at which energy is stored per unit length per unit time,
   (b) the rate at which energy is dissipated per unit length per unit time,
   (c) the maximum potential energy per unit length per cycle ($\max[\mathscr{P}]$),
   (d) the energy dissipated per unit length per cycle ($\Delta\mathcal{E}/cycle$),
   (e) the reciprocal quality factor (Q^{-1}), and
   (f) the damping ratio (ζ).
   (*Hint*: Substitute solution (1.4.15) into appropriate expressions in (1.2.5) through (1.2.10)).

(4) Use the expressions derived in problems 3(a) through 3(f) to find the corresponding expressions for a viscoelastic rod that responds as
   (a) an elastic solid
   (b) a Maxwell solid, and
   (c) a Voight solid.
   (*Hint*: Use appropriate expressions for the modulus as specified in terms of specific material parameters and frequency in Table (1.3.30).)

2

Three-Dimensional Viscoelasticity

Consideration of viscoelastic behavior in three dimensions allows consideration of stress distributions whose amplitude varies spatially over the surface to which it is applied. As a result energy flow and resultant displacements may occur in directions other than those of the maximum applied stress as they do for problems involving only one spatial dimension. These differences in physical phenomena will be shown to imply physical characteristics for radiation fields in three-dimensional viscoelastic media that do not occur in one dimension.

2.1 Constitutive Law

A homogeneous linear viscoelastic continuum is defined mathematically as one for which a causal fourth-order tensorial relaxation function r_{ijkl} exists independent of the spatial coordinates such that the constitutive equation relating stress and strain may be expressed as

$$p_{ij}(t) = \int_{-\infty}^{t} r_{ijkl}(t-\tau)\,de_{kl}(\tau), \tag{2.1.1}$$

where p_{ij} and e_{ij} denote the second-order time-dependent stress and strain tensors, respectively (Gurtin and Sternberg, 1962). Spatial independence of the tensorial relaxation function r_{ijkl} characterizes the continuum and corresponding medium as homogeneous.

The physical principle of causality requires that the relaxation function is zero for negative time, hence the constitutive relation may be written in the form

$$p_{ij}(t) = \int_{-\infty}^{\infty} r_{ijkl}(t-\tau)\,de_{kl}(\tau) \tag{2.1.2}$$

or more compactly with the definition of a Riemann–Stieltjes convolution operator as

$$p_{ij} = r_{ijkl} * de_{kl}. \tag{2.1.3}$$

Another function often used to characterize the behavior of a linear viscoelastic (LV) continuum or material is the so-called tensorial creep function $c_{ijkl}(t)$, which relates stress to strain by the following relation,

$$e_{ij} = c_{ijkl} * dp_{kl}. \tag{2.1.4}$$

A Homogeneous Isotropic Linear Viscoelastic (HILV) continuum is defined to be one for which the tensorial relaxation function $r_{ijkl}(t)$ is invariant with respect to the rotation of Cartesian coordinates. For such a material, scalar functions r_S and r_K exist, such that

$$rijkl = \frac{1}{3}(r_K - r_S)\delta_{ij}\,\delta_{kl} + \frac{1}{2}r_S(\delta_{ik}\,\delta_{jl} + \delta_{il}\,\delta_{jk}), \tag{2.1.5}$$

where the Kronecker delta δ_{ij} is defined as

$$\delta_{ij} \equiv \begin{Bmatrix} 1 & \text{if } i = j \\ 0 & \text{if } i \neq j \end{Bmatrix}. \tag{2.1.6}$$

For HILV materials, the constitutive relation simplifies to

$$p_{ij}' = r_S * de_{ij}' \qquad \text{for} \quad i \neq j \tag{2.1.7}$$

and

$$p_{kk} = r_K * de_{kk}, \tag{2.1.8}$$

where r_S and r_K are the spatially independent relaxation functions characteristic of the shear and bulk behaviors of the material, a repeated index *kk* indicates summation over index values 1, 2, and 3, and the prime denotes the deviatoric components of stress and strain defined by

$$p_{ij}' \equiv p_{ij} - \frac{1}{3}\delta_{ij}\,p_{kk} \tag{2.1.9}$$

and

$$e_{ij}' \equiv e_{ij} - \frac{1}{3}\delta_{ij}\,e_{kk}. \tag{2.1.10}$$

2.2 Stress–Strain Notation

Interpretation of the notation for components of the stress and strain tensors as presented in the previous sections for viscoelastic media is similar to that for elastic

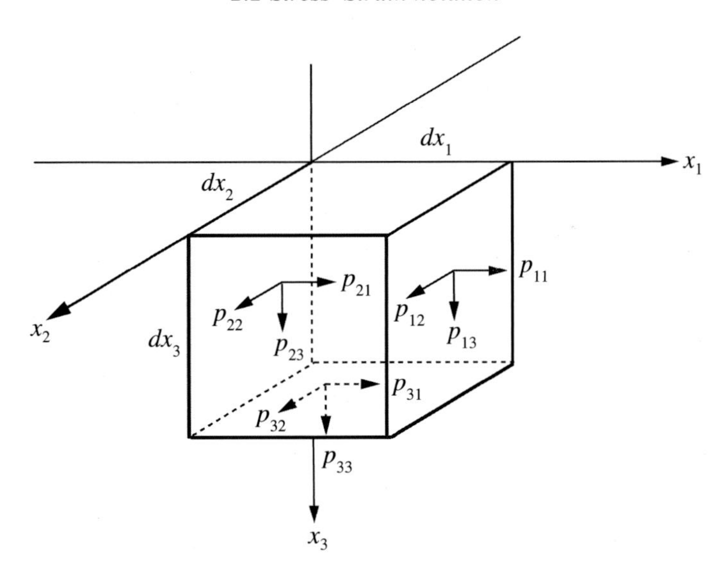

Figure (2.2.1). Notation for components of the stress tensor p_{ij} as illustrated at a fixed but arbitrary time t on the faces of an elemental cube.

media, if the tensors are interpreted at an arbitrary, but fixed time t. Diagrams illustrating the notation are shown here to aid the student not familiar with the notation. Formal derivations of the stress and strain tensors for elastic media are available in several texts on classical continuum mechanics.

Notation for components of the stress tensor can be visualized by considering an elemental cube located at the origin of a rectangular coordinate system (x_1, x_2, x_3) of dimension dx_1 by dx_2 by dx_3 (Figure (2.2.1)). Designating the plane perpendicular to a coordinate axis by the corresponding subscript, the stress acting on the ith plane in the direction of the jth coordinate axis at a fixed, but arbitrary, time t is designated as p_{ij} for $i = 1, 2, 3$ and $j = 1, 2, 3$. This notation is illustrated for the components of the stress tensor acting on faces of an elemental cube (Figure (2.2.1)). Stress components with the same subscripts, p_{ii} for $i = 1, 2, 3$, represent stresses normal to the indicated plane. Stress components with subscripts that are not the same, p_{ij} for $i \neq j$, represent shear stresses on the indicated plane.

Notation for components of the strain tensor may be visualized by considering undeformed and deformed states of the continuum. Deformation at a fixed but arbitrary time in the solid is described by a displacement vector $\vec{u} = (u_1, u_2, u_3)$, which specifies the position of each point in the continuum in the deformed state with respect to its position in an undeformed state. Uniaxial extension or strain at a point in the continuum as illustrated along the x_1 axis normal to the x_1 plane in Figure (2.2.2) is denoted by $e_{11} = \partial u_1 / \partial x_1$. Uniaxial extensions or compressions are denoted by components of the strain tensor with the same subscripts, namely

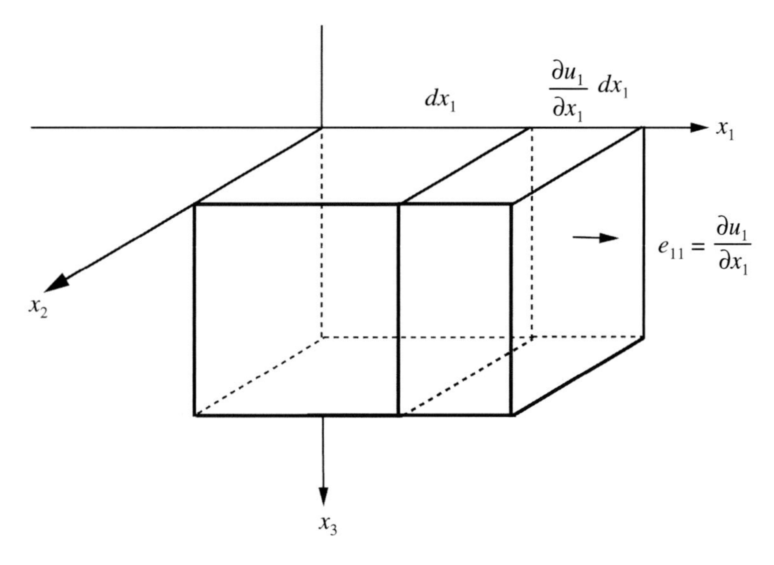

Figure (2.2.2). Notation for strain component e_{11} that designates uniaxial extension normal to the plane perpendicular to the x_1 axis at a fixed but arbitrary time t

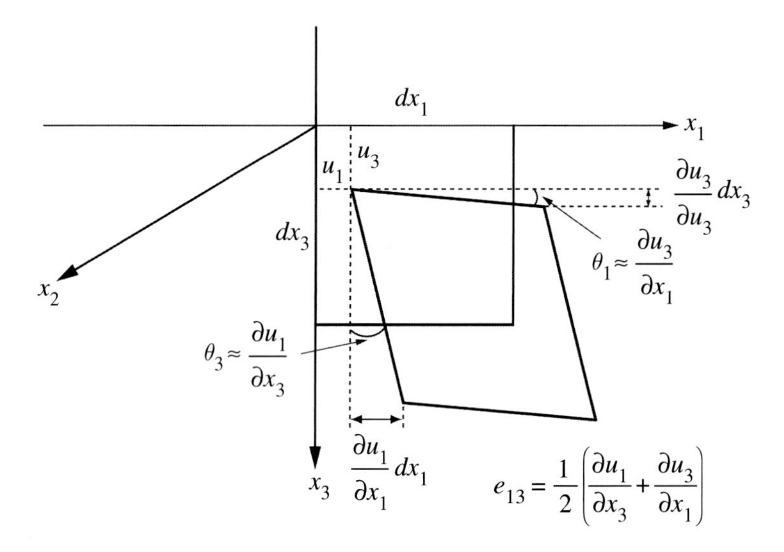

Figure (2.2.3). Notation for strain component e_{13} corresponding to average shear strain of a plane square in the x_1x_3 plane.

$e_{ii} = \partial u_i/\partial x_i$ for $i = 1, 2, 3$. These components of the strain tensor represent strains normal to the ith plane in the direction of the ith coordinate axis.

Shearing strains at a point in the continuum are illustrated by considering a planar elemental square deformed in one of the coordinate planes as illustrated for the x_1x_3 plane in Figure (2.2.3). The deformed square shows translation, shearing, and rotation specified with respect to the reference square by $\vec{u} = (u_1, u_2, u_3)$, where $u_2 = 0$ for the illustration considered. For small amounts of shear the tangents of the angles θ_1 and θ_3 are approximately equal to the angles and hence the average shear strain of the deformed square is given approximately by

$$e_{13} = \frac{1}{2}\left(\frac{\partial u_1}{\partial x_3} + \frac{\partial u_3}{\partial x_1}\right).$$

Similarly, each of the components of the strain tensor for which the subscripts are not the same, that is

$$e_{ij} = \frac{1}{2}\left(\frac{\partial u_i}{\partial x_j} + \frac{\partial u_j}{\partial x_i}\right),$$

can be visualized as the average shear strain in the $x_i\, x_j$ plane for $i \neq j$.

2.3 Equation of Motion

The motion of a material continuum with a specified set of boundary conditions is governed by the conservation laws of mass and momentum together with a constitutive relation characteristic of the material. The law of conservation of linear momentum implies the following equation of motion,

$$p_{ij,j} + f_i = \rho \ddot{u}_i, \tag{2.3.1}$$

where $f_i \equiv$ body force per unit volume, $\rho \equiv$ mass density, and $u_i \equiv i^{th}$ Cartesian component of the displacement vector.

To linearize the relation between the components of the stress tensor and the components of the displacement it is necessary to assume the displacements are infinitesimal. This assumption yields the following linear relation for the strain

$$e_{ij} = \frac{1}{2}(u_{i,j} + u_{j,i}), \tag{2.3.2}$$

where the notation $u_{i,j}$ denotes the partial derivative of the "i" component of displacement with respect to the "j" variable.

The equation of motion (2.3.1) together with the linearized strain displacement relation (2.3.2) and the constitutive relations (2.1.7) through (2.1.10) are sufficient to describe the infinitesimal motion of a homogeneous isotropic linear viscoelastic (HILV) continuum.

To find steady-state solutions of these equations, assume sufficient time has elapsed for the effect of initial conditions to be negligible. Employing the usual convention for steady state exhibited by

$$u_i \equiv U_i\, e^{i\omega t}, \tag{2.3.3}$$

where U_i is a complex function of the spatial coordinates and the physical displacement is given by the real part of u_i denoted by u_{i_R}, the equations governing the steady state motion of a HILV continuum in the absence of body forces are

$$P_{ij,j} + \rho\omega^2 U_i = 0, \tag{2.3.4}$$

$$E_{ij} = \frac{1}{2}\left(U_{i,j} + U_{j,i}\right), \tag{2.3.5}$$

$$P'_{ij} = i\omega R_S E'_{ij}, \tag{2.3.6}$$

$$P_{kk} = i\omega R_K E_{kk}, \tag{2.3.7}$$

where R_S and R_K denote the Fourier transforms of the relaxation functions r_S and r_K, respectively. When the complex shear modulus is defined by

$$M \equiv \frac{1}{2} i\omega R_S \tag{2.3.8}$$

and the complex bulk modulus is defined by

$$K \equiv \frac{1}{3} i\omega R_K \tag{2.3.9}$$

the constitutive relations (2.3.6) and (2.3.7) may be written as

$$P'_{ij} = 2ME'_{ij} \qquad \text{for} \quad i \neq j \tag{2.3.10}$$

and

$$P_{kk} = 3KE_{kk} \tag{2.3.11}$$

or as

$$P_{ij} = \delta_{ij}\left(K - \frac{2}{3}M\right)E_{kk} + 2ME_{ij}. \tag{2.3.12}$$

Other parameters familiar from elasticity theory are Lamé's parameters Λ and M, Young's modulus E, and Poisson's ratio N. Definitions of these parameters extended to viscoelastic media are

$$\Lambda \equiv K - \frac{2}{3}M, \tag{2.3.13}$$

$$E \equiv M\frac{3\Lambda + 2M}{\Lambda + M} \tag{2.3.14}$$

$$N \equiv \frac{\Lambda}{2(\Lambda + M)}, \tag{2.3.15}$$

where each of the parameters is in general a complex function of frequency.

In terms of Lamé's parameters the constitutive relation (2.3.12) becomes

$$P_{ij} = \delta_{ij}\,\Lambda E_{kk} + 2ME_{ij}. \tag{2.3.16}$$

The equation of motion (2.3.4) may be written entirely in terms of U_i by substituting (2.3.16) into (2.3.4) and simplifying with (2.3.5). The resulting desired equation of motion in vector notation is

$$\left(K+\frac{M}{3}\right)\nabla\theta + M\nabla^2\vec{u} = \rho\ddot{\vec{u}}, \tag{2.3.17}$$

where

$$\vec{u} \equiv u_k\hat{x}_k \tag{2.3.18}$$

and

$$\theta \equiv \nabla\bullet\vec{u}, \tag{2.3.19}$$

with the divergence of $\vec{u}$ defined by

$$\nabla\bullet\vec{u} \equiv \left(\frac{\partial}{\partial x_1}\hat{x}_1 + \frac{\partial}{\partial x_2}\hat{x}_2 + \frac{\partial}{\partial x_3}\hat{x}_3\right)\bullet(u_1\hat{x}_1 + u_2\hat{x}_2 + u_3\hat{x}_3) \equiv u_{i,i}. \tag{2.3.20}$$

2.4 Correspondence Principle

Three sets of fundamental equations are needed to derive the equation of motion for a HILV medium (2.3.17) under steady-state motion. The first set of equations is that implied by the conservation of momentum, namely (2.3.1) with body forces included or (2.3.4) in the absence of body forces. The second set is that which specifies that the problem is linear. It specifies, to first order, the components of strain in terms of the spatial derivatives of the displacement components, namely (2.3.2) or (2.3.5). The third set is the constitutive relations which specify the relationship between the components of stress and strain, namely (2.1.7) through (2.1.10) or (2.3.12). These equations are similar to those for elastic media, except that the parameters used to characterize the response of viscoelastic media are complex numbers instead of real numbers. This correspondence between equations for some problems in linear viscoelastic media and the equations for elastic problems has given rise to the term correspondence principle.

The concept of the correspondence principle is that solutions to certain steady-state problems in viscoelasticity can be inferred from the solutions to corresponding problems in elastic media upon replacement of real material parameters by complex parameters. This replacement of material parameters yields valid results for some problems, but for other problems the solution cannot be found using the correspondence principle. Some problems in viscoelastic behavior do not have a counterpart in elastic behavior.

Bland (1960, p. 67) states that the correspondence principle can only be used to obtain a solution for a steady-state problem in viscoelasticity if (1) a solution for a corresponding problem in elastic media exists, (2) no operation in obtaining the elastic solution would have a corresponding operation in the viscoelastic solution which would involve separating the complex modulus into real and imaginary parts, and (3) the boundary conditions for the two problems are identical. Bland defines a "corresponding problem" as an identical problem except that the body concerned is viscoelastic instead of elastic.

One of the previously mentioned examples for which the replacement of real material parameters with complex ones yields a correct equation is the equation of motion for HILV media as specified in one dimension by (1.4.9). This equation for viscoelastic media is similar to that for elastic media except that the material modulus (M) is in general a complex number instead of a real number.

Examples of problems for which the correspondence principle as stated by Bland (1960) cannot be used to derive solutions concern the dissipation and storage of energy in HILV media. These problems require separation of material moduli into their real and imaginary parts (see, for example, (1.2.7) through (1.2.11)). Energy balance equations that account for the dissipation and attenuation of energy due to intrinsic absorption in viscoelastic media have no corresponding counterpart in elastic media.

Problems and solutions derived herein are formulated from first principles for general linear viscoelastic media with the classic results for elastic media being readily implied upon specification of material parameters that are real numbers. This approach provides a rigorous framework from which the results for elastic media and an infinite number of linear anelastic models are readily deduced as special cases. This approach is needed to rigorously account for energy propagation and dissipation in viscoelastic media. It provides a framework for solution of reflection–refraction and surface-wave problems for general viscoelastic media some aspects of which have no counterpart in elastic media, especially those aspects pertaining to the physical characteristics of reflected, refracted, and surface waves required to satisfy the boundary conditions.

2.5 Energy Balance

Expressions for the various energy characteristics of harmonic displacement fields governed by the equation of motion provide the basis for understanding the nature of wave propagation in three-dimensional HILV media. Explicit energy expressions together with their physical interpretation are derived in this section.

Energy characteristics of viscoelastic radiation fields can be determined from an equation of mechanical energy balance which states that

$$\begin{aligned} \textit{rate of mechanical work} &= \textit{rate of change of internal energies} \\ &\quad + \textit{rate of energy dissipated}. \end{aligned} \tag{2.5.1}$$

Considering the internal energies per unit volume as the sum of the kinetic energy, $\mathscr{K}$, and stored strain, $\mathscr{P}$, energies per unit volume and denoting the rate of energy dissipated per unit volume by $\mathscr{D}$, and the energy intensity by $\vec{\mathscr{I}}$, equation (2.5.1) can be written as

$$-\int_V \nabla \cdot \vec{\mathscr{I}} dv = \int_V \left(\frac{\partial}{\partial t} (\mathscr{K} + \mathscr{P}) + \mathscr{D} \right) dV. \tag{2.5.2}$$

A mechanical energy-balance equation of the form (2.5.2) is derived by forming the dot product of the physical velocity field, $\vec{\dot{u}}_R$, with the steady-state equation of motion, where the ith component is given by

$$\begin{aligned} &\left(K_R - \frac{2}{3}M_R\right) u_{Rj,ji} + M_R(u_{Ri,jj} + u_{Rj,jj}) + \frac{1}{\omega}\left(\left(K_I - \frac{2}{3}M_I\right)\dot{u}_{Rj,ji}\right. \\ &\quad \left. + M_I(\dot{u}_{Ri,jj} + \dot{u}_{Rj,jj})\right) = \rho \ddot{u}_{Ri}. \end{aligned} \tag{2.5.3}$$

Upon applying the chain rule for differentiation, integrating over an arbitrary volume V, and rearranging terms, an equation of the form (2.5.2) as initially derived by Borcherdt (1971, 1973) and extended with physical interpretation of the terms by Borcherdt and Wennerberg (1985) is obtained, where

$$\mathscr{K} = \frac{1}{2}\rho \dot{u}_{Ri}^2, \tag{2.5.4}$$

$$\mathscr{P} = \frac{1}{2}\left(\left(K_R - \frac{2}{3}M_R\right) u_{Rj,j}^2 + M_R(u_{Ri,j}^2 + u_{Ri,j} u_{Rj,i})\right), \tag{2.5.5}$$

$$\mathscr{D} = \frac{1}{\omega}\left(K_I - \frac{2}{3}M_I\right)\dot{u}_{Rj,j}^2 + \frac{M_I}{\omega}(\dot{u}_{Ri,j}^2 + \dot{u}_{Ri,j}\dot{u}_{Rj,i}), \tag{2.5.6}$$

and

$$\begin{aligned} \mathscr{I}_j = &-\left(\left(K_R - \frac{2}{3}M_R\right) u_{Ri,i}\dot{u}_{Rj} + M_R(u_{Ri,j} + u_{Rj,i})\,\dot{u}_{Ri}\right. \\ &\left. + \frac{1}{\omega}\left(K_I - \frac{2}{3}M_I\right)\dot{u}_{Ri,i}\dot{u}_{Rj} + \frac{M_I}{\omega}(\dot{u}_{Ri,j} + \dot{u}_{Rj,i})\,\dot{u}_{Ri}.\right) \end{aligned} \tag{2.5.7}$$

In terms of the physical displacement field, u_R, the various terms (2.5.4) through (2.5.7) in the mechanical energy balance equation (2.5.2) also may be written in vector notation as

$$\mathscr{K} = \frac{1}{2}\rho \dot{\vec{u}}_R^2, \tag{2.5.8}$$

$$\mathscr{P} = \frac{1}{2}\left(K_R - \frac{2}{3}M_R\right)\theta_R^2 + M_R(\nabla \cdot \nabla \vec{u}_R \cdot \vec{u}_R - \nabla^2 \vec{u}_R \cdot \vec{u}_R), \tag{2.5.9}$$

$$\mathscr{D} = \frac{1}{\omega}\left(K_I - \frac{2}{3}M_I\right)\dot{\theta}_R^2 + \frac{M_I}{\omega}(\nabla \cdot \nabla \dot{\vec{u}}_R \cdot \dot{\vec{u}}_R - \nabla^2 \dot{\vec{u}}_R \cdot \dot{\vec{u}}_R), \tag{2.5.10}$$

$$\begin{aligned}\vec{\mathscr{J}} &= \left(K_R - \frac{2}{3}M_R\right)\theta_R \dot{\vec{u}}_R + M_R(\nabla \vec{u}_R \cdot \dot{\vec{u}}_R + \dot{\vec{u}}_R \cdot \nabla \vec{u}_R) \\ &\quad + \frac{1}{\omega}\left(K_I - \frac{2}{3}M_I\right)\dot{\theta}_R \dot{\vec{u}}_R + \frac{M_I}{\omega}(\nabla \dot{\vec{u}}_R \cdot \dot{\vec{u}}_R + \dot{\vec{u}}_R \cdot \nabla \dot{\vec{u}}_R),\end{aligned} \tag{2.5.11}$$

where $\theta_R = \nabla \cdot \vec{u}_R$.

To physically interpret the terms given in (2.5.4) through (2.5.7), the rate of working of the stresses or the mechanical work on an arbitrary volume, V, with surface element $d\vec{s} = \hat{n}\, ds$ is defined by (Aki and Richards, 1980; Hudson, 1980)

$$-\int_S p_{R\,ij}\, \dot{u}_{R\,i}\, \hat{n}\, ds. \tag{2.5.12}$$

Neglecting body forces and using the real components of (2.3.1) to represent the physical equation of motion, where the ith component is

$$p_{R\,ij,j} = \rho \ddot{u}_{R\,i}, \tag{2.5.13}$$

then forming the dot product of (2.5.13) with the velocity field $\dot{\vec{u}}_R$ yields

$$p_{R\,ij,j}\, \dot{u}_{R\,i} = \rho\, \ddot{u}_{R\,i}\, \dot{u}_{R\,i}. \tag{2.5.14}$$

The chain rule for differentiation implies

$$p_{R\,ij,j}\, \dot{u}_{R\,i} = \left(p_{R\,ij}\, \dot{u}_{R\,i}\right)_{,j} - p_{R\,ij}\, \dot{u}_{R\,i,j}, \tag{2.5.15}$$

so upon substitution of (2.5.15) into (2.5.14) the resulting equation can be written in terms of the components of the strain tensor as

$$\left(p_{R\,ij}\,\dot{u}_{R\,i}\right)_{,j} = \frac{1}{2}\frac{\partial}{\partial t}\left(\rho\,\dot{u}_{R\,i}^{2}\right) + p_{R\,ij}\,\dot{e}_{R\,ij}. \tag{2.5.16}$$

Integration of both sides of (2.5.16) over an arbitrary but fixed volume V with outward surface element $\hat{n}\,ds$ yields

$$\int_{V}\left(p_{R\,ij}\,\dot{u}_{R\,i}\right)_{,j} dV = \int_{V}\left(\frac{1}{2}\frac{\partial}{\partial t}\left(\rho\,\dot{u}_{R\,i}^{2}\right) + p_{R\,ij}\,\dot{e}_{R\,ij}\right) dV, \tag{2.5.17}$$

which upon application of Gauss' theorem becomes

$$-\int_{S} p_{R\,ij}\,\dot{u}_{R\,i}\,\hat{n}\,ds = \int_{V}\left(\frac{1}{2}\frac{\partial}{\partial t}\left(\rho\,\dot{u}_{R\,i}^{2}\right) + p_{R\,ij}\,\dot{e}_{R\,ij}\right) dV. \tag{2.5.18}$$

Substituting the constitutive equations for steady state and rewriting with Gauss' theorem (2.5.18) becomes

$$\begin{aligned} -\int_{V} \nabla\cdot\left(p_{R\,ij}\,\dot{u}_{R\,i}\right) dV = {} & \frac{\partial}{\partial t}\frac{1}{2}\int_{V}\left(\rho\,\dot{u}_{R\,i}^{2}\right) + \left(K_{R} - \frac{2}{3}M_{R}\right)e_{R\,jj}^{2} + 2M_{\mathrm{R}}e_{R\,ij}^{2}\,dV \\ & + \frac{1}{\omega}\int_{V}\left(\left(K_{I} - \frac{2}{3}M_{I}\right)\dot{e}_{R\,jj}^{2} + 2M_{I}\,\dot{e}_{R\,ij}^{2}\right) dV. \end{aligned} \tag{2.5.19}$$

Substitution of definitions (2.3.2) permits the right-hand side of (2.5.19) to be easily expressed in terms of the physical displacement field with the resultant equation being composed of terms of the form (2.5.4) through (2.5.7).

To physically interpret the terms on the right-side side of (2.5.19), the average over one cycle implies that the average rate of mechanical work over one cycle equals the average rate of energy dissipated; hence, the second term on the right-hand side can be physically interpreted as the rate of energy dissipated; hence, the rate of energy dissipated per unit volume, $\mathscr{D}$ is as given by (2.5.6). Physically interpreting the kinetic energy per unit volume, $\mathscr{K}$ as $\frac{1}{2}\rho\dot{\vec{u}}_{R}^{2}$ indicates that the remaining integrand in (2.5.19) can be interpreted as the stored (strain) energy per unit volume, $\mathscr{P}$, which is given by (2.5.5). Simplification of the integrand on the left-hand side of (2.5.19) with the constitutive law for viscoelastic media yields the desired expression for the components of the energy flux vector $\vec{\mathscr{I}}$ as given by (2.5.7). Hence, the desired physical interpretation and corresponding explicit description have been derived for each of the various energy characteristics ($\mathscr{K}, \mathscr{P}, \mathscr{D}$, and $\vec{\mathscr{I}}$) as specified in terms of the components of a harmonic displacement by (2.5.4) through (2.5.11).

2.6 Problems

(1) Derive the steady-state equation of motion (2.3.17) for a viscoelastic medium with complex shear and bulk moduli M and K and density ρ.

(2) Provide examples of equations that can and cannot be derived using the *Correspondence Principle.*

(3) Using the steady-state equation of motion (2.5.3) as expressed in terms of the components of the displacement field, derive expressions for the
   (a) kinetic energy density $\mathscr{K}$,
   (b) potential energy density $\mathscr{P}$,
   (c) dissipated energy density $\mathscr{D}$, and
   (d) energy flux $\vec{\mathscr{I}}$.

4) Show several energy conservation relations of the form

$$\int_V \nabla \bullet (\,) dV = \int_V \frac{\partial}{\partial t}\Big((\,) + (\,)\Big) dV \tag{2.6.1}$$

can be derived from the steady-state equation of motion (2.5.3) with the desired relation being that chosen based on physical interpretation of each of the terms. *Hint*: For illustration purposes consider (2.5.3) for elastic media, in which case three different identities can be derived by interchanging the order of partial differentiation, namely

$$\left(K_R + \frac{M_R}{3}\right) u_{Rj,ji} + M_R\, u_{Ri,jj} = \rho\, \ddot{u}_{Ri}, \tag{2.6.2}$$

$$\left(K_R + \frac{M_R}{3}\right) u_{Ri,ij} + M_R\, u_{Ri,jj} = \rho\, \ddot{u}_{Ri}, \tag{2.6.3}$$

and

$$\left(K_R - \frac{2}{3} M_R\right) u_{Rj,ji} + M_R(u_{Ri,jj} + u_{Rj,ij}) = \rho \ddot{u}_{Ri}. \tag{2.6.4}$$

The dot product of the vectors for which these identities specify the ith component with the velocity field yields three additional energy-conservation relations, the first of which is implied by

$$\begin{aligned} &\left(\left(K_R + \frac{M_R}{3}\right) u_{Ri,i}\, \dot{u}_{Rj} + M_R u_{Ri,j}\, \dot{u}_{Ri}\right)_{,j} \\ &= \frac{1}{2}\frac{\partial}{\partial t}\left(\rho\, \dot{u}_{Ri}^2 + \left(K_R + \frac{M_R}{3}\right) u_{Rj,j}^2 + M_R\, u_{Ri,j}\, u_{Ri,j}\right). \end{aligned} \tag{2.6.5}$$

Similarly, sixteen energy-conservation relations may be derived for an arbitrary viscoelastic solid. The preferred conservation relation as defined by the terms in (2.5.4) through (2.5.11) and the subsequent physical interpretation of the terms may be obtained from that implied by (2.6.5) upon subtracting the following identity from each side, where $M_I = 0$ for elastic media,

$$\int_V \left(\left(M_R \left(u_{Ri,j} \dot{u}_{Ri} - u_{Rj,j} \dot{u}_{Ri} \right) + \frac{M_I}{\omega} \left(\dot{u}_{Ri,j} \dot{u}_{Ri} - \dot{u}_{Rj,i} \dot{u}_{Ri} \right) \right)_{,j} \right) dV$$
$$= \frac{\partial}{\partial t} \frac{M_R}{2} \int_V \left(u_{Ri,j} u_{Rj,i} - u_{Rj,j} u_{Ri,i} \right) dV + \frac{M_I}{\omega} \int_V \left(\dot{u}_{Ri,j} \dot{u}_{Rj,i} - \dot{u}_{Rj,j} \dot{u}_{Ri,i} \right) dV. \tag{2.6.6}$$

As a final note, the fact that different energy-conservation relations can be derived from equations of the form (2.5.3) has led to confusion in the literature as to the proper expressions for the various energy characteristics. For example, for elastic media the energy characteristics defined by (2.5.4) through (2.5.11) for elastic media with $M_I = 0$ correspond to those derived for elastic media by Morse and Feshback (1953, p. 151), while (2.6.5) corresponds to that derived by Lindsay (1960, p. 154). Physical interpretation of the terms for the energy characteristics as given following (2.5.12) and specialized for elastic media implies the expressions derived by Morse and Feshback for elastic media are preferable.

3

Viscoelastic P, SI, and SII Waves

The equation of motion (2.3.17) governs wave propagation in a HILV continuum. Solutions of the wave equation and physical characteristics of the corresponding wave fields are presented in this chapter.

3.1 Solutions of Equation of Motion

A theorem due to Helmholtz implies that there exists a scalar potential ϕ and a vector potential $\vec{\psi}$ such that the displacement field may be expressed as

$$\vec{u} = \nabla\phi + \nabla \times \vec{\psi}, \tag{3.1.1}$$

where

$$\nabla \bullet \vec{\psi} = 0. \tag{3.1.2}$$

Employing the notational convention (2.3.3) for steady-state considerations and substituting (3.1.1) into (2.3.17) shows that the equation of motion (2.3.17) is satisfied if

$$\nabla^2\Phi + k_P^2 = 0 \tag{3.1.3}$$

and or

$$\nabla^2\vec{\Psi} + k_S^2 = 0, \tag{3.1.4}$$

where the complex wave numbers are defined in terms of the material parameters by

$$k_P \equiv \frac{\omega}{\alpha} \tag{3.1.5}$$

and

$$k_S \equiv \frac{\omega}{\beta} \tag{3.1.6}$$

with complex velocities defined by

$$\alpha \equiv \sqrt{\frac{K+\frac{4}{3}M}{\rho}} \tag{3.1.7}$$

and

$$\beta \equiv \sqrt{\frac{M}{\rho}}. \tag{3.1.8}$$

One solution of the equation of motion (2.3.17) is given by the solution of (3.1.3) with $\nabla \times \vec{\Psi} = 0$. For such a solution (3.1.1) implies that the corresponding displacement field is irrotational. This solution of the equation of motion states that irrotational or P waves may propagate independently in a HILV continuum. Another solution of the equation of motion (2.3.17) is given by the solution of (3.1.4) with $\nabla \Phi = 0$. For such a solution (3.1.1) implies that

$$\nabla \cdot \vec{u} = \nabla \cdot (\nabla \times \vec{\psi}) = 0 \tag{3.1.9}$$

and hence the volumetric strain as denoted by Δ or θ, namely

$$\Delta \equiv \theta \equiv \nabla \cdot \vec{u} = 0, \tag{3.1.10}$$

vanishes and the corresponding displacement field is termed equivoluminal. This solution states that an equivoluminal or S wave may propagate independently in a HILV continuum.

To consider general characteristics of the P and S wave solutions specified by (3.1.3) and (3.1.4), it is sufficient to consider general solutions to a general Helmholtz equation of the form

$$\nabla^2 \vec{G} + k^2 \vec{G} = 0, \tag{3.1.11}$$

where for P solutions $k = k_P$ and $\vec{G}$ is interpreted as the complex scalar potential Φ and for S solutions $k = k_S$ and $\vec{G}$ is interpreted as the complex vector potential $\vec{\Psi}$. Consideration of solutions of this general equation allows common characteristics of any wave (P, S, or surface wave) as governed by a Helmholtz equation of the form (3.1.11) to be derived.

A general solution of (3.1.11) for an assumed steady-state condition is given by a general expression of the form

$$\vec{G} = \vec{G}_0 \exp\left[-\vec{A} \cdot \vec{r}\right] \exp\left[-i\vec{P} \cdot \vec{r}\right], \tag{3.1.12}$$

where (1) $\vec{P}$ represents the propagation vector, which is perpendicular to surfaces of constant phase defined by $\vec{P} \cdot \vec{r} = constant$, (2) $\vec{A}$ represents the attenuation vector, which is perpendicular to surfaces of constant amplitude defined by $\vec{A} \bullet \vec{r} = constant$, and (3) $\vec{P}$ and $\vec{A}$ must be related to the material parameters through the wave number k by identities,

$$\vec{P} \bullet \vec{P} - \vec{A} \bullet \vec{A} = \text{Re}[k^2] \tag{3.1.13}$$

and

$$\vec{P} \bullet \vec{A} = |\vec{P}||\vec{A}| \cos(\gamma) = -\frac{1}{2} \text{Im}[k^2], \tag{3.1.14}$$

where γ represents the angle between $\vec{P}$ and $\vec{A}$ and "Re[z]" and "Im[z]" denote the real and imaginary parts of a complex number "z". The physical requirement that the amplitude of the wave not increase in the direction of propagation is satisfied by requiring that the angle between the propagation vector $\vec{P}$ and the attenuation vector $\vec{A}$ not exceed $\pi/2$, that is $0 \leq \gamma \leq \pi/2$.

Expression (3.1.12) is readily shown to be a solution of the Helmholtz equation (3.1.11) if conditions (3.1.13) and (3.1.14) are satisfied by substituting the solution (3.1.12) into (3.1.11) and simplifying with the vector identities in Appendix 2. Solutions of this general form also may be derived from first principles using separation of variables as shown in Appendix 3.

An *inhomogeneous* wave is defined as a wave for which surfaces of constant phase are not parallel to surfaces of constant amplitude, that is, $\vec{A}$ is not parallel to $\vec{P}$. If $\vec{A}$ is parallel to $\vec{P}$, then the wave is defined to be *homogeneous*. The term *general* will often be used to describe a wave that can be considered either homogeneous or inhomogeneous in the context of the problem under consideration.

The velocity of surfaces of constant phase for the general solution (3.1.12) is given by

$$\vec{v} = \omega \frac{\vec{P}}{|\vec{P}|^2} \tag{3.1.15}$$

with maximum attenuation for the general solution given by

$$|\vec{A}|. \tag{3.1.16}$$

Equations (3.1.13) and (3.1.14) provide considerable insight into the nature of wave fields that can be solutions of the general Helmholtz equation (3.1.11) and hence the equation of motion (2.3.17) for a HILV medium. It will be shown that they imply that the type of inhomogeneous waves that propagate in elastic media cannot propagate in anelastic HILV media and vice versa. This important result will be shown to imply that the characteristics of P, S, and surface waves in layered anelastic media are distinctly different from those in elastic media. The results specifying the types of inhomogeneous waves that may propagate in elastic and anelastic media are stated formally in the following theorems (Borcherdt, 1971, 1973a).

Theorem (3.1.17). A wave governed by the Helmholtz equation (3.1.11) for HILV media may propagate in non-dissipative or elastic media if and only if

(1) $|\vec{A}| = 0$ *and the wave is homogeneous*

or

(2) $|\vec{A}| \neq 0$ *and the direction of maximum attenuation for the inhomogeneous wave is perpendicular to the direction of phase propagation.*

To prove the first part of this theorem assume the medium is elastic. Interpreting k as k_P for P wave solutions as specified by (3.1.5) or as k_S for S wave solutions as specified by (3.1.8) in terms of the material parameters, the assumption of elastic media implies $\text{Im}[k^2] = \text{Im}[k_P^2] = \text{Im}[k_S^2] = 0$, which upon substitution into (3.1.14) implies that either (1) $|\vec{A}| = 0$ and the wave is homogeneous or (2) $\gamma = \pi/2$ and the direction of maximum attenuation, $\vec{A}/|\vec{A}|$, is perpendicular to the direction of phase propagation, $\vec{P}/|\vec{P}|$. Conversely, (3.1.14) shows that if $|\vec{A}| = 0$ or $\gamma = \pi/2$, then $\text{Im}[k^2] = 0$, which depending upon whether one is considering a P or S solution implies $\text{Im}[k_P^2] = 0$ or $\text{Im}[k_S^2] = 0$, which together with (3.1.5) through (3.1.8) implies the medium is elastic.

This theorem establishes that the only type of inhomogeneous P and S solutions for elastic media of the Helmholtz equations (3.1.3) and (3.1.4) as generalized to (3.1.12) is one for which the direction of propagation is perpendicular to the direction of maximum attenuation. It also establishes that the only type of homogeneous wave in elastic media is one for which there is no attenuation.

The corresponding result for dissipative or anelastic HILV media is stated formally in the following theorem.

Theorem (3.1.18). A wave governed by the Helmholtz equation (3.1.11) for HILV media propagates in dissipative or anelastic HILV media if and only if

(1) the maximum attenuation of the wave is non-zero, that is, $|\vec{A}| \neq 0$,

and

(2) the direction of maximum attenuation of the wave is not perpendicular to the direction of phase propagation, that is, $\gamma \neq \pi/2$.

The proof of this theorem is entirely analogous to Theorem (3.1.17). If the medium is assumed to be anelastic, then (3.1.5) through (3.1.8) imply $\text{Im}[k^2] \neq 0$, hence, (3.1.14) implies $|\vec{A}| \neq 0$, that is, the maximum attenuation is non-zero and

$\gamma \neq \pi/2$, that is the direction of maximum attenuation is not perpendicular to the direction of phase propagation. Conversely, if $|\vec{A}| \neq 0$ and $\gamma \neq \pi/2$, then (3.1.14) implies $\text{Im}[k^2] \neq 0$, from which it follows the medium is anelastic.

This theorem establishes that the only type of inhomogeneous P and S solutions for anelastic HILV media of the Helmholtz equations (3.1.3) and (3.1.4) as generalized to (3.1.12) is one for which the direction of propagation is not perpendicular to the direction of maximum attenuation. It also establishes that the only type of homogeneous wave in anelastic media is one for which the maximum attenuation does not vanish.

The results pertaining to inhomogeneous waves as proved for Theorems (3.1.17) and (3.1.18) also may be stated as

Theorem (3.1.19). For HILV media the only type of inhomogeneous wave that propagates in anelastic media, namely a wave for which $\gamma \neq \pi/2$, cannot propagate in elastic media and vice versa.

These results emphasize an important difference in the nature of inhomogeneous wave fields in elastic versus anelastic media. This difference is the basis for significant differences in the physical characteristics of two- and three-dimensional wave fields in layered anelastic media versus those in layered elastic media. These differences and the characteristics of the wave fields will be fully explored in subsequent sections and chapters.

For $\gamma \neq \pi/2$ (3.1.13) and (3.1.14) when solved simultaneously yield

$$|\vec{P}| = \sqrt{\frac{1}{2}\left(\text{Re}[k^2] + \sqrt{(\text{Re}[k^2])^2 + \frac{(\text{Im}[k^2])^2}{\cos^2\gamma}}\right)} \tag{3.1.20}$$

and

$$|\vec{A}| = -\frac{1}{2}\frac{\text{Im}[k^2]}{|\vec{P}|\cos\gamma} = \sqrt{\frac{1}{2}\left(-\text{Re}[k^2] + \sqrt{(\text{Re}[k^2])^2 + \frac{(\text{Im}[k^2])^2}{\cos^2\gamma}}\right)}. \tag{3.1.21}$$

Equations (3.1.20) and (3.1.21) show that for a specified angle $\gamma \neq \pi/2$, the magnitudes of the propagation and attenuation vectors are uniquely determined in terms of the complex wave number k. Equations (3.1.5) through (3.1.8) show that the complex wave number is uniquely determined in terms of the parameters of the material. These results indicate that, once the angle between the propagation and attenuation vectors is specified for an inhomogeneous wave in anelastic media, the velocity (3.1.15) and maximum attenuation (3.1.16) of the wave are uniquely determined in terms of the wave number k and in turn the parameters of the material. This is not true, however, for elastic media or non-dissipative media.

For an inhomogeneous wave in elastic media with $\gamma=\pi/2$, (3.1.13) and (3.1.14) imply that a unique solution does not exist for the propagation and attenuation vectors. In other words, for elastic media with the only type of permissible inhomogeneous solution being $\gamma=\pi/2$, the solution for the attenuation and propagation vectors is not unique in terms of the parameters of the material. For those familiar with elastic reflection–refraction problems, examples of this non-uniqueness for elastic media are the solutions for inhomogeneous P waves generated by SV waves incident on a plane boundary at angles beyond critical. The speed of the reflected solutions is the apparent phase speed along the boundary of the incident wave field, and hence must vary with the angle of incidence and is not uniquely determined once $\gamma=\pi/2$ is specified.

For a homogeneous wave with $\gamma=0$, the magnitude of the phase velocity expressed in terms of the real part of the wave number, which in turn can be expressed in terms of the parameters of the material, is given by

$$|\vec{v}_H| = \frac{\omega}{k_R} \tag{3.1.22}$$

and the magnitude of the attenuation vector by

$$|\vec{A}_H| = -k_I. \tag{3.1.23}$$

Equations (3.1.13), (3.1.14), (3.1.20), and (3.1.21) imply that for all γ the magnitude of the phase velocity for an inhomogeneous wave, denoted here by $|\vec{v}_{iH}|$, is less than that for a corresponding homogeneous wave in the same medium, that is,

$$|\vec{v}_{iH}| < |\vec{v}_H|. \tag{3.1.24}$$

Similarly, the maximum attenuation for an inhomogeneous wave, denoted by $|\vec{A}_{iH}|$, can be shown to be greater than that for a corresponding homogeneous wave in the same material, that is,

$$|\vec{A}_H| < |\vec{A}_{iH}|. \tag{3.1.25}$$

3.2 Particle Motion for P Waves

The solution of the equation of motion corresponding to the Helmholtz equation, (3.1.3) or (3.1.11), and in turn the equation of motion (2.3.17) for a general (homogeneous or inhomogeneous) P wave, is given by

$$\phi = G_0 \exp\left[-\vec{A}_P \bullet \vec{r}\right] \exp\left[i\left(\omega t - \vec{P}_P \bullet \vec{r}\right)\right] \tag{3.2.1}$$

with

$$\nabla \times \vec{\psi} = 0, \tag{3.2.2}$$

where $\vec{P}_P$ and $\vec{A}_P$ are given by (3.1.20) and (3.1.21) with $k=k_P$. Introduction of a complex wave vector, $\vec{K}_P$, defined in terms of the propagation and attenuation vectors as

$$\vec{K}_P \equiv \vec{P}_P - i\vec{A}_P \tag{3.2.3}$$

permits the solution for a general P wave to be written in a slightly more compact form as

$$\phi = G_0 \exp[i(\omega t - \vec{K}_P \bullet \vec{r})], \tag{3.2.4}$$

where the condition for propagation, namely (3.1.13) and (3.1.14) rewritten in terms of $\vec{K}_P$, is

$$\vec{K}_P \bullet \vec{K}_P = \left(\vec{P}_P \bullet \vec{P}_P - \vec{A}_P \bullet \vec{A}_P\right) - i\,2\vec{P}_P \bullet \vec{A}_P = k_P^2 = \left(\frac{\omega}{\alpha}\right)^2 = \frac{\omega^2 \rho}{K + \frac{4}{3}M}. \tag{3.2.5}$$

The complex displacement field for the P wave solution (3.2.1) is, from (3.1.1) and identity (11.2.6), given by

$$\vec{u} = \nabla\phi = -i(\vec{P}_P - i\vec{A}_P)\phi = -i\vec{K}_P\,\phi, \tag{3.2.6}$$

from which it follows that the complex velocity field for a P wave is given by

$$\dot{\vec{u}} = \omega(\vec{P}_P - i\vec{A}_P)\phi = \omega\vec{K}_P\,\phi. \tag{3.2.7}$$

The complex volumetric strain, $\Delta = \theta = \nabla \bullet \vec{u}$ associated with the displacement field $\vec{u}$ may be written in terms of the complex scalar potential ϕ for P waves using (11.2.7) as

$$\Delta = \nabla^2\phi = -\vec{K}_P \bullet \vec{K}_P\,\phi = -k_P^2\,\phi \tag{3.2.8}$$

or in terms of the complex displacement field as

$$\Delta = -i(\vec{P}_P - i\vec{A}_P) \bullet \vec{u} = -i\vec{K}_P \bullet \vec{u}. \tag{3.2.9}$$

The actual particle motion for a wave field is specified by the real part of the displacement field. The real part of the displacement field for a P wave is

$$\vec{u}_R = \mathrm{Re}[\nabla\phi] = \exp[-\vec{A}_P \bullet \vec{r}\,]\mathrm{Re}[-(\vec{A}_P + i\vec{P}_P)G_0 \exp[i(\omega t - \vec{P}_P \bullet \vec{r})]]. \tag{3.2.10}$$

Simplification of (3.2.10) allows the physical displacement field describing the particle motion for a general P wave (Borcherdt, 1973a) to be written as

$$\vec{u}_R = |G_0||k_P| \exp[-\vec{A}_P \bullet \vec{r}\,]\left(\vec{\xi}_{1P} \cos\zeta_P(t) + \vec{\xi}_{2P} \sin\zeta_P(t)\right) \tag{3.2.11}$$

where

$$\vec{\xi}_{1P} \equiv \frac{k_{P_R}\vec{P}_P - k_{P_I}\vec{A}_P}{|k_P|^2}, \tag{3.2.12}$$

$$\vec{\xi}_{2P} \equiv \frac{k_{P_I}\vec{P}_P + k_{P_R}\vec{A}_P}{|k_P|^2}, \tag{3.2.13}$$

and

$$\zeta_P(t) \equiv \omega t - \vec{P}_P \bullet \vec{r} + \arg[G_0 k_P] - \pi/2. \tag{3.2.14}$$

The identities required for propagation, namely (3.1.13) and (3.1.14), imply that

$$\vec{\xi}_{1P} \bullet \vec{\xi}_{2P} = 0 \tag{3.2.15}$$

and

$$\vec{\xi}_{1P} \bullet \vec{\xi}_{1P} - \vec{\xi}_{2P} \bullet \vec{\xi}_{2P} = 1. \tag{3.2.16}$$

Equations (3.2.11) through (3.2.16) indicate that the particle motion for a general P wave is elliptical. To see this result more clearly, consider the transformation defined by

$$u_1' \equiv \frac{\vec{u}_R \bullet \vec{\xi}_{1P}}{C|\vec{\xi}_{1P}|} \tag{3.2.17}$$

and

$$u_2' \equiv \frac{\vec{u}_R \bullet \vec{\xi}_{2P}}{C|\vec{\xi}_{2P}|}, \tag{3.2.18}$$

where

$$C \equiv |k_P||G_0| \exp\left[-\vec{A}_P \bullet \vec{r}\right]. \tag{3.2.19}$$

Substitution of (3.2.11) for $\vec{u}_R$ into (3.2.17) and (3.2.18) for the transformation shows that

$$\cos \zeta_P(t) = \frac{u_1'}{|\vec{\xi}_{1P}|} \tag{3.2.20}$$

and

$$\sin \zeta_P(t) = \frac{u_2'}{|\vec{\xi}_{2P}|}. \tag{3.2.21}$$

Substitution of these expressions into the trigonometric identity

$$\sin^2 \zeta_P(t) + \cos^2 \zeta_P(t) = 1 \tag{3.2.22}$$

yields the desired equation of an ellipse, namely

$$\frac{u_1'^2}{|\vec{\xi}_{1P}|^2} + \frac{u_2'^2}{|\vec{\xi}_{2P}|^2} = 1, \tag{3.2.23}$$

with major axis $|\vec{\xi}_{1P}|$, minor axis $|\vec{\xi}_{2P}|$, and eccentricity $e = 1/|\vec{\xi}_{1P}|$. The direction of motion is from $\vec{P}_P$ to $\vec{A}_P$ with the angle between the major axis of the ellipse and the direction of propagation given by

$$\cos \eta_P = \frac{\vec{\xi}_{1P} \bullet \vec{P}_P}{|\vec{\xi}_{1P}||\vec{P}_P|} = \frac{k_{P_R}}{|k_P|}\sqrt{1 + \frac{k_{P_I}^2}{|\vec{P}_P|^2}}. \tag{3.2.24}$$

For a homogeneous P wave,

$$|\vec{P}_{PH}| = k_{P_R} \tag{3.2.25}$$

and

$$|\vec{A}_{PH}| = -k_{P_I} \tag{3.2.26}$$

so (3.2.13) implies $\vec{\xi}_{2P} = 0$ and (3.2.24) implies $\eta_P = 0$; hence the elliptical motion for an inhomogeneous P wave degenerates to linear motion parallel to the direction of propagation for a homogeneous wave.

The parameters describing the elliptical particle motion for a general P wave are illustrated in Figure (3.2.27).

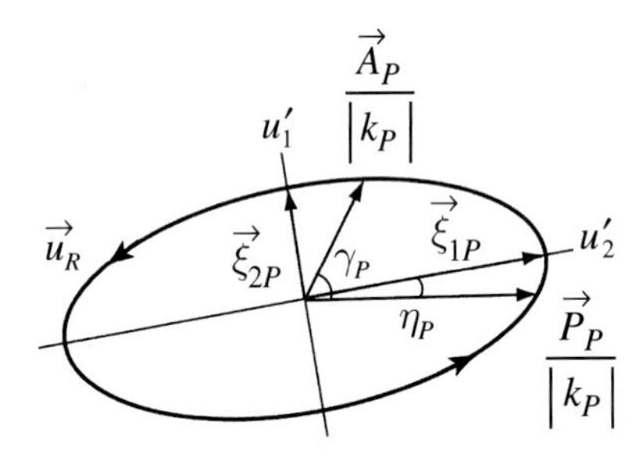

Figure (3.2.27). Diagram illustrating the parameters describing the elliptical particle motion orbit of a P wave in HILV media. If the wave is homogeneous, then the elliptical particle motion degenerates to linear motion parallel to the direction of phase propagation.

3.3 Particle Motion For Elliptical and Linear S Waves

The solution of the equation of motion corresponding to the Helmholtz equation, (3.1.3) or (3.1.12), and in turn the equation of motion (2.3.17) for a general (homogeneous or inhomogeneous) S wave, is given by

$$\vec{\psi} = \vec{G}_0 \exp\left[-\vec{A}_S \bullet \vec{r}\right] \exp\left[i\left(\omega t - \vec{P}_S \bullet \vec{r}\right)\right] \tag{3.3.1}$$

with

$$\nabla \phi = 0 \tag{3.3.2}$$

and

$$\nabla \bullet \vec{\psi} = 0, \tag{3.3.3}$$

where $\vec{P}_S$ and $\vec{A}_S$ are given by (3.1.20) and (3.1.21) with $k = k_S$. The most general form of the solution for an S wave is given by (3.3.1) with $\vec{G}_0$ being a complex vector of the form

$$\vec{G}_0 = z_1 \hat{x}_1 + z_2 \hat{x}_2 + z_3 \hat{x}_3, \tag{3.3.4}$$

where z_1, z_2, and z_3 are arbitrary, but fixed, complex numbers and $\hat{x}_1$, $\hat{x}_2$, and $\hat{x}_3$ are orthogonal real unit vectors for a Cartesian coordinate system. The form of the complex vector $\vec{G}_0$ and its relation to the plane defined by the directions of propagation and maximum attenuation determines the type and corresponding characteristics of the S wave solution as initially shown by Borcherdt (1973a, 1977).

Introduction of a complex wave vector, $\vec{K}_S$, defined in terms of the propagation and attenuation vectors as

$$\vec{K}_S \equiv \vec{P}_S - i\vec{A}_S \tag{3.3.5}$$

permits the solution for a general S wave to be written in a slightly more compact form as

$$\vec{\psi} = \vec{G}_0 \exp\left[i\left(\omega t - \vec{K}_S \bullet \vec{r}\right)\right], \tag{3.3.6}$$

where the condition for propagation, namely (3.1.13) and (3.1.14) rewritten in terms of $\vec{K}_S$, is

$$\vec{K}_S \bullet \vec{K}_S = \left(\vec{P}_S \bullet \vec{P}_S - \vec{A}_S \bullet \vec{A}_S\right) - i2\vec{P}_S \bullet \vec{A}_S = k_S^2 = \left(\frac{\omega}{\beta}\right)^2 = \frac{\omega^2 \rho}{M}. \tag{3.3.7}$$

Condition (3.3.3) and identity (11.2.9) imply

$$\nabla \bullet \vec{\psi} = -i\left(\vec{P}_S - i\vec{A}_S\right) \bullet \vec{G}_0 = -i\vec{K}_S \bullet \vec{G}_0 = 0. \tag{3.3.8}$$

The complex displacement field for the S wave solution (3.3.1) is, from (3.1.1) and (11.3.10), given by

$$\vec{u} = \nabla \times \vec{\psi} = -i\left(\vec{P}_S - i\vec{A}_S\right) \times \vec{\psi} = -i\vec{K}_S \times \vec{\psi} \tag{3.3.9}$$

and the complex velocity field by

$$\dot{\vec{u}} = \omega\left(\vec{P}_S - i\vec{A}_S\right) \times \vec{\psi} = \omega\vec{K}_S \times \vec{\psi}. \tag{3.3.10}$$

Introduction of a complex vector displacement coefficient $\vec{D}$, defined as

$$\vec{D} \equiv -i\vec{K}_S \times \vec{G}_0, \tag{3.3.11}$$

allows the complex displacement field to be written conveniently as

$$\vec{u} = \vec{D}\exp\left[i\left(\omega t - \vec{K}_S \bullet \vec{r}\right)\right]. \tag{3.3.12}$$

The vector identity for a quadruple vector product, (11.2.4), implies that

$$\vec{D} \bullet \vec{D} = -\left(\vec{K}_S \times \vec{G}_0\right) \bullet \left(\vec{K}_S \times \vec{G}_0\right) = -\vec{K}_S^2 \vec{G}_0{}^2 + \left(\vec{K}_S \bullet \vec{G}_0\right)^2, \tag{3.3.13}$$

hence conditions (3.3.8) and (3.3.7) show that the dot product of the vector displacement coefficient with itself, i.e. $\vec{D} \bullet \vec{D}$, is related to the corresponding dot product of the vector displacement potential coefficient for S waves by

$$\vec{D} \bullet \vec{D} = -\vec{K}_S^2 \vec{G}_0 \bullet \vec{G}_0 = -k_S^2 \vec{G}_0 \bullet \vec{G}_0. \tag{3.3.14}$$

Substitution of (3.3.12) into (2.3.17) shows that the equation of motion for a general S wave reduces to

$$\nabla^2 \vec{u} = \frac{\rho}{M}\ddot{\vec{u}}, \tag{3.3.15}$$

with solutions of the form (3.3.12), where $\nabla \bullet \vec{u} = 0$ and the corresponding Helmholtz equation is

$$\nabla^2 \vec{U} + k_S^2 \vec{U} = 0. \tag{3.3.16}$$

3.3.1 Type-I or Elliptical S (SI) Wave

If the complex vector $\vec{G}_0$ in the general solution (3.3.1) for the displacement potential for an S wave is a simple vector multiplied by an arbitrary, but fixed complex number, that is, if

$$\vec{G}_0 = z\hat{n}, \tag{3.3.17}$$

then the corresponding S wave is defined as a Type-I S wave (Borcherdt, 1973a). For brevity, this type of S wave often will be referred to as an SI wave. It sometimes will be called an Elliptical S wave, because it will be shown that the particle motion for an inhomogeneous SI wave is elliptical as opposed to linear for an inhomogeneous Type-II S wave, to be defined in the next section.

For S wave solutions of the form (3.3.17), the equivoluminal property of the wave as specified by (3.3.8) implies that

$$\vec{P}_S \cdot \hat{n} = 0 \tag{3.3.18}$$

and

$$\vec{A}_S \cdot \hat{n} = 0. \tag{3.3.19}$$

For an inhomogeneous SI wave solution in elastic or anelastic media (3.1.17) and (3.1.18) imply $\vec{A}_S \neq 0$ and $\vec{P}_S \neq 0$, so the equivoluminal property of the solution implies that $\hat{n}$ must be perpendicular to the plane defined by $\vec{P}_S$ and $\vec{A}_S$, in order that (3.3.1) represent a solution of the equation of motion.

For solutions with complex vector amplitude of the form (3.3.17), the complex vector displacement (3.3.11) may be written as

$$\vec{D}_{SI} = -i\vec{K}_S \times \vec{G}_0 = -iz\left(\vec{P}_S \times \hat{n} - i\vec{A}_S \times \hat{n}\right), \tag{3.3.20}$$

where $\vec{D}_{SI} \cdot \vec{D}_{SI}$ is given from (3.3.14) by

$$\vec{D}_{SI} \cdot \vec{D}_{SI} = -k_S^2 \vec{G}_0 \cdot \vec{G}_0 = -k_S^2 z^2. \tag{3.3.21}$$

The particle motion for a Type-I S wave is specified by the real part of the displacement field, namely

$$\begin{aligned} \vec{u}_R &= \mathrm{Re}[\nabla \times \vec{\psi}] \\ &= \exp\left[-\vec{A}_S \cdot \vec{r}\right] \mathrm{Re}\left[-i\left(\vec{P}_S - i\vec{A}_S\right) \times \vec{G}_0 \exp\left[i\left(\omega t - \vec{P}_S \cdot \vec{r}\right)\right]\right]. \end{aligned} \tag{3.3.22}$$

Simplification of (3.3.22) allows the physical displacement field for a Type-I S wave (Borcherdt, 1973a) to be written as

$$\vec{u}_R = |\vec{G}_0||k_S| \exp\left[-\vec{A}_S \cdot \vec{r}\right](-1)\left(\vec{\xi}_{1SI} \cos\zeta_{SI}(t) + \vec{\xi}_{2SI} \sin\zeta_{SI}(t)\right), \tag{3.3.23}$$

where

$$\vec{\xi}_{1SI} \equiv \frac{k_{S_R}\vec{P}_S - k_{S_I}\vec{A}_S}{|k_S|^2} \times \hat{n}, \tag{3.3.24}$$

$$\vec{\xi}_{2SI} \equiv \frac{k_{S_I}\vec{P}_S + k_{S_R}\vec{A}_S}{|k_S|^2} \times \hat{n}, \tag{3.3.25}$$

with $\hat{n} = \left(\vec{P}_S \times \vec{A}_S\right) / \left|\vec{P}_S \times \vec{A}_S\right|$ and

$$\zeta_{SI}(t) \equiv \omega t - \vec{P}_S \cdot \vec{r} + \arg[(\vec{G}_0 \cdot \hat{n})k_S] + \pi/2. \tag{3.3.26}$$

The identities required for propagation, namely (3.1.13) and (3.1.14), imply that

$$\vec{\xi}_{1SI} \bullet \vec{\xi}_{2SI} = 0 \tag{3.3.27}$$

and

$$\vec{\xi}_{1SI} \bullet \vec{\xi}_{1SI} - \vec{\xi}_{2SI} \bullet \vec{\xi}_{2SI} = 1. \tag{3.3.28}$$

Analogous to the derivation for a P wave, equations (3.3.23) through (3.3.28) indicate that the particle motion for a general inhomogeneous Type-I S wave is elliptical in the plane of the propagation and attenuation vectors, $\vec{P}_S$ and $\vec{A}_S$. The plane of the elliptical particle motion is perpendicular to $\vec{G}_0 = z\hat{n}$. The major axis of the particle motion ellipse is $|\vec{\xi}_{1SI}|$. Its minor axis is $|\vec{\xi}_{2SI}|$ and its eccentricity is $e = 1/|\vec{\xi}_{1SI}|$. The direction of motion is from $\vec{P}_S$ to $\vec{A}_S$ with the angle between the major axis of the ellipse and the direction of propagation given by

$$\cos \eta_{SI} = \frac{\left(\vec{P}_S \times \hat{n}\right) \bullet \vec{\xi}_{1SI}}{|\vec{P}_S \times \hat{n}||\vec{\xi}_{1SI}|} = \frac{k_{S_R}}{|k_S|}\sqrt{1 + \frac{k_{S_I}^2}{|\vec{P}_S|^2}}. \tag{3.3.29}$$

The parameters describing the elliptical particle motion of a general Type-I S wave are illustrated in Figure (3.3.30).

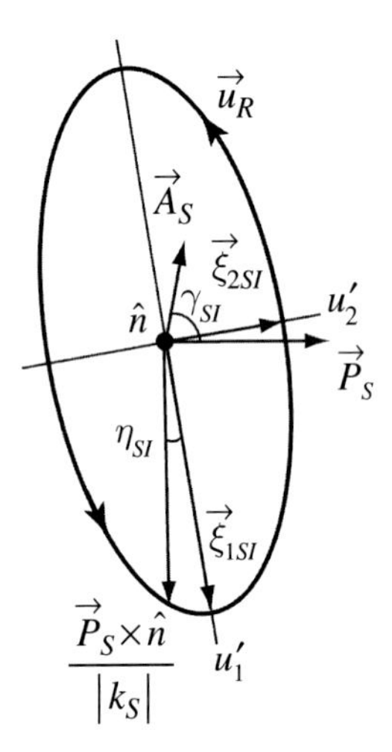

Figure (3.3.30). Diagram illustrating the parameters describing the elliptical particle motion orbit of a Type-I S (SI) wave in HILV media. If the wave is homogeneous, then the elliptical particle motion degenerates to linear motion perpendicular to the direction of phase propagation.

For a homogeneous Type-I S wave,

$$\left|\vec{P}_{SH}\right| = k_{S_R} \tag{3.3.31}$$

and

$$|\vec{A}_{SH}| = -k_{S_I}, \tag{3.3.32}$$

so (3.3.25) implies $\vec{\xi}_{2SI} = 0$ and (3.3.29) implies $\eta_{SI} = 0$. Hence, the elliptical motion for an inhomogeneous Type-I S wave degenerates to linear motion in the plane of $\vec{P}_S$ and $\vec{A}_S$ perpendicular to the direction of propagation if the wave is homogeneous.

3.3.2 Type-II or Linear S (SII) Wave

If the complex vector $\vec{G}_0$ in the general solution (3.3.1) for an S wave is not a simple vector, $\vec{G}_0 = z\hat{n}$, but a more general complex vector, which upon introduction of a rectangular Cartesian coordinate system may be written as

$$\vec{G}_0 = z_1\hat{x}_1 + z_2\hat{x}_2 + z_3\hat{x}_3, \tag{3.3.33}$$

where z_1, z_2, and z_3 are arbitrary, but fixed, complex numbers and $\hat{x}_1$, $\hat{x}_2$, and $\hat{x}_3$ are orthogonal real unit vectors for a Cartesian coordinate system, then a special case of interest is that for which $z_2 = 0$ and the propagation and attenuation vectors are in the x_1x_3 plane. This type of S wave is defined as a Type-II S wave (Borcherdt, 1977). For brevity, it shall be referred to as an SII wave and sometimes called a Linear S wave, because the particle motion for both inhomogeneous and homogeneous Type-II S waves will be shown to be linear.

For expression (3.3.1) to represent a solution of the equation of motion with

$$\vec{G}_0 = z_1\hat{x}_1 + z_3\hat{x}_3, \tag{3.3.34}$$

condition (3.3.8) implies that the complex constants z_1 and z_3 for the wave must be chosen in relation to the directions of propagation such that

$$\vec{K}_S \bullet \vec{G}_0 = 0. \tag{3.3.35}$$

For S wave solutions of the form (3.3.34), the definition of the cross product of two vectors implies that the complex displacement coefficient, $\vec{D}$, for a Type-II S wave is given from (3.3.11) by

$$\vec{D}_{SII} = D_{SII}\hat{x}_2 = i\vec{K}_S \bullet (z_3\hat{x}_1 - z_1\hat{x}_3)\hat{x}_2. \tag{3.3.36}$$

Expression (3.3.36) implies that the complex displacement field is given by

$$\vec{u} = i\left(\vec{K}_S\right) \bullet (z_3\hat{x}_1 - z_1\hat{x}_3)\exp\left[i\left(\omega t - \vec{K}_S \bullet \vec{r}\right)\right]\hat{x}_2 \tag{3.3.37}$$

and the corresponding physical displacement field by

$$\vec{u}_R = \left|\vec{D}_{SII}\right| \exp\left[-\vec{A}_S \bullet \vec{r}\right] \cos\left[\omega t - \vec{P}_S \bullet \vec{r} + \arg[D_{SII}]\right]\hat{x}_2. \tag{3.3.38}$$

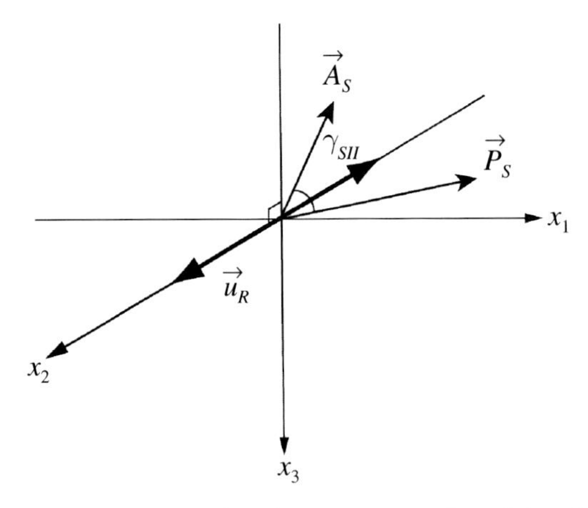

Figure (3.3.39). Diagram illustrating the parameters describing the linear particle motion orbit of a Type-II S wave in HILV media. The particle motion is linear for both homogeneous and inhomogeneous Type-II S waves in contrast to elliptical particle motion for inhomogeneous Type-I S waves.

Equation (3.3.38) shows that the particle motion for a Type-II S wave is not elliptical as it is for a Type-I S wave, but that it is linear perpendicular to the directions of propagation and attenuation. The particle motion for a Type-II S wave is linear parallel to the x_2 axis for both homogeneous and inhomogeneous waves. Parameters for the particle motion of a Type-II S wave are illustrated in Figure (3.3.39). This important difference in the nature of the particle motions for the two types of S waves, which may propagate independently in a HILV medium, has significant implications for interpretation and measurement of S waves in a HILV medium.

3.4 Energy Characteristics of P, SI, and SII Waves

Energy characteristics of a viscoelastic radiation field are expressed in terms of the corresponding physical displacement field by the scalar equations (2.5.4) through (2.5.7) or by the vector equations (2.5.8) through (2.5.11). Substitution of the expressions for the physical displacement fields for P, SI, and SII waves into these expressions provides explicit expressions for the energy characteristics of each wave type in terms of the wave's basic parameters, namely amplitude, propagation vector, attenuation vector, frequency, and the material parameters These energy expressions for each wave type are derived in subsequent sections (Borcherdt, 1973a, 1977; Borcherdt and Wennerberg, 1985).

3.4.1 Mean Energy Flux (Mean Intensity)

The energy flux or intensity for a steady-state wave field is specified in terms of the physical displacement field for the wave by (2.5.7) or (2.5.11). Substitution of expressions (3.2.11) and (3.3.22) for the physical displacement fields for P and Type-I S waves, respectively, into (2.5.7) or (2.5.11) yields after some algebra the following expressions for the energy flux time averaged over one cycle for a P wave as

$$\left\langle \vec{\mathcal{J}}_P \right\rangle = |\vec{G}_0|^2 \exp\left[-2\vec{A}_P \cdot \vec{r}\right] \frac{\omega}{2} \left(\rho\omega^2 \vec{P}_P + 4\left(\vec{P}_P \times \vec{A}_P\right) \times \left(M_I \vec{P}_P - M_R \vec{A}_P\right)\right) \tag{3.4.1}$$

and for a Type-I S wave as

$$\left\langle \vec{\mathcal{J}}_{SI} \right\rangle = |\vec{G}_0|^2 \exp\left[-2\vec{A}_S \cdot \vec{r}\right] \frac{\omega}{2} \left(\rho\omega^2 \vec{P}_S + 4\left(\vec{P}_S \times \vec{A}_S\right) \times \left(M_I \vec{P}_S - M_R \vec{A}_S\right)\right), \tag{3.4.2}$$

where the time average of a function $f(t)$ over one cycle of period $T = 2\pi/\omega$ is denoted by "$\langle\ \rangle$" and defined by

$$\langle f(t) \rangle \equiv \frac{1}{T} \int_0^T f(t)\, dt. \tag{3.4.3}$$

Equations (3.4.1) and (3.4.2) express the time-averaged energy flux for P and Type-I S waves in terms of the propagation and attenuation vectors for the respective waves. The expressions for P and Type-I S waves are of the same form with the expression for one wave derivable from that for the other by substitution of the correct propagation and attenuation vector and the appropriate interpretation of $\vec{G}_0$.

Similarly, substitution of expression (3.3.38) for the physical displacement field of a Type-II S wave into (2.5.7) or (2.5.11) yields the expression for the mean energy flux of a Type-II S wave, namely

$$\left\langle \vec{\mathcal{J}}_{SII} \right\rangle = \frac{\left|\vec{D}_{SII}\right|^2}{h_S} \exp\left[-2\vec{A}_S \cdot \vec{r}\right] \frac{\omega}{2} \left(\rho\omega^2 \vec{P}_S + 2\left(\vec{P}_S \times \vec{A}_S\right) \times \left(M_I \vec{P}_S - M_R \vec{A}_S\right)\right), \tag{3.4.4}$$

where

$$h_S \equiv \left|\vec{P}_S\right|^2 + \left|\vec{A}_S\right|^2. \tag{3.4.5}$$

This expression for a Type-II S wave may be rewritten as

$$\left\langle \vec{\mathcal{J}}_{SII} \right\rangle = \left|\vec{D}_{SII}\right|^2 \exp\left[-2\vec{A}_S \cdot \vec{r}\right] \frac{\omega}{2} \left(M_R \vec{P}_S + M_I \vec{A}_S\right). \tag{3.4.6}$$

Comparison of equations (3.4.2) and (3.4.4) shows that the mean energy flux for an inhomogeneous Type-I S wave is not equal to that of a corresponding inhomogeneous Type-II S wave. For inhomogeneous waves, examination of the expressions for the mean energy flux shows that the direction of maximum energy flow does not coincide with the direction of phase propagation or the direction of maximum attenuation, but instead is at some intermediate direction between those of phase propagation and maximum attenuation for each wave type.

For homogeneous waves, $\vec{P} \times \vec{A} = 0$, so the expressions for the mean energy flux simplify for

a P wave,

$$\left\langle \vec{\mathcal{I}}_{HP} \right\rangle = |G_0|^2 \exp\left[-2\vec{A}_P \bullet \vec{r}\right] \frac{\omega}{2} \left(\rho\omega^2 \vec{P}_P\right), \tag{3.4.7}$$

an SI wave,

$$\left\langle \vec{\mathcal{I}}_{HSI} \right\rangle = |\vec{G}_0|^2 \exp\left[-2\vec{A}_S \bullet \vec{r}\right] \frac{\omega}{2} \left(\rho\omega^2 \vec{P}_S\right), \tag{3.4.8}$$

and an SII wave,

$$\left\langle \vec{\mathcal{I}}_{HSII} \right\rangle = \frac{\left|\vec{D}_{SII}\right|^2}{h_S} \exp\left[-2\vec{A}_S \bullet \vec{r}\right] \frac{\omega}{2} \left(\rho\omega^2 \vec{P}_S\right). \tag{3.4.9}$$

For homogeneous waves equations (3.4.7) through (3.4.9) show that the direction of mean energy flux coincides with the direction of phase propagation.

For homogeneous S waves

$$h_{HS} = \left|\vec{P}_S\right|^2 + |\vec{A}_S|^2 = k_{S_R}^2 + k_{S_I}^2 = |k_S|^2. \tag{3.4.10}$$

Hence, (3.3.14) implies for homogeneous SII waves that

$$\frac{\left|\vec{D}_{SII}\right|^2}{h_{HS}} = |\vec{G}_0|^2 \tag{3.4.11}$$

showing that the mean energy flux of a homogeneous Type-II S wave equals that of a corresponding homogeneous Type-I S wave as expected. The equations for the mean energy flux in terms of the propagation and attenuation vectors for each of the respective homogeneous wave types are of the same form.

The simple linear particle motion of an SII wave allows some simple expressions for the energy characteristics of the wave to be readily derived. The steady-state displacement field for a Type-II S wave as specified by (3.3.38) may be expressed simply as

$$\vec{u}_R = u_R\,\hat{x}_2, \tag{3.4.12}$$

where

$$\frac{\partial u_R}{\partial x_2} = u_{R,2} = 0. \tag{3.4.13}$$

The general expressions, which describe the energy characteristics of steady-state viscoelastic radiation fields (2.5.4) through (2.5.7), simplify for a Type-II S wave as specified by (3.4.12) to

$$\mathscr{K}_{SII} = \frac{1}{2}\rho \dot{u}_R^2, \tag{3.4.14}$$

$$\mathscr{P}_{SII} = \frac{1}{2} M_R \left(u_{R,1}^2 + u_{R,3}^2 \right), \tag{3.4.15}$$

$$\mathscr{D}_{SII} = \frac{1}{\omega} M_I \left(\dot{u}_{R,1}^2 + \dot{u}_{R,3}^2 \right), \tag{3.4.16}$$

and

$$\mathscr{I}_{SII} = -\dot{u}_R \left(M_R \left(u_{R,1}\, \hat{x}_1 + u_{R,3}\, \hat{x}_3 \right) + \frac{1}{\omega} M_I \left(\dot{u}_{R,1}\, \hat{x}_1 + \dot{u}_{R,3}\, \hat{x}_3 \right) \right). \tag{3.4.17}$$

Components of the physical stress tensor for a displacement field of the form (3.4.12) are given from (2.3.5) and (2.3.12) by

$$p_{R12} = M_R\, u_{R,1} + \frac{1}{\omega} M_I\, \dot{u}_{R,1} \tag{3.4.18}$$

and

$$p_{R32} = M_R\, u_{R,3} + \frac{1}{\omega} M_I\, \dot{u}_{R,3}, \tag{3.4.19}$$

where the physical stress tensor is given by the real part denoted by p_{Rij}.

Hence, the mean energy flux for a Type-II S wave may be written in terms of the velocity field and the components of the stress tensor as

$$\langle \mathscr{I}_{SII} \rangle = -(\langle \dot{u}_R\, p_{R12} \rangle \hat{x}_1 + \langle \dot{u}_R\, p_{R32} \rangle \hat{x}_3). \tag{3.4.20}$$

Substitution of the physical displacement field as specified by (3.3.38) into (3.4.18) and (3.4.19) implies that the components of the physical stress tensor may be written in terms of the parameters of an SII wave as

$$p_{R12} = |D_{SII}| \exp\left[-\vec{A}_S \bullet \vec{r} \right] \left(G \vec{P}_S \bullet \hat{x}_1 + H \vec{A}_S \bullet \hat{x}_1 \right) \tag{3.4.21}$$

and

$$p_{R32} = |D_{SII}| \exp\left[-\vec{A}_S \bullet \vec{r} \right] \left(G \vec{P}_S \bullet \hat{x}_3 + H \vec{A}_S \bullet \hat{x}_3 \right), \tag{3.4.22}$$

where

$$G \equiv M_R \sin \Omega + M_I \cos \Omega, \tag{3.4.23}$$

$$H \equiv -M_R \cos \Omega + M_I \sin \Omega, \tag{3.4.24}$$

and

$$\Omega \equiv \omega t - \vec{P}_S \bullet \vec{r} + \arg[D_{SII}]. \tag{3.4.25}$$

Substitution of (3.4.21) through (3.4.25) and (3.3.38) into (3.4.20) and taking time averages over a cycle yields (3.4.6). This independent derivation of the expression for the mean energy flux of a SII wave provides explicit expressions for the components of the physical stress tensor for a SII wave. These equations will be useful in later derivations concerning the nature of energy flow at viscoelastic boundaries.

3.4.2 Mean Energy Densities

The kinetic and potential energy densities per unit time for a steady-state viscoelastic radiation field are given by (2.5.4) and (2.5.5) or by (2.5.8) and (2.5.9). Substitution of expressions (3.2.11), (3.3.22), and (3.3.38) for the physical displacement fields for P, Type-I S, and Type-II S waves, respectively yields after some algebra the expression for the kinetic energy density time averaged over one cycle for a

P wave,

$$\langle \mathcal{K}_P \rangle = |G_0|^2 \exp\left[-2\vec{A}_P \bullet \vec{r}\right] \left(\frac{\rho\omega^2}{4}\left(\left|\vec{P}_P\right|^2 + \left|\vec{A}_P\right|^2\right)\right), \tag{3.4.26}$$

SI wave,

$$\langle \mathcal{K}_{SI} \rangle = |\vec{G}_0|^2 \exp[-2\vec{A}_S \bullet \vec{r}] \left(\frac{\rho\omega^2}{4}\left(\left|\vec{P}_S\right|^2 + \left|\vec{A}_S\right|^2\right)\right), \tag{3.4.27}$$

and SII wave

$$\langle \mathcal{K}_{SII} \rangle = \frac{\left|\vec{D}_{SII}\right|^2}{h_S} \exp\left[-2\vec{A}_S \bullet \vec{r}\right] \left(\frac{\rho\omega^2}{4}\left(\left|\vec{P}_S\right|^2 + \left|\vec{A}_S\right|^2\right)\right). \tag{3.4.28}$$

Similarly, substitution of the physical displacement fields for each wave into the expressions for the potential energy density per unit time for a steady-state wave field yields the expression for the time-averaged potential energy density for a

P wave,

$$\langle \mathscr{P}_P \rangle = |G_0|^2 \exp\left[-2\vec{A}_P \bullet \vec{r}\right] \frac{1}{4} \left(\rho\omega^2 \left(|\vec{P}_P|^2 - |\vec{A}_P|^2\right) + 8M_R |\vec{P}_P \times \vec{A}_P|^2\right), \tag{3.4.29}$$

SI wave,

$$\langle \mathscr{P}_{SI} \rangle = |\vec{G}_0|^2 \exp\left[-2\vec{A}_S \bullet \vec{r}\right] \frac{1}{4} \left(\rho\omega^2 \left(|\vec{P}_S|^2 - |\vec{A}_S|^2\right) + 8M_R |\vec{P}_S \times \vec{A}_S|^2\right), \tag{3.4.30}$$

and SII wave,

$$\langle \mathscr{P}_{SII} \rangle = \frac{|\vec{D}_{SII}|^2}{h_S} \exp\left[-2\vec{A}_S \bullet \vec{r}\right] \frac{1}{4} \left(\rho\omega^2 \left(|\vec{P}_S|^2 - |\vec{A}_S|^2\right) + 4M_R |\vec{P}_S \times \vec{A}_S|^2\right). \tag{3.4.31}$$

For inhomogeneous waves in elastic and anelastic media and homogeneous waves in anelastic media, $|\vec{A}| \neq 0$, so comparison of expressions (3.4.26) through (3.4.28) with corresponding expressions (3.4.29) through (3.4.31) shows that for such waves the mean kinetic energy density is not equal to the mean potential energy density. For homogeneous waves in elastic media, however, $|\vec{A}| = 0$, so the above expressions do imply the familiar result that for elastic homogeneous waves the mean kinetic energy density equals the mean potential energy density.

Expressions (3.4.30) and (3.4.31) show that the mean potential energy density for an inhomogeneous Type-I S wave is not equal to that of a corresponding Type-II S wave.

For homogeneous waves $\vec{P} \times \vec{A} = 0$, so these expressions for each wave type simplify to

$$\langle \mathscr{P}_{HP} \rangle = |G_0|^2 \exp\left[-2\vec{A}_P \bullet \vec{r}\right] \frac{1}{4} \left(\rho\omega^2 \left(|\vec{P}_P|^2 - |\vec{A}_P|^2\right)\right), \tag{3.4.32}$$

$$\langle \mathscr{P}_{HSI} \rangle = |\vec{G}_0|^2 \exp\left[-2\vec{A}_S \bullet \vec{r}\right] \frac{1}{4} \left(\rho\omega^2 \left(|\vec{P}_S|^2 - |\vec{A}_S|^2\right)\right), \tag{3.4.33}$$

and

$$\langle \mathscr{P}_{HSII} \rangle = |\vec{G}_0|^2 \exp\left[-2\vec{A}_S \bullet \vec{r}\right] \frac{1}{4} \left(\rho\omega^2 \left(|\vec{P}_S|^2 - |\vec{A}_S|^2\right)\right), \tag{3.4.34}$$

where (3.4.34) follows from (3.4.10) and (3.4.11). Expressions (3.4.33) and (3.4.34) show that the mean potential energy density for a homogeneous Type-I S wave is equal to that of a corresponding homogeneous Type-II S wave.

The mean total energy density for a wave field, denoted by $\langle \mathcal{E} \rangle$, is defined as the sum of the mean kinetic and potential energy densities, that is,

$$\langle \mathcal{E} \rangle \equiv \langle \mathcal{K} \rangle + \langle \mathcal{P} \rangle. \tag{3.4.35}$$

Substitution of the corresponding expressions for the mean kinetic and potential energy densities for each wave type into (3.4.35) yields for a

P wave,

$$\langle \mathcal{E}_P \rangle = |G_0|^2 \exp\left[-2\vec{A}_P \bullet \vec{r}\right] \frac{1}{2} \left(\rho\omega^2 \left|\vec{P}_P\right|^2 + 4M_R \left|\vec{P}_P \times \vec{A}_P\right|^2 \right), \tag{3.4.36}$$

SI wave,

$$\langle \mathcal{E}_{SI} \rangle = \left|\vec{G}_0\right|^2 \exp\left[-2\vec{A}_S \bullet \vec{r}\right] \frac{1}{2} \left(\rho\omega^2 \left|\vec{P}_S\right|^2 + 4M_R \left|\vec{P}_S \times \vec{A}_S\right|^2 \right), \tag{3.4.37}$$

and SII wave

$$\langle \mathcal{E}_{SII} \rangle = \frac{\left|\vec{D}_{SII}\right|^2}{h_S} \exp\left[-2\vec{A}_S \bullet \vec{r}\right] \frac{1}{2} \left(\rho\omega^2 \left|\vec{P}_S\right|^2 + 2M_R \left|\vec{P}_S \times \vec{A}_S\right|^2 \right). \tag{3.4.38}$$

Comparison of the equations for the mean energy flux (3.4.1) through (3.4.4) with those for the mean total energy density (3.4.36) through (3.4.38) shows that for each wave type

$$\langle \mathcal{E} \rangle = \frac{\vec{P} \bullet \left\langle \vec{\mathcal{I}} \right\rangle}{\omega}. \tag{3.4.39}$$

Hence, the total energy density for each wave type depends on the component of the mean energy flux in the direction of phase propagation.

The corresponding expressions for homogeneous waves with $\vec{P} \times \vec{A} = 0$ are for a

P wave,

$$\langle \mathcal{E}_{HP} \rangle = |G_0|^2 \exp\left[-2\vec{A}_P \bullet \vec{r}\right] \frac{1}{2} \left(\rho\omega^2 \left|\vec{P}_P\right|^2 \right), \tag{3.4.40}$$

SI wave,

$$\langle \mathcal{E}_{HSI} \rangle = \left|\vec{G}_0\right|^2 \exp\left[-2\vec{A}_S \bullet \vec{r}\right] \frac{1}{2} \left(\rho\omega^2 \left|\vec{P}_S\right|^2 \right), \tag{3.4.41}$$

and SII wave,

$$\langle \mathcal{E}_{HSII} \rangle = \left|\vec{G}_0\right|^2 \exp\left[-2\vec{A}_S \bullet \vec{r}\right] \frac{1}{2} \left(\rho\omega^2 \left|\vec{P}_S\right|^2 \right). \tag{3.4.42}$$

The expressions for the total energy density for homogeneous waves are of the same form for each wave type. They show that the total energy density of a homogeneous Type-I S wave is equal to that of a homogeneous Type-II S wave.

3.4.3 Energy Velocity

Energy velocity, denoted by $\vec{v}_E$, for a viscoelastic radiation field is defined as the ratio of the mean energy flux to the mean total energy density, that is,

$$\vec{v}_E \equiv \frac{\left\langle \vec{\mathcal{J}} \right\rangle}{\langle \mathcal{E} \rangle}. \tag{3.4.43}$$

Substitution of the corresponding expressions for the mean energy flux (3.4.1) through (3.4.4) and mean total energy density (3.4.36) through (3.4.38) yields the energy velocity for a

P wave,

$$\vec{v}_{EP} = \frac{\omega\left(\rho\omega^2 \vec{P}_P + 4\left(\vec{P}_P \times \vec{A}_P\right) \times \left(M_I \vec{P}_P - M_R \vec{A}_P\right)\right)}{\left(\rho\omega^2 \left|\vec{P}_P\right|^2 + 4M_R \left|\vec{P}_P \times \vec{A}_P\right|^2\right)}, \tag{3.4.44}$$

SI wave,

$$\vec{v}_{ESI} = \frac{\omega\left(\rho\omega^2 \vec{P}_S + 4\left(\vec{P}_S \times \vec{A}_S\right) \times \left(M_I \vec{P}_S - M_R \vec{A}_S\right)\right)}{\left(\rho\omega^2 \left|\vec{P}_S\right|^2 + 4M_R \left|\vec{P}_S \times \vec{A}_S\right|^2\right)}, \tag{3.4.45}$$

and SII wave,

$$\vec{v}_{ESII} = \frac{\omega\left(\rho\omega^2 \vec{P}_S + 2\left(\vec{P}_S \times \vec{A}_S\right) \times \left(M_I \vec{P}_S - M_R \vec{A}_S\right)\right)}{\left(\rho\omega^2 \left|\vec{P}_S\right|^2 + 2M_R \left|\vec{P}_S \times \vec{A}_S\right|^2\right)}. \tag{3.4.46}$$

Comparison of these expressions for the energy velocity with the corresponding expressions for phase velocity indicated by (3.1.15) shows that the energy velocity is not equal to the phase velocity in either direction or amplitude for inhomogeneous anelastic waves. In addition, equations (3.4.45) and (3.4.46) show that the energy velocity for an inhomogeneous anelastic Type-I S wave is not equal to that of a corresponding Type-II S wave.

For homogeneous waves, $\vec{P} \times \vec{A} = 0$ and the expression for energy velocity for each wave type reduces to

$$\vec{v}_{EP} = \frac{\omega \vec{P}_P}{\left|\vec{P}_P\right|^2} = \vec{v}_P, \tag{3.4.47}$$

$$\vec{v}_{ESI} = \frac{\omega \vec{P}_S}{\left|\vec{P}_S\right|^2} = \vec{v}_S, \tag{3.4.48}$$

and

$$\vec{v}_{ESII} = \frac{\omega \vec{P}_S}{\left|\vec{P}_S\right|^2} = \vec{v}_S, \tag{3.4.49}$$

showing that for each type of homogeneous wave the energy velocity equals the phase velocity.

3.4.4 Mean Rate of Energy Dissipation

The rate of energy dissipation for a harmonic radiation field is given by (2.5.6) or (2.5.10). Substitution of the expression for the physical displacement field for each wave type into these expressions and taking time averages yields the following expressions for the time-averaged rate of energy dissipation for a

P wave,

$$\langle \mathscr{D}_P \rangle = \left|G_0\right|^2 \exp\left[-2\vec{A}_P \cdot \vec{r}\right] \omega \left(\rho\omega^2 \vec{P}_P \cdot \vec{A}_P + 4M_I \left|\vec{P}_P \times \vec{A}_P\right|^2\right), \tag{3.4.50}$$

SI wave,

$$\langle \mathscr{D}_{SI} \rangle = \left|\vec{G}_0\right|^2 \exp\left[-2\vec{A}_S \cdot \vec{r}\right] \omega \left(\rho\omega^2 \vec{P}_S \cdot \vec{A}_S + 4M_I \left|\vec{P}_S \times \vec{A}_S\right|^2\right), \tag{3.4.51}$$

and SII wave,

$$\langle \mathscr{D}_{SII} \rangle = \frac{\left|\vec{D}_{SII}\right|^2}{h_S} \exp\left[-2\vec{A}_S \cdot \vec{r}\right] \omega \left(\rho\omega^2 \vec{P}_S \cdot \vec{A}_S + 2M_I \left|\vec{P}_S \times \vec{A}_S\right|^2\right). \tag{3.4.52}$$

Comparison of the equations for the mean energy flux (3.4.1) through (3.4.4) with those for the mean rate of energy dissipation (3.4.50) through (3.4.52) shows that for each wave type

$$\langle \mathscr{D} \rangle = -\nabla \cdot \langle \vec{\mathscr{J}} \rangle = 2\vec{A} \cdot \langle \vec{\mathscr{J}} \rangle. \tag{3.4.53}$$

Hence, the mean rate of energy dissipation for each wave type depends on the component of the mean intensity in the direction of maximum attenuation.

Equations (3.4.51) and (3.4.52) show that the mean rate of energy dissipation of inhomogeneous Type-I S waves in anelastic media is not equal to that of a corresponding Type-II S wave.

For homogeneous waves with $\vec{P} \times \vec{A} = 0$ the expressions simplify to

$$\langle \mathcal{D}_{HP} \rangle = |G_0|^2 \exp\left[-2\vec{A}_P \bullet \vec{r}\right] \rho\omega^3 |\vec{P}_P||\vec{A}_P|, \tag{3.4.54}$$

$$\langle \mathcal{D}_{HSI} \rangle = |\vec{G}_0|^2 \exp\left[-2\vec{A}_S \bullet \vec{r}\right] \rho\omega^3 |\vec{P}_S||\vec{A}_S|, \tag{3.4.55}$$

and

$$\langle \mathcal{D}_{HSII} \rangle = |\vec{G}_0|^2 \exp\left[-2\vec{A}_S \bullet \vec{r}\right] \rho\omega^3 |\vec{P}_S||\vec{A}_S|. \tag{3.4.56}$$

Equations (3.4.55) and (3.4.56) show that the mean rates of energy dissipation for the two types of S wave are equal if the waves are homogeneous.

3.4.5 Reciprocal Quality Factor, Q^{-1}

A dimensionless parameter useful for describing energy loss is the reciprocal quality factor denoted by Q^{-1}. It is defined here as the ratio of the loss in energy density per cycle of forced oscillation to the peak energy density stored during the cycle with the ratio normalized by 2π. Substitution of the expression for the physical displacement field for each wave type into (2.5.6) or (2.5.10) shows that the energy loss per cycle of forced oscillation for each wave type is

$$\frac{\Delta\mathcal{E}_P}{cycle} = \frac{4\pi\vec{A}_P \bullet \langle \vec{\mathcal{I}}_P \rangle}{\omega}, \tag{3.4.57}$$

$$\frac{\Delta\mathcal{E}_{SI}}{cycle} = \frac{4\pi\vec{A}_S \bullet \langle \vec{\mathcal{I}}_{SI} \rangle}{\omega}, \tag{3.4.58}$$

and

$$\frac{\Delta\mathcal{E}_{SII}}{cycle} = \frac{4\pi\vec{A}_S \bullet \langle \vec{\mathcal{I}}_{SII} \rangle}{\omega}. \tag{3.4.59}$$

The peak energy density stored during a cycle is given by the maximum of the potential energy density during the cycle. Substitution of the physical displacement field for each wave type into (2.5.5) or (2.5.9) yields the following expressions for the peak energy density stored during a cycle for each wave type:

$$\max[\mathscr{P}_P] = \frac{1}{2}|G_0|^2 \exp\left[-2\vec{A}_P \bullet \vec{r}\right] g_P, \tag{3.4.60}$$

$$\max[\mathscr{P}_{SI}] = \frac{1}{2}\left|\vec{G}_0\right|^2 \exp\left[-2\vec{A}_S \bullet \vec{r}\right] g_{SI}, \tag{3.4.61}$$

and

$$\max[\mathscr{P}_{SII}] = \frac{\left|\vec{D}_{SII}\right|^2}{4h_S} \exp\left[-2\vec{A}_S \bullet \vec{r}\right]\left(g_{SII} + \sqrt{g_{SII}\rho\omega^2\left(\left|\vec{P}_S\right|^2 - \left|\vec{A}_S\right|^2\right)}\right), \tag{3.4.62}$$

where

$$g_P \equiv \rho\omega^2\left(\left|\vec{P}_P\right|^2 - |A_P|^2 + 4M_R\left|\vec{P}_P \times \vec{A}_P\right|^2\right), \tag{3.4.63}$$

$$g_{SI} \equiv \rho\omega^2\left(\left|\vec{P}_S\right|^2 - |A_S|^2 + 4M_R\left|\vec{P}_S \times \vec{A}_S\right|^2\right), \tag{3.4.64}$$

and

$$g_{SII} \equiv \rho\omega^2\left(\left|\vec{P}_S\right|^2 - |A_S|^2 + 2M_R\left|\vec{P}_S \times \vec{A}_S\right|^2\right). \tag{3.4.65}$$

Using the definition for the reciprocal quality factor, namely

$$Q^{-1} \equiv \frac{1}{2\pi}\frac{\Delta\mathcal{E}/cycle}{\max[\mathscr{P}]}, \tag{3.4.66}$$

the corresponding expressions for the reciprocal quality factor for each wave type are

$$Q_P^{-1} = \frac{1}{g_P}\left(\rho\omega^2 2\vec{P}_P \bullet \vec{A}_P + 8M_I\left|\vec{P}_P \times \vec{A}_P\right|^2\right), \tag{3.4.67}$$

$$Q_{SI}^{-1} = \frac{1}{g_{SI}}\left(\rho\omega^2 2\vec{P}_S \bullet \vec{A}_S + 8M_I\left|\vec{P}_S \times \vec{A}_S\right|^2\right), \tag{3.4.68}$$

and

$$Q_{SII}^{-1} = \frac{4\left(\rho\omega^2\vec{P}_S \bullet \vec{A}_S + 2M_I\left|\vec{P}_S \times \vec{A}_S\right|^2\right)}{g_{SII} + \sqrt{g_{SII}\rho\omega^2\left(\left|\vec{P}_S\right|^2 - \left|\vec{A}_S\right|^2\right)}}. \tag{3.4.69}$$

For inhomogeneous waves in anelastic media expressions (3.4.68) and (3.4.69) show that the reciprocal quality factors for the two types of S waves are not equal, that is

$$Q_{SI}^{-1} \neq Q_{SII}^{-1}. \tag{3.4.70}$$

For homogeneous wave fields with $\vec{P} \times \vec{A} = 0$ the expressions for the reciprocal quality factor reduce to

$$Q_{HP}^{-1} = \frac{2|\vec{P}_P||\vec{A}_P|}{|\vec{P}_P|^2 - |\vec{A}_P|^2}, \tag{3.4.71}$$

$$Q_{HSI}^{-1} = \frac{2|\vec{P}_S||\vec{A}_S|}{|\vec{P}_S|^2 - |\vec{A}_S|^2}, \tag{3.4.72}$$

and

$$Q_{HSII}^{-1} = \frac{2|\vec{P}_S||\vec{A}_S|}{|\vec{P}_S|^2 - |\vec{A}_S|^2}. \tag{3.4.73}$$

Equations (3.4.72) and (3.4.73) show that the reciprocal quality factors for homogeneous Type-I and Type-II S waves are equal and hence the subscript distinguishing the quantities for the two types of homogeneous S waves can be dropped. Henceforth the reciprocal quality factor for homogeneous S waves shall be denoted by Q_{HS}^{-1}.

Expressions (3.1.11) though (3.4.73) provide a through description of the physical characteristics of harmonic P, Type-I S, and Type-II S waves in a general HILV medium in terms of their respective propagation and attenuation vectors whose characteristics are determined by properties of the material as characterized by complex bulk and shear moduli. These expressions will be used in subsequent sections to consider alternative characterizations of material response in terms of characteristics of homogeneous waves, to derive expressions for the physical characteristics of inhomogeneous waves in terms of those for homogeneous waves, and to derive explicit expressions for the physical characteristics of general waves in HILV media with small amounts of intrinsic absorption.

3.5 Viscoelasticity Characterized by Parameters for Homogeneous P and S Waves

Physical characteristics of P, SI, and SII waves as derived in previous sections are based on characterization of the material response using complex bulk and shear moduli, K and M. Additional insight is provided by deriving here the expressions showing that the phase speeds and reciprocal quality factors for homogeneous P and S waves also can be used to characterize the response of the material (Borcherdt and Wennerberg, 1985).

The reciprocal quality factors for homogeneous P and S waves as given by (3.4.71) through (3.4.73) may be expressed in terms of the real and imaginary parts of the complex moduli using (3.1.13), (3.1.14), and (3.1.5) through (3.1.8) as

$$Q_{HP}^{-1} = \frac{-\text{Im}\left[k_P^2\right]}{\text{Re}\left[k_P^2\right]} = \frac{K_I + \frac{4}{3}M_I}{K_R + \frac{4}{3}M_R} \tag{3.5.1}$$

and

$$Q_{HS}^{-1} = \frac{-\text{Im}\left[k_S^2\right]}{\text{Re}\left[k_S^2\right]} = \frac{M_I}{M_R}. \tag{3.5.2}$$

The phase speed for a homogeneous P wave as specified from (3.1.15) and (3.1.20) is

$$v_{HP} = \frac{\omega}{\left|\vec{P}_P\right|} = 2\omega\Bigg/\sqrt{\text{Re}\left[k_P^2\right]\left(1 + \sqrt{1 + \frac{\text{Im}\left[k_P^2\right]}{\text{Re}\left[k_P^2\right]}}\right)}. \tag{3.5.3}$$

Using the definition (3.1.5) of k_P and substitution of (3.5.1) the wave speed for a homogeneous P wave may be written as

$$v_{HP} = \sqrt{\left(\frac{K_R + \frac{4}{3}M_R}{\rho}\right)\frac{2\left(1 + Q_{HP}^{-2}\right)}{1 + \sqrt{1 + Q_{HP}^{-2}}}}. \tag{3.5.4}$$

Similarly, the wave speed for a homogeneous S wave may be written as

$$v_{HS} = \sqrt{\left(\frac{M_R}{\rho}\right)\frac{2\left(1 + Q_{HS}^{-2}\right)}{1 + \sqrt{1 + Q_{HS}^{-2}}}}. \tag{3.5.5}$$

For elastic media, $K_I = M_I = Q_{HP}^{-1} = Q_{HS}^{-1} = 0$, the expressions for the wave speeds reduce to the familiar expressions for the wave speeds of elastic P and S waves, namely

$$v_{HPe} \equiv \sqrt{\frac{K_R + \frac{4}{3}M_R}{\rho}} \tag{3.5.6}$$

and

$$v_{HSe} \equiv \sqrt{\frac{M_R}{\rho}}. \tag{3.5.7}$$

The complex wave number for P and S waves defined in terms of the complex bulk and shear moduli by (3.1.5) through (3.1.8) may be written in terms of wave speed and reciprocal quality factor for the corresponding homogeneous wave using expressions (3.5.1) through (3.5.5) as

$$k_P^2 = \frac{2\omega^2}{v_{HP}^2} \frac{1 - iQ_{HP}^{-1}}{1 + \sqrt{1 + Q_{HP}^{-2}}} \tag{3.5.8}$$

and

$$k_S^2 = \frac{2\omega^2}{v_{HS}^2} \frac{1 - iQ_{HS}^{-1}}{1 + \sqrt{1 + Q_{HS}^{-2}}}. \tag{3.5.9}$$

With the complex bulk and shear moduli expressed in terms of the complex wave number for homogeneous P and S waves using (3.1.13), (3.1.14), and (3.1.5) through (3.1.8) by

$$K + \frac{4}{3}M = \frac{\rho\omega^2}{k_P^2} \tag{3.5.10}$$

and

$$M = \frac{\rho\omega^2}{k_S^2} \tag{3.5.11}$$

(3.5.8) and (3.5.9) imply that specification of v_{HP}, v_{HS}, Q_{HP}^{-1}, and Q_{HS}^{-1} specify the real and imaginary parts of the complex moduli. Hence, specification of these parameters for homogeneous P and S waves together with density specifies the steady-state response of a HILV medium for a given circular frequency ω. Specification of the response of HILV media in terms of these parameters for homogeneous P and S waves will be shown in subsequent sections to afford a number of analytic simplifications for problems regarding wave propagation in layered viscoelastic media.

3.6 Characteristics of Inhomogeneous Waves in Terms of Characteristics of Homogeneous Waves

The predominant type of P, SI, and SII waves that propagate through a stack of linear anelastic layers will be shown in later chapters to be inhomogeneous. Hence, to better understand the nature of body waves in layered anelastic media, it is of interest to describe the physical characteristics of inhomogeneous waves in more detail. This section will use the characterization of material response in terms of the parameters for homogeneous P and S waves as just presented to express the physical characteristics of inhomogeneous waves in terms of those for corresponding homogeneous waves. These expressions will afford simplifications in formulae and prove convenient for solution of subsequent problems. They will be used to derive closed-form expressions herein for the physical characteristics of inhomogeneous P, SI, and SII waves in media with small amounts of intrinsic absorption (Borcherdt and Wennerberg, 1985). The results will be shown to be of interest for wave-propagation problems in layered low-loss anelastic Earth materials.

3.6.1 Wave Speed and Maximum Attenuation

The wave speed for a general P wave as implied by (3.1.15) may be written in terms of the wave speed, the reciprocal quality factor for a homogeneous P wave, and the angle between the directions of propagation and attenuation, γ, as

$$|\vec{v}_P| = v_{HP}\sqrt{\frac{\left(1+\sqrt{1+Q_{HP}^{-2}}\right)}{1+\sqrt{1+Q_{HP}^{-2}\sec^2\gamma}}}. \tag{3.6.1}$$

Similarly, the wave speed for a general Type-I or Type-II S wave may be written in terms of the wave speed, reciprocal quality factor for a homogeneous S wave, and γ as

$$|\vec{v}_S| = v_{HS}\sqrt{\frac{\left(1+\sqrt{1+Q_{HS}^{-2}}\right)}{1+\sqrt{1+Q_{HS}^{-2}\sec^2\gamma}}}, \tag{3.6.2}$$

where (3.6.2) shows that for equivalent degrees of inhomogeneity, γ, for SI and SII waves the wave speed of an SI wave equals that of an SII wave. Hence, with this understanding the introduction of subscripts to distinguish the wave speeds for the two types of S waves is not necessary. This convention simplifies the notation and is applied to other characteristics for which equivalency of inhomogeneity implies the characteristics for SI and SII waves are equal.

The maximum attenuation for P and both types of S waves may be written in terms of that for corresponding homogeneous P and S waves as

$$\left|\vec{A}_P\right| = \left|\vec{A}_{HP}\right|\sqrt{\frac{\left(-1+\sqrt{1+Q_{HP}^{-2}\sec^2\gamma}\right)}{-1+\sqrt{1+Q_{HP}^{-2}}}} \tag{3.6.3}$$

and

$$\left|\vec{A}_S\right| = \left|\vec{A}_{HS}\right| \sqrt{\frac{\left(-1 + \sqrt{1 + Q_{HS}^{-2} \sec^2 \gamma}\right)}{-1 + \sqrt{1 + Q_{HS}^{-2}}}}, \tag{3.6.4}$$

where the corresponding maximum attenuation for homogeneous waves is given by

$$\left|\vec{A}_{HP}\right| = \frac{\omega}{v_{HP}} \frac{Q_{HP}^{-1}}{1 + \sqrt{1 + Q_{HP}^{-2}}} \tag{3.6.5}$$

and

$$\left|\vec{A}_{HS}\right| = \frac{\omega}{v_{HS}} \frac{Q_{HS}^{-1}}{1 + \sqrt{1 + Q_{HS}^{-2}}}. \tag{3.6.6}$$

Expressions (3.6.1) through (3.6.6) express the phase speed and maximum attenuation of inhomogeneous anelastic P and S waves in terms of the corresponding quantities for homogeneous waves. They show that as the degree of inhomogeneity, γ, approaches its physical limit of $\pi/2$, the phase speed decreases to zero and the maximum attenuation becomes infinite. As the amount of absorption becomes small, the phase speed of a P wave is given approximately by

$$\left|\vec{v}_P\right| \approx v_{HPe} \sqrt{\frac{2}{1 + \sqrt{1 + Q_{HP}^{-2} \sec^2 \gamma}}}, \tag{3.6.7}$$

showing that the phase speed for an inhomogeneous wave field may differ significantly from that of a homogeneous wave in materials with small amounts of absorption provided that the inhomogeneity of the wave field is sufficiently close to $\pi/2$. A similar expression may be readily derived for S waves.

The term $\sqrt{1 + Q_H^{-2} \sec^2 \gamma}$ which appears in the preceding equations is a function of the amount of inhomogeneity γ and Q_H^{-1} for the respective homogeneous wave field. Its magnitude plays a central role in determining the magnitude of the phase speed and maximum attenuation of P, SI and SII waves as specified by (3.6.1) through (3.6.4). The term plays a similar role for other characteristics of the waves. Hence, it is expedient to introduce the definitions

$$\chi_P \equiv \sqrt{1 + Q_{HP}^{-2} \sec^2 \gamma} \tag{3.6.8}$$

and

$$\chi_S \equiv \sqrt{1 + Q_{HS}^{-2} \sec^2 \gamma}. \tag{3.6.9}$$

Corresponding definitions for homogeneous P and S wave solutions are

$$\chi_{HP} \equiv \sqrt{1 + Q_{HP}^{-2}} \tag{3.6.10}$$

and

$$\chi_{HS} \equiv \sqrt{1 + Q_{HS}^{-2}}. \tag{3.6.11}$$

The above definitions show that for each wave type χ increases as either the inhomogeneity γ increases or the intrinsic material absorption increases. For elastic media, $\chi_{HP} = \chi_{HS} = 1$. For media with small amounts of absorption, which are often referred to as low-loss media, Q_{HP}^{-1} and Q_{HS}^{-1} are assumed to be much less than unity, that is, $Q_{HP}^{-1} \ll 1$ and $Q_{HS}^{-1} \ll 1$. For such materials $\chi_{HP} \approx \chi_{HS} \approx 1$ for both P and S wave solutions.

Identities for P, SI, and SII waves, involving v_{HP}, v_{HS}, Q_{HP}^{-1}, Q_{HS}^{-1}, χ_{HP}, and χ_{HS}, that are useful for subsequent derivations are

$$k_P = \frac{\omega}{v_{HP}}\left(1 - i\frac{Q_{HP}^{-1}}{1+\chi_{HP}}\right), \qquad k_S = \frac{\omega}{v_{HS}}\left(1 - i\frac{Q_{HS}^{-1}}{1+\chi_{HS}}\right), \tag{3.6.12}$$

$$\mathrm{Re}[k_P] = k_{P_R} = \frac{\omega}{v_{HP}}, \qquad \mathrm{Re}[k_S] = k_{S_R} = \frac{\omega}{v_{HS}}, \tag{3.6.13}$$

$$\mathrm{Im}[k_P] = k_{P_I} = \frac{-\omega}{v_{HP}}\frac{Q_{HP}^{-1}}{1+\chi_{HP}}, \qquad \mathrm{Im}[k_S] = k_{S_I} = \frac{-\omega}{v_{HS}}\frac{Q_{HS}^{-1}}{1+\chi_{HS}}, \tag{3.6.14}$$

$$k_P^2 = \frac{2\omega^2}{v_{HP}^2}\frac{1 - iQ_{HP}^{-1}}{1+\chi_{HP}}, \qquad k_S^2 = \frac{2\omega^2}{v_{HS}^2}\frac{1 - iQ_{HS}^{-1}}{1+\chi_{HS}} \tag{3.6.15}$$

$$\left|k_P^2\right| = \frac{2\omega^2}{v_{HP}^2}\frac{\chi_{HP}}{1+\chi_{HP}}, \qquad \left|k_S^2\right| = \frac{2\omega^2}{v_{HS}^2}\frac{\chi_{HS}}{1+\chi_{HS}}, \tag{3.6.16}$$

$$\left|\vec{P}_P\right|^2 = \frac{\omega^2}{v_{HP}^2}\frac{1+\chi_P}{1+\chi_{HP}}, \qquad \left|\vec{P}_S\right|^2 = \frac{\omega^2}{v_{HS}^2}\frac{1+\chi_S}{1+\chi_{HS}}, \tag{3.6.17}$$

$$\left|\vec{A}_P\right|^2 = \frac{\omega^2}{v_{HP}^2}\frac{\chi_P - 1}{1+\chi_{HP}}, \qquad \left|\vec{A}_S\right|^2 = \frac{\omega^2}{v_{HS}^2}\frac{\chi_S - 1}{1+\chi_{HS}}, \tag{3.6.18}$$

$$\vec{P}_P \bullet \vec{A}_P = \frac{\omega^2}{v_{HP}^2}\frac{Q_{HP}^{-1}}{1+\chi_{HP}}, \qquad \vec{P}_S \bullet \vec{A}_S = \frac{\omega^2}{v_{HS}^2}\frac{Q_{HS}^{-1}}{1+\chi_{HS}}, \tag{3.6.19}$$

$$\left|\vec{P}_P \times \vec{A}_P\right| = \frac{\omega^2}{v_{HP}^2}\frac{Q_{HP}^{-1}}{1+\chi_{HP}}\left|\tan\gamma_P\right|, \qquad \left|\vec{P}_S \times \vec{A}_S\right| = \frac{\omega^2}{v_{HS}^2}\frac{Q_{HS}^{-1}}{1+\chi_{HS}}\left|\tan\gamma_S\right|, \tag{3.6.20}$$

$$|\vec{v}_P| \equiv v_P = v_{HP}\sqrt{\frac{1+\chi_{HP}}{1+\chi_P}}, \qquad |\vec{v}_S| \equiv v_S = v_{HS}\sqrt{\frac{1+\chi_{HS}}{1+\chi_S}}, \tag{3.6.21}$$

$$|\vec{A}_{HP}| = \frac{\omega}{v_{HP}} \frac{Q_{HP}^{-1}}{1+\chi_{HP}}, \qquad |\vec{A}_{HS}| = \frac{\omega}{v_{HS}} \frac{Q_{HS}^{-1}}{1+\chi_{HS}}, \tag{3.6.22}$$

$$\begin{aligned} |\vec{A}_P| &= |\vec{A}_{HP}|\sqrt{\frac{-1+\chi_P}{-1+\chi_{HP}}} = |\vec{A}_{HP}|\frac{v_P}{v_{HP}} \sec\gamma \quad \text{for } Q_{HP}^{-1} \neq 0, \\ |\vec{A}_S| &= |\vec{A}_{HS}|\sqrt{\frac{-1+\chi_S}{-1+\chi_{HS}}} = |\vec{A}_{HS}|\frac{v_S}{v_{HS}} \sec\gamma \quad \text{for } Q_{HS}^{-1} \neq 0, \end{aligned} \tag{3.6.23}$$

and

$$K_R + \frac{4}{3}M_R = \frac{\rho v_{HP}^2}{2} \frac{1+\chi_{HP}}{\chi_{HP}^2}\left(1+iQ_{HP}^{-1}\right), \qquad M_R = \frac{\rho v_{HS}^2}{2} \frac{1+\chi_{HS}}{\chi_{HS}^2}\left(1+iQ_{HS}^{-1}\right). \tag{3.6.24}$$

Proofs of these useful identities are left to the reader. Equations (3.6.12) through (3.6.14) show the following additional relations useful for later reference

$$\text{Re}\left[k_P^2\right] = k_{P_R}^2 - k_{P_I}^2 \geq 0, \qquad \text{Re}\left[k_S^2\right] = k_{S_R}^2 - k_{S_I}^2 \geq 0, \tag{3.6.25}$$

$$k_{P_R} \geq 0, \quad k_{S_R} \geq 0, \tag{3.6.26}$$

and

$$k_{P_I} \leq 0, \quad k_{S_I} \leq 0. \tag{3.6.27}$$

The expressions describing the physical characteristics of P and S waves in terms of those for homogeneous waves readily permit insight into the quantitative dependence of the characteristics on inhomogeneity and the intrinsic material absorption of the media. The phase speed and attenuation coefficient in the direction of phase propagation,$|\vec{A}|\cos\gamma$, as normalized by the corresponding quantity for a homogeneous P, SI, or SII wave are plotted as a function of the degree of inhomogeneity, γ, and various values of intrinsic material absorption as measured by Q_H^{-1} for the corresponding homogeneous wave in Figure (3.6.28). The curves show that the wave speed and attenuation coefficient in the direction of phase propagation of an inhomogeneous P, SI, or SII wave in anelastic media decreases to zero as the inhomogeneity of the wave increases toward its physical limit of $\pi/2$ regardless of the amount of intrinsic material absorption. They also show that the phase speed and attenuation coefficient of an inhomogeneous wave differ most significantly from those for homogeneous waves for materials with significant amounts of intrinsic absorption ($Q_H^{-1} > 0.1$) or for materials with small amounts of absorption,

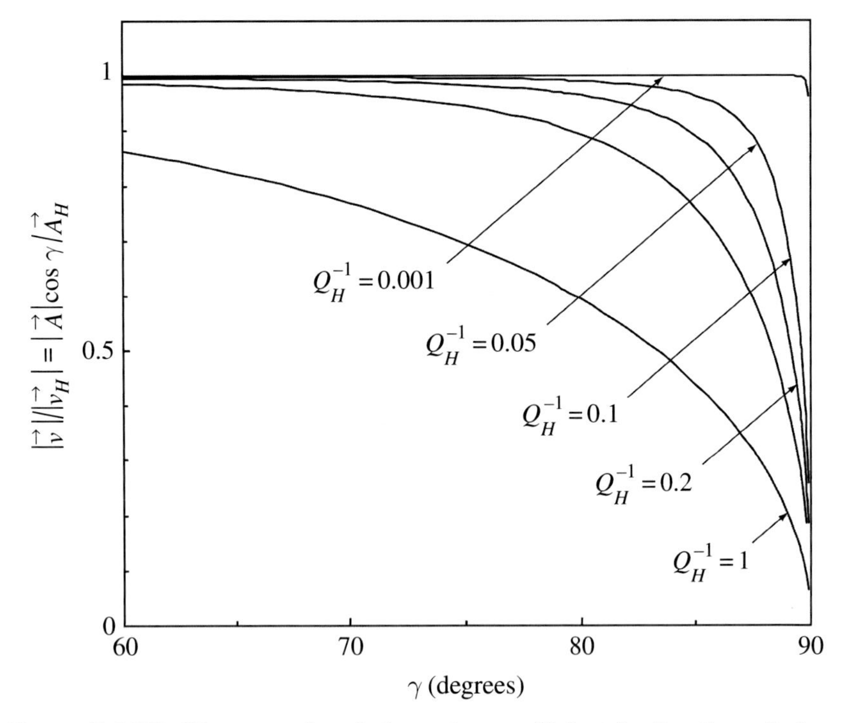

Figure (3.6.28). Phase speed and absorption coefficient in direction of phase propagation of an inhomogeneous P or S wave normalized by the corresponding quantity for a homogeneous P or S wave.

say $Q_H^{-1} > 0.1$, and larger degrees of inhomogeneity, say $\gamma > \sim 75°$. The curves show that for fixed amount of inhomogeneity the phase speed and attenuation coefficient in the direction of phase propagation decrease with increasing amounts of intrinsic material absorption.

3.6.2 Particle Motion for P and SI Waves

The particle motions for P and Type-I S waves are explicitly described by (3.2.11) and (3.3.23). The parameters describing the elliptical particle motion for inhomogeneous P and SI waves may be readily expressed in terms of χ and the velocity and intrinsic absorption for homogeneous waves using identities (3.6.12) through (3.6.19) together with expressions (3.2.12), (3.2.13), and (3.2.24) for P waves and expressions (3.3.24), (3.3.25), and (3.3.29) for SI waves.

The parameters describing the elliptical particle motion of a P wave may be rewritten for the

major axis,

$$\left|\vec{\xi}_{1P}\right|^2 = \frac{1}{2}\left(\frac{\chi_P}{\chi_{HP}} + 1\right), \tag{3.6.29}$$

minor axis,

$$\left|\vec{\xi}_{2P}\right|^2 = \frac{1}{2}\left(\frac{\chi_P}{\chi_{HP}} - 1\right), \tag{3.6.30}$$

tilt η_P,

$$\cos^2 \eta_P = \frac{1}{2}\left(\frac{\chi_{HP}+1}{\chi_{HP}}\right)\left(1 + \frac{\chi_{HP}-1}{\chi_P+1}\right), \tag{3.6.31}$$

and eccentricity,

$$e_P = \sqrt{\frac{2\chi_{HP}}{\chi_P + \chi_{HP}}}. \tag{3.6.32}$$

Similarly, the parameters describing the elliptical particle motion of an SI wave may be rewritten for the

major axis,

$$\left|\vec{\xi}_{1SI}\right|^2 = \frac{1}{2}\left(\frac{\chi_{SI}}{\chi_{HS}} + 1\right), \tag{3.6.33}$$

minor axis,

$$\left|\vec{\xi}_{2SI}\right|^2 = \frac{1}{2}\left(\frac{\chi_{SI}}{\chi_{HS}} - 1\right), \tag{3.6.34}$$

tilt η_{SI},

$$\cos^2 \eta_{SI} = \frac{1}{2}\left(\frac{\chi_{HS}+1}{\chi_{HS}}\right)\left(1 + \frac{\chi_{HS}-1}{\chi_{SI}+1}\right), \tag{3.6.35}$$

and eccentricity,

$$e_{SI} = \sqrt{\frac{2\chi_{HS}}{\chi_{SI} + \chi_{HS}}}. \tag{3.6.36}$$

Equations (3.6.29) through (3.6.36) show that the eccentricity and tilt of the particle motion ellipse for P and Type-I S waves vary with χ and hence are dependent only on the amount of intrinsic material absorption and the amount of inhomogeneity of the wave field. These equations and the corresponding calculations shown in Figures (3.6.37) and (3.6.38) show that as the amount of inhomogeneity, γ, approaches its physical limit of $\pi/2$, χ becomes infinite and the eccentricity approaches zero, indicating that the particle motion for P and SI waves becomes circular. As the amount of inhomogeneity, γ, approaches zero, χ approaches χ_H, indicating that the

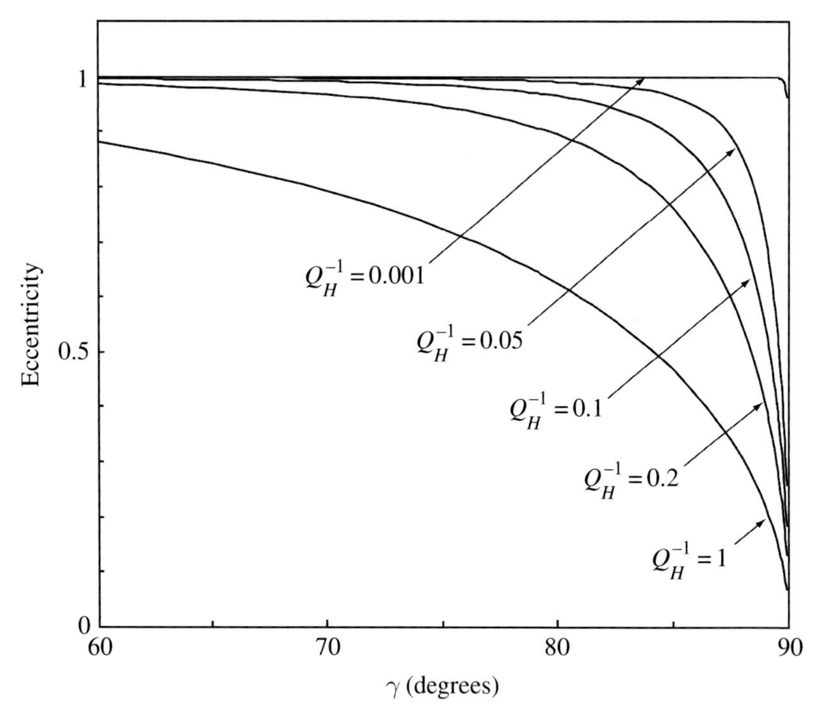

Figure (3.6.37). Eccentricity for particle motion ellipse for P or Type-I S waves as a function of inhomogeneity and intrinsic material absorption as measured by Q_H^{-1} for the corresponding homogeneous wave.

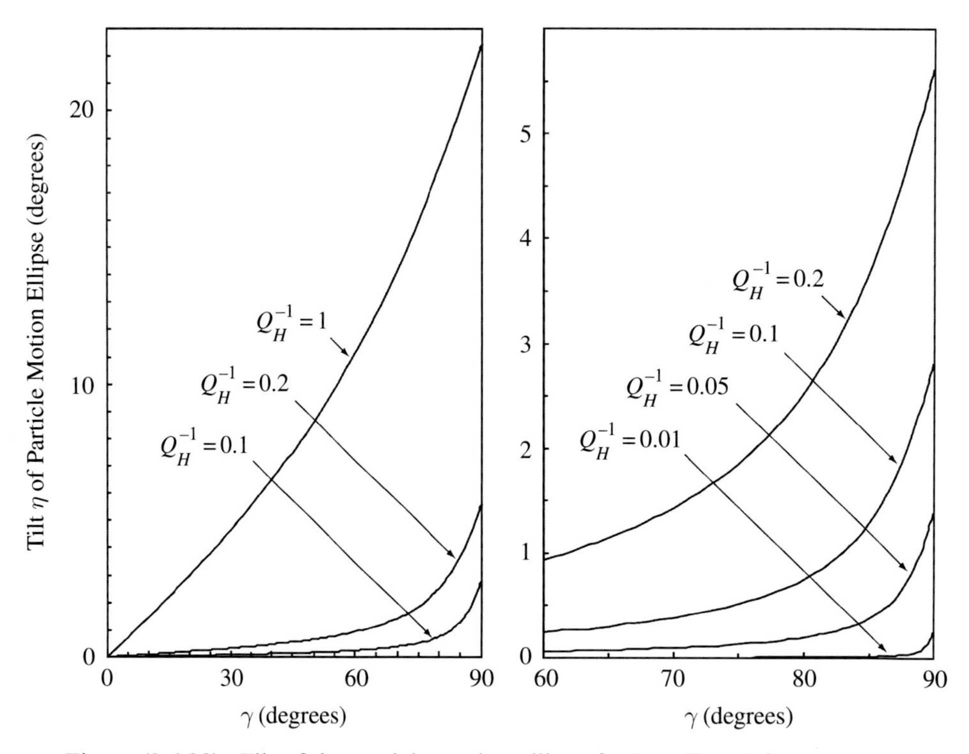

Figure (3.6.38). Tilt of the particle motion ellipse for P or Type-I S waves as a function of inhomogeneity and intrinsic material absorption as measured by Q_H^{-1} for the corresponding homogeneous wave.

elliptical motion for both wave types becomes linear with the eccentricity e approaching unity and the tilt η approaching zero. The figures show that the eccentricity and tilt of the particle motion ellipse deviate most significantly from those for corresponding homogeneous waves for materials with large amounts of intrinsic absorption ($Q_H^{-1} > 0.1$) or for waves in materials with smaller amounts of absorption (say $Q_H^{-1} < 0.1$) with larger degrees of inhomogeneity, say $\gamma > \sim 75°$.

3.6.3 Energy Characteristics for P, SI, and SII Waves

Towards developing expressions for the energy characteristics of P, SI, and SII waves in terms of corresponding expressions for homogeneous waves, the expressions for homogeneous waves as derived in preceding sections may be easily rewritten in terms of χ. The resulting expressions for homogeneous P and S waves are, respectively,

magnitude of mean intensity,

$$\left|\left\langle \vec{\mathcal{I}}_{HP}\right\rangle\right| = |G_0|^2 \exp\left[-2\vec{A}_{HP}\cdot\vec{r}\right]\frac{\rho\omega^4}{2v_{HP}},$$
$$\left|\left\langle \vec{\mathcal{I}}_{HS}\right\rangle\right| = \left|\vec{G}_0\right|^2 \exp\left[-2\vec{A}_{HS}\cdot\vec{r}\right]\frac{\rho\omega^4}{2v_{HS}}, \tag{3.6.39}$$

mean kinetic energy density,

$$\langle \mathcal{K}_{HP}\rangle = \left(\frac{\chi_{HP}}{v_{HP}(1+\chi_{HP})}\right)\left\langle \vec{\mathcal{I}}_{HP}\right\rangle, \qquad \langle \mathcal{K}_{HS}\rangle = \left(\frac{\chi_{HS}}{v_{HS}(1+\chi_{HS})}\right)\left\langle \vec{\mathcal{I}}_{HS}\right\rangle, \tag{3.6.40}$$

mean potential energy density,

$$\langle \mathcal{P}_{HP}\rangle = \left(\frac{1}{\chi_{HP}}\right)\langle \mathcal{K}_{HP}\rangle, \qquad \langle \mathcal{P}_{HS}\rangle = \left(\frac{1}{\chi_{HS}}\right)\langle \mathcal{K}_{HS}\rangle, \tag{3.6.41}$$

mean total energy density,

$$\langle \mathcal{E}_{HP}\rangle = \left(\frac{1+\chi_{HP}}{\chi_{HP}}\right)\langle \mathcal{K}_{HP}\rangle, \qquad \langle \mathcal{E}_{HS}\rangle = \left(\frac{1+\chi_{HS}}{\chi_{HS}}\right)\langle \mathcal{K}_{HS}\rangle, \tag{3.6.42}$$

mean energy density dissipated,

$$\langle \mathcal{D}_{HP}\rangle = \left(\frac{2\omega Q_{HP}^{-1}}{1+\chi_{HP}}\right)\langle \mathcal{E}_{HP}\rangle = \left(\frac{2\omega Q_{HP}^{-1}}{\chi_{HP}}\right)\langle \mathcal{K}_{HP}\rangle,$$
$$\langle \mathcal{D}_{HS}\rangle = \left(\frac{2\omega Q_{HS}^{-1}}{1+\chi_{HS}}\right)\langle \mathcal{E}_{HS}\rangle = \left(\frac{2\omega Q_{HS}^{-1}}{\chi_{HS}}\right)\langle \mathcal{K}_{HS}\rangle, \tag{3.6.43}$$

maximum potential energy density,

$$\max[\mathscr{P}_{HP}] = 2\langle\mathscr{P}_{HP}\rangle, \qquad \max[\mathscr{P}_{HS}] = 2\langle\mathscr{P}_{HS}\rangle, \tag{3.6.44}$$

and energy speed,

$$|\vec{v}_{EHP}| = v_{HP}, \qquad |\vec{v}_{EHS}| = v_{HS}, \tag{3.6.45}$$

where with the exception of the solution amplitudes ($|\vec{F}_0|^2$ is interpreted as $|\vec{G}_0|^2$ for SI waves and as $|\vec{D}_{SII}|^2/h_S$ for SII waves) the characteristics indicated for homogeneous SI waves equal those for homogeneous SII waves. Hence, subscripts distinguishing these characteristics for the two types of homogeneous S waves are not necessary.

Equations (3.6.39) through (3.6.45) provide explicit relations between the various energy characteristics of homogeneous waves in general viscoelastic media. As a check, they can be used to readily derive the familiar expressions for elastic media. Introducing the subscript "*e*" to designate terms for elastic media, the result that

$$\chi_{HPe} = \chi_{HSe} = 1 \tag{3.6.46}$$

implies for P and S waves in elastic media that

$$\left|\vec{\mathscr{I}}_{HPe}\right| = 2v_{HPe}\langle\mathscr{K}_{HPe}\rangle, \qquad \left|\vec{\mathscr{I}}_{HSe}\right| = 2v_{HSe}\langle\mathscr{K}_{HSe}\rangle, \tag{3.6.47}$$

$$\langle\mathscr{P}_{HPe}\rangle = \langle\mathscr{K}_{HPe}\rangle, \qquad \langle\mathscr{P}_{HSe}\rangle = \langle\mathscr{K}_{HSe}\rangle, \tag{3.6.48}$$

$$\langle\mathscr{E}_{HPe}\rangle = 2\langle\mathscr{K}_{HPe}\rangle, \qquad \langle\mathscr{E}_{HSe}\rangle = 2\langle K_{HSe}\rangle, \tag{3.6.49}$$

$$\langle\mathscr{D}_{HPe}\rangle = \langle\mathscr{D}_{HSe}\rangle = 0, \tag{3.6.50}$$

and

$$|\vec{v}_{HEPe}| = v_{HPe}, \qquad |\vec{v}_{HESe}| = v_{HSe}. \tag{3.6.51}$$

These expressions imply the familiar results for homogeneous P and S waves in elastic media, namely, the mean intensity of the wave equals twice the product of the wave speed and the mean kinetic energy density (3.6.47), the mean potential energy density is equal to the mean kinetic energy density (3.6.48), the total energy density is equal to twice the mean kinetic or potential energy density (3.6.49), the mean rate of energy dissipated per unit volume vanishes (3.6.50), and the energy speed is equal to the phase speed for both homogeneous P and S waves (3.6.51).

Energy characteristics for general P, SI, and SII waves as specified by equations (3.4.1) through (3.4.69) can be readily written in terms of those for homogeneous waves as specified by (3.6.39) through (3.6.45). The expressions for the magnitude of the mean energy flux for general P, SI, and SII waves are

$$\left|\langle \vec{\mathcal{J}}_P \rangle\right| = Y_P \sqrt{\frac{1+\chi_P+4F_P H_P}{1+\chi_{HP}}} \left|\langle \vec{\mathcal{J}}_{HP} \rangle\right|, \tag{3.6.52}$$

$$\left|\langle \vec{\mathcal{J}}_{SI} \rangle\right| = Y_{SI} \sqrt{\frac{1+\chi_{SI}+4F_{SI} H_{SI}}{1+\chi_{HS}}} \left|\langle \vec{\mathcal{J}}_{HS} \rangle\right|, \tag{3.6.53}$$

and

$$\left|\langle \vec{\mathcal{J}}_{SII} \rangle\right| = Y_{SII} \sqrt{\frac{1+\chi_{SII}+2F_{SII} H_{SII}}{1+\chi_{HS}}} \left|\langle \vec{\mathcal{J}}_{HS} \rangle\right|, \tag{3.6.54}$$

where Y, H, and F are defined as

$$Y_P \equiv \exp\left[-2\left(\vec{A}_P - \vec{A}_{HP}\right) \bullet \vec{r}\right], \tag{3.6.55}$$

$$Y_{SI} \equiv \exp\left[-2\left(\vec{A}_{SI} - \vec{A}_{HS}\right) \bullet \vec{r}\right], \tag{3.6.56}$$

$$Y_{SII} \equiv \exp\left[-2\left(\vec{A}_{SII} - \vec{A}_{HS}\right) \bullet \vec{r}\right], \tag{3.6.57}$$

$$H_P \equiv \frac{Q_{HP}^{-2}}{\chi_{HS}^2} \frac{v_{HS}^2}{v_{HP}^2} \frac{1+\chi_{HS}}{1+\chi_{HP}} \tan^2 \gamma, \tag{3.6.58}$$

$$H_S \equiv H_{SI} = H_{SII} = \frac{Q_{HS}^{-2}}{\chi_{HS}^2} \tan^2 \gamma, \tag{3.6.59}$$

$$F_P \equiv 1 + \frac{1}{\chi_{HS}^2} \frac{v_{HS}^2}{v_{HP}^2} \frac{1+\chi_{HS}}{1+\chi_{HP}} \left(\chi_{HS}^2 \chi_P - \left(2Q_{HS}^{-1} Q_{HP}^{-1} - Q_{HP}^{-2} + 1\right)\right), \tag{3.6.60}$$

$$\begin{aligned} F_{SI} &\equiv 1 + \frac{1}{\chi_{HS}^2} \frac{v_{HS}^2}{v_{HS}^2} \frac{1+\chi_{HS}}{1+\chi_{HS}} \left(\chi_{HS}^2 \chi_{SI} - \left(2Q_{HS}^{-1} Q_{HS}^{-1} - Q_{HS}^{-2} + 1\right)\right) \\ &= \chi_{SI}, \end{aligned} \tag{3.6.61}$$

$$\begin{aligned} F_{SII} &\equiv 1 + \frac{1}{2\chi_{HS}^2} \frac{v_{HS}^2}{v_{HS}^2} \frac{1+\chi_{HS}}{1+\chi_{HS}} \left(\chi_{HS}^2 \chi_{SII} - \left(2Q_{HS}^{-1} Q_{HS}^{-1} - Q_{HS}^{-2} + 1\right)\right) \\ &= \frac{1}{2}\left(\chi_{SII} + 1\right). \end{aligned} \tag{3.6.62}$$

Expressions for the other energy characteristics written respectively, for general P, SI, and SII waves in terms of corresponding expressions for homogeneous waves are

mean kinetic energy density,

$$\langle \mathcal{K}_P \rangle = Y_P \frac{\chi_P}{\chi_{HP}} \langle \mathcal{K}_{HP} \rangle, \tag{3.6.63}$$

$$\langle \mathcal{K}_{SI} \rangle = Y_{SI} \frac{\chi_{SI}}{\chi_{HS}} \langle \mathcal{K}_{HS} \rangle, \tag{3.6.64}$$

$$\langle \mathcal{K}_{SII} \rangle = Y_{SII} \frac{\chi_{SII}}{\chi_{HS}} \langle \mathcal{K}_{HS} \rangle, \tag{3.6.65}$$

mean potential energy density,

$$\langle \mathcal{P}_P \rangle = Y_P(1 + 2H_P) \langle \mathcal{P}_{HP} \rangle, \tag{3.6.66}$$

$$\langle \mathcal{P}_{SI} \rangle = Y_{SI}(1 + 2H_S) \langle \mathcal{P}_{HS} \rangle, \tag{3.6.67}$$

$$\langle \mathcal{P}_{SII} \rangle = Y_{SII}(1 + H_S) \langle \mathcal{P}_{HS} \rangle, \tag{3.6.68}$$

mean total energy density,

$$\langle \mathcal{E}_P \rangle = Y_P \left(\frac{1 + \chi_P + 2H_P}{1 + \chi_{HP}} \right) \langle \mathcal{E}_{HP} \rangle, \tag{3.6.69}$$

$$\langle \mathcal{E}_{SI} \rangle = Y_{SI} \left(\frac{1 + \chi_{SI} + 2H_S}{1 + \chi_{HS}} \right) \langle \mathcal{E}_{HS} \rangle, \tag{3.6.70}$$

$$\langle \mathcal{E}_{SII} \rangle = Y_{SII} \left(\frac{1 + \chi_{SII} + H_S}{1 + \chi_{HS}} \right) \langle \mathcal{E}_{HS} \rangle, \tag{3.6.71}$$

mean rate of energy dissipation per unit volume,

$$\langle \mathcal{D}_P \rangle = Y_P \left(1 + 2 \frac{Q_{HS}^{-1}}{Q_{HP}^{-1}} H_P \right) \langle \mathcal{D}_{HP} \rangle, \tag{3.6.72}$$

$$\langle \mathcal{D}_{SI} \rangle = Y_{SI} \left(1 + 2 \frac{Q_{HS}^{-1}}{Q_{HS}^{-1}} H_S \right) \langle \mathcal{D}_{HS} \rangle = Y_{SI}(1 + 2H_S) \langle \mathcal{D}_{HS} \rangle, \tag{3.6.73}$$

$$\langle \mathcal{D}_{SII} \rangle = Y_{SII} \left(1 + \frac{Q_{HS}^{-1}}{Q_{HS}^{-1}} H_S \right) \langle \mathcal{D}_{HS} \rangle = Y_{SII}(1 + H_S) \langle \mathcal{D}_{HS} \rangle, \tag{3.6.74}$$

maximum strain energy density,

$$\max[\mathcal{P}_P] = Y_P(1 + H_P) \max[\mathcal{P}_{HP}], \tag{3.6.75}$$

$$\max[\mathcal{P}_{SI}] = Y_{SI}(1 + H_S) \max[\mathcal{P}_{HS}], \tag{3.6.76}$$

$$\max[\mathcal{P}_{SII}] = \frac{Y_{SII}}{2} \left(1 + H_S + \sqrt{1 + H_S} \right) \max[\mathcal{P}_{HS}], \tag{3.6.77}$$

Q^{-1},

$$Q_P^{-1} = \left(\frac{\left(1 + 2Q_{HS}^{-1}/Q_{HP}^{-1}\right) H_P}{1 + H_P} \right) Q_{HP}^{-1}, \tag{3.6.78}$$

$$Q_{SI}^{-1} = \left(\frac{\left(1 + 2Q_{HS}^{-1}/Q_{HS}^{-1}\right) H_S}{1 + H_S} \right) Q_{HS}^{-1} = \left(\frac{1 + 2H_S}{1 + H_S} \right) Q_{HS}^{-1}, \tag{3.6.79}$$

$$Q_{SII}^{-1} = \left(\frac{2(1 + H_S)}{1 + H_S + \sqrt{1 + H_S}} \right) Q_{HS}^{-1}, \tag{3.6.80}$$

energy velocity,

$$\vec{v}_{EP} = \frac{\langle \vec{\mathcal{J}}_P \rangle}{\langle \mathcal{E}_P \rangle} = \frac{\sqrt{1 + \chi_P + 4F_P H_P}}{1 + \chi_P + 2H_P} \vec{v}_{EHP}, \tag{3.6.81}$$

$$\vec{v}_{ESI} = \frac{\langle \vec{\mathcal{J}}_{SI} \rangle}{\langle \mathcal{E}_{SI} \rangle} = \frac{\sqrt{1 + \chi_{SI} + 4F_{SI} H_S}}{1 + \chi_{SI} + 2H_S} \vec{v}_{EHSI}, \tag{3.6.82}$$

$$\vec{v}_{ESII} = \frac{\langle \vec{\mathcal{J}}_{SII} \rangle}{\langle \mathcal{E}_{SII} \rangle} = \frac{\sqrt{1 + \chi_{SII} + 2F_{SII} H_S}}{1 + \chi_{SII} + 2H_S} \vec{v}_{EHSII}. \tag{3.6.83}$$

For homogeneous waves $\gamma = 0$, so $H_P = H_S = 0$, $Y_P = Y_{SI} = Y_{SII} = 1$, and $\chi = \chi_H$ for each wave type. Substituting these identities into the preceding energy expressions confirms that each of the expressions reduces to that for the corresponding homogeneous wave.

The expressions for the energy density characteristics of general P, SI and SII waves, namely equations (3.6.63) through (3.6.74), readily permit comparison of the magnitudes of various energy characteristics of inhomogeneous waves with those of corresponding homogeneous waves. Choosing the position vector $\vec{r}$ as a unit vector in the direction of wave propagation, that is letting $\vec{r} = \vec{P}/|\vec{P}|$ for each wave type, together with (3.6.19) and (3.6.22) implies

$$Y = \exp\left[-2\left(\vec{A} - \vec{A}_H\right) \cdot \vec{r}\right] = \exp\left[-2 \frac{\omega}{v_H} \frac{Q_H^{-1}}{1 + \chi_H} \left(\frac{\sqrt{1 + \chi_H}}{\sqrt{1 + \chi}} - 1 \right)\right] \tag{3.6.84}$$

and hence $Y \geq 1$ for each wave type. This result, together with (3.6.55) through (3.6.74) implies that the mean kinetic energy density, mean potential energy density, mean energy dissipated per unit volume, and total energy density for harmonic inhomogeneous P, SI, and SII waves in HILV are greater than the corresponding energy density quantity for homogeneous waves, that is,

$$\langle \mathcal{K}_{HP} \rangle \leq \langle \mathcal{K}_P \rangle, \qquad \langle \mathcal{K}_{HSI} \rangle \leq \langle \mathcal{K}_{SI} \rangle, \qquad \langle \mathcal{K}_{HSII} \rangle \leq \langle \mathcal{K}_{SII} \rangle, \tag{3.6.85}$$

$$\langle \mathcal{P}_{HP} \rangle \leq \langle \mathcal{P}_P \rangle, \qquad \langle \mathcal{P}_{HSI} \rangle \leq \langle \mathcal{P}_{SI} \rangle, \qquad \langle \mathcal{P}_{HSII} \rangle \leq \langle \mathcal{P}_{SII} \rangle, \tag{3.6.86}$$

$$\langle \mathcal{D}_{HP} \rangle \leq \langle \mathcal{D}_P \rangle, \qquad \langle \mathcal{D}_{HSI} \rangle \leq \langle \mathcal{D}_{SI} \rangle, \qquad \langle \mathcal{D}_{HSII} \rangle \leq \langle \mathcal{D}_{SII} \rangle, \tag{3.6.87}$$

and

$$\langle \mathcal{E}_{HP} \rangle \leq \langle \mathcal{E}_P \rangle, \qquad \langle \mathcal{E}_{HSI} \rangle \leq \langle \mathcal{E}_{SI} \rangle, \qquad \langle \mathcal{E}_{HSII} \rangle \leq \langle \mathcal{E}_{SII} \rangle. \tag{3.6.88}$$

Similarly, (3.6.79) and (3.6.80) imply the following inequalities for the reciprocal quality factors for SI and SII waves, namely

$$Q_{HS}^{-1} < Q_{SII}^{-1} < Q_{SI}^{-1} < 2Q_{HS}^{-1}. \tag{3.6.89}$$

These inequalities indicate that regardless of the amount of inhomogeneity of either type of S wave, the reciprocal quality factor for the wave is less than

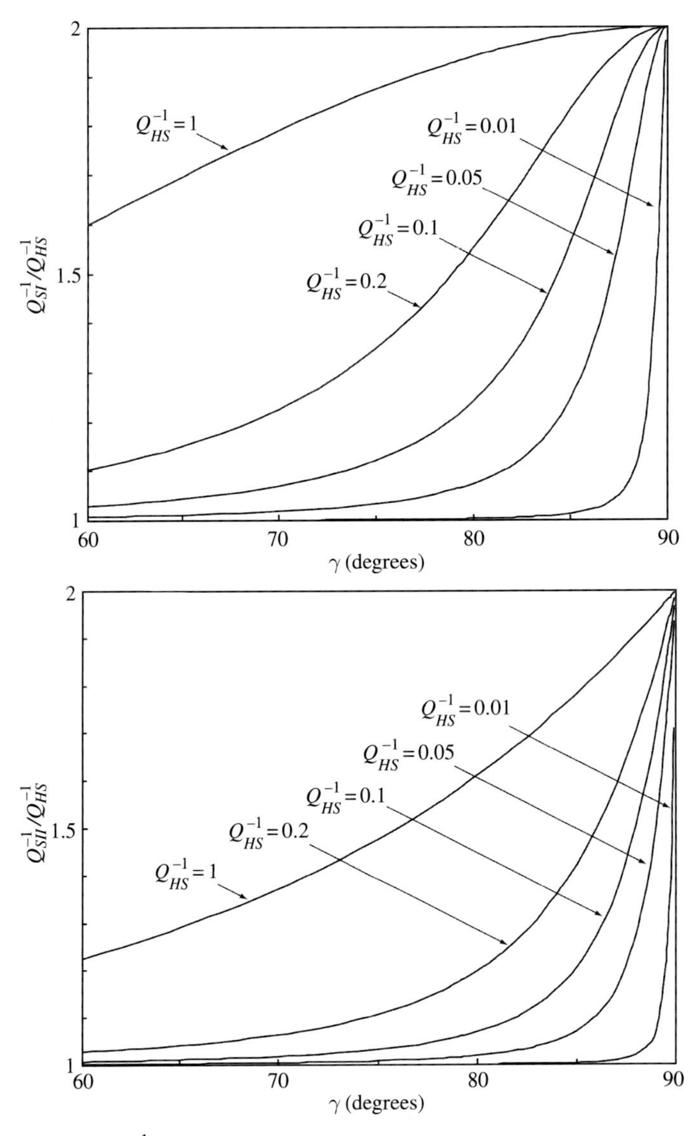

Figure (3.6.90). Q^{-1} for inhomogeneous Type-I (a) and Type-II (b) S waves normalized by that for a homogeneous S wave (i.e. Q_{SI}^{-1}/Q_{HS}^{-1}, Q_{SII}^{-1}/Q_{HS}^{-1}) computed as a function of inhomogeneity, γ, and the amount of intrinsic material absorption in shear as measured by Q_{HS}^{-1}.

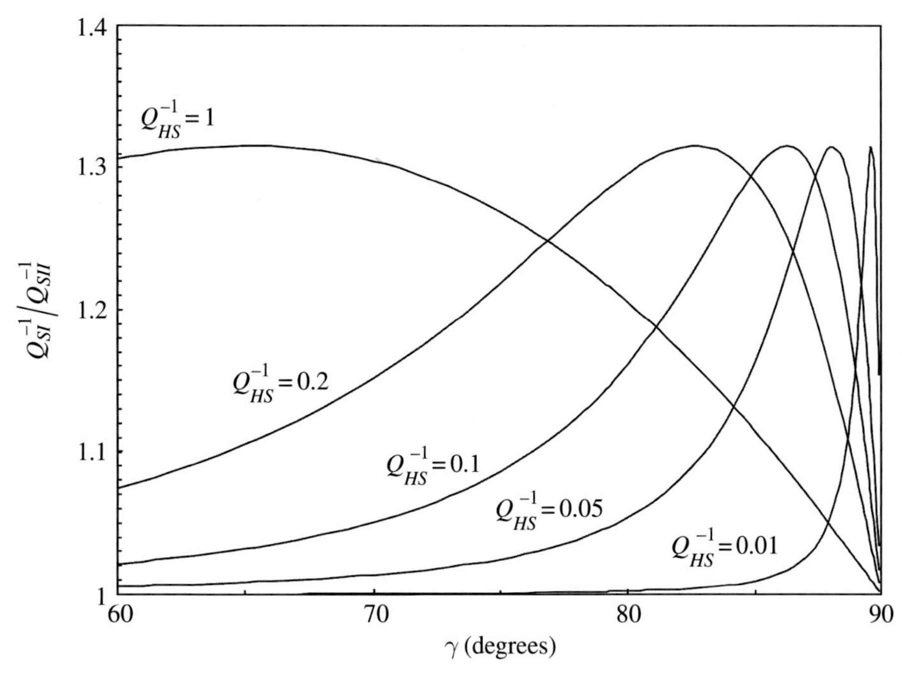

Figure (3.6.91). Ratio of Q^{-1} for an inhomogeneous Type-I S wave to that for a corresponding inhomogeneous Type-II S wave (i.e. Q_{SI}^{-1}/Q_{SII}^{-1}) computed as a function of inhomogeneity, γ, and the amount of intrinsic material absorption in shear as measured by Q_{HS}^{-1}.

twice that of a corresponding homogeneous S wave. They indicate that Q_{SII}^{-1} for an inhomogeneous SII wave is less than Q_{SI}^{-1} for an inhomogeneous SI wave with the same degree of inhomogeneity and that both are greater than that for a corresponding homogeneous wave. They indicate that Q_{SII}^{-1} and Q_{SI}^{-1} for inhomogeneous SII and SI waves are bounded and contained within the half-open interval $[Q_{HS}^{-1}, 2Q_{HS}^{-1})$.

Calculations showing the reciprocal quality factor for inhomogeneous SI and SII waves normalized by the corresponding factor for a homogeneous S wave and their ratio are shown in Figures (3.6.90) and (3.6.91). The calculations indicate that the amount the reciprocal quality factors for the two types of inhomogeneous S waves differ from that for a corresponding homogeneous wave increases with increasing inhomogeneity of the wave and the amount of intrinsic material absorption. For fixed, but arbitrary, amounts of intrinsic absorption and inhomogeneity, the curves shown in Figures (3.6.90) and (3.6.91) are consistent with inequalities (3.6.89). The curves in Figure (3.6.91) show that the amount by which Q_{SI}^{-1} exceeds Q_{SII}^{-1} increases with the intrinsic absorption of the media to a maximum amount of 31 percent. They indicate that this maximum deviation occurs at degrees of inhomogeneity of the waves which increase as the amount of intrinsic absorption decreases.

An important result for Earth materials establishes limits on the magnitude of Q^{-1} for an inhomogeneous P wave in terms of those for homogeneous P and S waves. This result can be stated most efficiently as

Theorem (3.6.92). If $Q_{HP}^{-1} < 2Q_{HS}^{-1}$ for a HILV medium, then $Q_{HP}^{-1} < Q_{\bar{P}}^{-1} < 2Q_{HS}^{-1}$.

The proof of this result follows readily from (3.6.78).

For most Earth materials $Q_{HP}^{-1} < Q_{HS}^{-1}$, hence Theorem (3.6.92) indicates that $Q_{\bar{P}}^{-1}$ for an inhomogeneous P wave, regardless of the amount of inhomogeneity, is contained within the half-open interval $\left[Q_{HP}^{-1}, 2Q_{HS}^{-1}\right)$ and hence less than twice that of a homogeneous S wave.

Curves showing $Q_{\bar{P}}^{-1}/Q_{HP}^{-1}$ for a set of chosen material parameters (Figure (3.6.93)) show that the amount $Q_{\bar{P}}^{-1}$ for an inhomogeneous P wave differs from

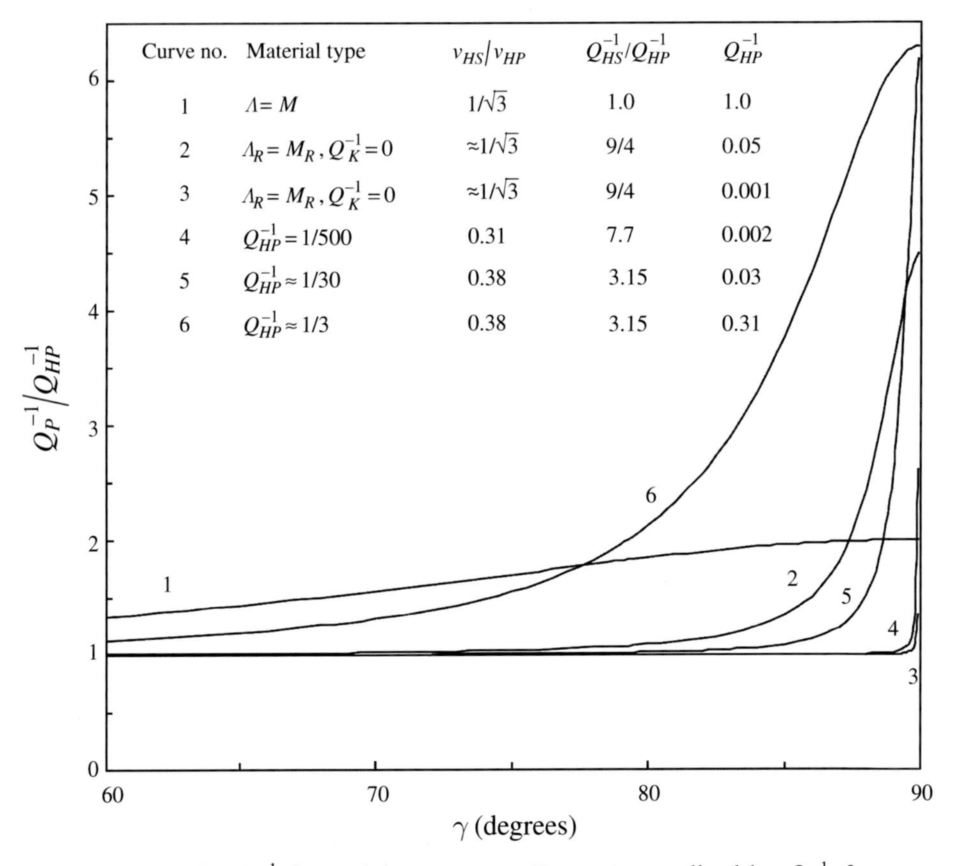

Figure (3.6.93). $Q_{\bar{P}}^{-1}$ for an inhomogeneous P wave normalized by Q_{HP}^{-1} for a homogeneous P wave computed as a function of inhomogeneity, γ, for six types of material with varying amounts of intrinsic material absorption.

that for a homogeneous wave increases with increasing inhomogeneity of the wave towards a value equal to twice that for a homogeneous S wave as indicated by Theorem (3.6.92). The curves show that $Q_{\overline{P}}^{-1}$ depends on not only the material parameters chosen for homogeneous P waves, but also those chosen for homogeneous S waves.

3.7 P, SI, and SII Waves in Low-Loss Viscoelastic Media

Many wave-propagation problems are concerned with media and wave frequencies for which the amount of wave energy absorbed by the media as the wave passes is small. Many seismological problems involving wave propagation in the core, mantle, and crust of the Earth are best modeled using materials with small amounts of intrinsic absorption. Consequently, it is of considerable interest to develop explicit expressions for the physical characteristics of P, SI, and SII waves in such materials.

Materials with small amounts of absorption can be quantitatively characterized by the assumptions that Q_{HS}^{-1} and Q_{HP}^{-1} are much less than unity, as denoted by

$$Q_{HS}^{-1} \ll 1 \qquad \text{and} \qquad Q_{HP}^{-1} \ll 1. \tag{3.7.1}$$

Applying these assumptions to the preceding equations for the characteristics of P and both types of S waves readily permits the derivation of explicit expressions for the physical characteristics of P, SI, and SII waves in low-loss anelastic media as initially derived by the author (Borcherdt and Wennerberg, 1985). Parameters describing the physical characteristics for waves in low-loss media are distinguished from those in media for which the assumption is not necessarily valid by the subscript "L".

Equations (3.5.4) and (3.5.5) together with the low-loss assumption (3.7.1) imply the wave speeds for homogeneous P and S waves in low-loss media are given by

$$v_{HP} \approx v_{HP_L} = \sqrt{\frac{K_R + \frac{4}{3}M_R}{\rho}} \tag{3.7.2}$$

and

$$v_{HS} \approx v_{HS_L} = \sqrt{\frac{M_R}{\rho}}. \tag{3.7.3}$$

The maximum attenuation for a homogeneous P wave is given from (3.6.5) by the familiar expression often used to characterize attenuation for waves in the Earth, namely

$$\left|\vec{A}_{HP}\right| \approx \left|\vec{A}_{HP}\right|_L = \frac{\omega Q_{HP}^{-1}}{2 v_{HP_L}} \tag{3.7.4}$$

and for homogeneous S waves from (3.6.6) by

$$\left|\vec{A}_{HS}\right| \approx \left|\vec{A}_{HS}\right|_L = \frac{\omega Q_{HS}^{-1}}{2 v_{HS_L}}. \tag{3.7.5}$$

Expressions for the wave speeds of inhomogeneous P, SI, and SII waves in low-loss media from (3.6.1) and (3.6.2) are

$$|\vec{v}_P| \approx |\vec{v}_P|_L = |\vec{v}_{HP}|_L \sqrt{\frac{2}{1+\chi_P}} \tag{3.7.6}$$

and

$$|\vec{v}_{SI}| = |\vec{v}_{SII}| \approx |\vec{v}_S|_L = |\vec{v}_{HS}|_L \sqrt{\frac{2}{1+\chi_S}}, \tag{3.7.7}$$

where for brevity it is assumed $\gamma_{SI}=\gamma_{SII}$, so $\chi_{SI}=\chi_{SII}=\chi_S$ and $\left|\vec{v}_{SI}\right| = \left|\vec{v}_{SII}\right|$. Corresponding expressions for the maximum attenuation from (3.6.3) and (3.6.4) are

$$\left|\vec{A}_P\right| \approx \left|\vec{A}_P\right|_L = \sqrt{\frac{2}{1+\chi_P}} \left(\frac{1}{\cos\gamma}\right) \left|\vec{A}_{HP}\right|_L \tag{3.7.8}$$

and

$$\left|\vec{A}_{SI}\right| = \left|\vec{A}_{SII}\right| \approx \left|\vec{A}_S\right|_L = \sqrt{\frac{2}{1+\chi_S}} \left(\frac{1}{\cos\gamma}\right) \left|\vec{A}_{HS}\right|_L, \tag{3.7.9}$$

where for brevity it is assumed $\gamma_{SI}=\gamma_{SII}$, so $\chi_{SI}=\chi_{SII}=\chi_S$ and $\left|\vec{A}_{SI}\right| = \left|\vec{A}_{SII}\right|$.

Particle motion characteristics for P waves in low-loss media from (3.6.29) through (3.6.32) are given by

$$\left|\vec{\xi}_{1P}\right|^2 \approx \left|\vec{\xi}_{1P}\right|_L^2 = \frac{1}{2}(\chi_P + 1), \tag{3.7.10}$$

$$\left|\vec{\xi}_{2P}\right|^2 \approx \left|\vec{\xi}_{2P}\right|_L^2 = \frac{1}{2}(\chi_P - 1), \tag{3.7.11}$$

$$e_P \approx e_{P_L} = \sqrt{\frac{2}{\chi_P + 1}}, \tag{3.7.12}$$

and

$$\cos^2 \eta_P \approx 1. \tag{3.7.13}$$

Corresponding particle motion characteristics for SI waves from (3.6.33) through (3.6.36) are given by

$$\left|\vec{\xi}_{1SI}\right|^2 \approx \left|\vec{\xi}_{1SI}\right|_L^2 = \frac{1}{2}(\chi_{SI}+1), \tag{3.7.14}$$

$$\left|\vec{\xi}_{2SI}\right|^2 \approx \left|\vec{\xi}_{2SI}\right|_L^2 = \frac{1}{2}(\chi_{SI}-1), \tag{3.7.15}$$

$$e_{SI} \approx e_{SI_L} = \sqrt{\frac{2}{\chi_{SI}+1}}, \tag{3.7.16}$$

and

$$\cos^2\eta_{SI} \approx 1. \tag{3.7.17}$$

The mean energy flux for inhomogeneous waves in low-loss media in terms of the corresponding expressions for homogeneous waves from equations (3.6.52) through (3.6.54) are given

for P waves by

$$\left|\langle\vec{\mathcal{J}}_P\rangle\right| \approx \left|\langle\vec{\mathcal{J}}_P\rangle\right|_L = Y_P\sqrt{\frac{1+\chi_P+4F_{P_L}H_{P_L}}{2}}\left|\langle\vec{\mathcal{J}}_{HP}\rangle_L\right|, \tag{3.7.18}$$

for SI waves by

$$\left|\langle\vec{\mathcal{J}}_{SI}\rangle\right| \approx \left|\langle\vec{\mathcal{J}}_{SI}\rangle\right|_L = Y_S\sqrt{\frac{1+\chi_{SI}+4F_{SI_L}H_{S_L}}{2}}\left|\langle\vec{\mathcal{J}}_{HS}\rangle_L\right|, \tag{3.7.19}$$

and for SII waves by

$$\left|\langle\vec{\mathcal{J}}_{SII}\rangle\right| \approx \left|\langle\vec{\mathcal{J}}_{SII}\rangle\right|_L = Y_S\sqrt{\frac{1+\chi_{SII}+2F_{SII_L}H_{S_L}}{2}}\left|\langle\vec{\mathcal{J}}_{HS}\rangle_L\right|, \tag{3.7.20}$$

where

$$H_P \approx H_{P_L} = Q_{HP}^{-2}\frac{v_{HS}^2}{v_{HP}^2}\tan^2\gamma, \tag{3.7.21}$$

$$H_S = H_{SI} = H_{SII} \approx H_{S_L} = Q_{HS}^{-2}\tan^2\gamma, \tag{3.7.22}$$

$$F_P \approx F_{P_L} = 1+\frac{v_{HS}^2}{v_{HP}^2}(\chi_P-1), \tag{3.7.23}$$

$$F_{SI} \approx F_{SI_L} = 1+\frac{v_{HS}^2}{v_{HS}^2}(\chi_{SI}-1) = \chi_{SI}, \tag{3.7.24}$$

$$F_{SII} \approx F_{SII_L} = 1+\frac{1}{2}\frac{v_{HS}^2}{v_{HS}^2}(\chi_{SII}-1) = \frac{1}{2}(\chi_{SII}+1). \tag{3.7.25}$$

The corresponding approximate expressions for the other energy characteristics for low-loss media from equations (3.6.63) through (3.6.83) for P, SI, and SII waves are

mean kinetic energy density,

$$\langle \mathcal{K}_P \rangle \approx \langle \mathcal{K}_P \rangle_L = Y_P \chi_P \langle \mathcal{K}_{HP} \rangle_L, \tag{3.7.26}$$

$$\langle \mathcal{K}_{SI} \rangle \approx \langle \mathcal{K}_{SI} \rangle_L = Y_{SI} \chi_{SI} \langle \mathcal{K}_{HS} \rangle_L, \tag{3.7.27}$$

$$\langle \mathcal{K}_{SII} \rangle \approx \langle \mathcal{K}_{SII} \rangle_L = Y_{SII} \chi_{SII} \langle \mathcal{K}_{HS} \rangle_L, \tag{3.7.28}$$

mean potential energy density,

$$\langle \mathcal{P}_P \rangle \approx \langle \mathcal{P}_P \rangle_L = Y_P \left(1 + 2H_{P_L}\right) \langle \mathcal{P}_{HP} \rangle_L, \tag{3.7.29}$$

$$\langle \mathcal{P}_{SI} \rangle \approx \langle \mathcal{P}_{SI} \rangle_L = Y_S \left(1 + 2H_{S_L}\right) \langle \mathcal{P}_{HS} \rangle_L, \tag{3.7.30}$$

$$\langle \mathcal{P}_{SII} \rangle \approx \langle \mathcal{P}_{SII} \rangle_L = Y_S \left(1 + H_{S_L}\right) \langle \mathcal{P}_{HS} \rangle_L, \tag{3.7.31}$$

total energy density,

$$\langle \mathcal{E}_P \rangle \approx \langle \mathcal{E}_P \rangle_L = \frac{Y_P}{2} \left(1 + \chi_P + 2H_{P_L}\right) \langle \mathcal{E}_{HP} \rangle_L, \tag{3.7.32}$$

$$\langle \mathcal{E}_{SI} \rangle \approx \langle \mathcal{E}_{SI} \rangle_L = \frac{Y_{SI}}{2} \left(1 + \chi_{SI} + 2H_{S_L}\right) \langle \mathcal{E}_{HSI} \rangle_L, \tag{3.7.33}$$

$$\langle \mathcal{E}_{SII} \rangle \approx \langle \mathcal{E}_{SII} \rangle_L = \frac{Y_{SII}}{2} \left(1 + \chi_{SII} + H_{S_L}\right) \langle \mathcal{E}_{HSII} \rangle_L, \tag{3.7.34}$$

mean rate of energy dissipation per unit volume,

$$\langle \mathcal{D}_P \rangle \approx \langle \mathcal{D}_P \rangle_L = Y_P \left(1 + 2\frac{Q_{HS}^{-1}}{Q_{HP}^{-1}} H_{P_L}\right) \langle \mathcal{D}_{HP} \rangle_L, \tag{3.7.35}$$

$$\langle \mathcal{D}_{SI} \rangle \approx \langle \mathcal{D}_{SI} \rangle_L = Y_{SI} \left(1 + 2H_{S_L}\right) \langle \mathcal{D}_{HSI} \rangle_L, \tag{3.7.36}$$

$$\langle \mathcal{D}_{SII} \rangle \approx \langle \mathcal{D}_{SII} \rangle_L = Y_{SII} \left(1 + H_{S_L}\right) \langle \mathcal{D}_{HSII} \rangle_L, \tag{3.7.37}$$

maximum potential energy density per cycle,

$$\max[\mathcal{P}_P] \approx \max[\mathcal{P}_P]_L = Y_P \left(1 + H_{P_L}\right) \max[\mathcal{P}_{HP}]_L, \tag{3.7.38}$$

$$\max[\mathcal{P}_{SI}] \approx \max[\mathcal{P}_{SI}]_L = Y_{SI} \left(1 + H_{S_L}\right) \max[\mathcal{P}_{HS}]_L, \tag{3.7.39}$$

$$\max[\mathcal{P}_{SII}] \approx \max[\mathcal{P}_{SII}]_L = \frac{Y_{SII}}{2} \left(1 + \sqrt{1 + H_{S_L}} + H_{S_L}\right) \max[\mathcal{P}_{HS}]_L, \tag{3.7.40}$$

and Q^{-1}

$$Q_P^{-1} \approx Q_{P_L}^{-1} = \left(\frac{1 + \left(2Q_{HS}^{-1}/Q_{HP}^{-1}\right) H_{P_L}}{1 + H_{P_L}}\right) Q_{HP}^{-1}, \tag{3.7.41}$$

$$Q_{SI}^{-1} \approx Q_{SI_L}^{-1} = \left(\frac{1+2H_{S_L}}{1+H_{S_L}}\right) Q_{HS}^{-1}, \tag{3.7.42}$$

$$Q_{SII}^{-1} \approx Q_{SII_L}^{-1} = \left(\frac{2(1+H_{S_L})}{1+\sqrt{1+H_{S_L}}+H_{S_L}}\right) Q_{HS}^{-1}. \tag{3.7.43}$$

Equations (3.7.2) through (3.7.43) provide an explicit description of the physical characteristics of P, SI, and SII waves in low-loss general viscoelastic media. Inspection of the expressions for inhomogeneous waves, namely (3.7.6) through (3.7.43), shows that each of the characteristics with the exception of the tilt of the particle motion ellipse for P and SI waves exhibits an explicit dependence on the amount of inhomogeneity, γ, of the wave field. The expressions show that the amount that the characteristics differ from those for homogeneous waves increases as the inhomogeneity of the wave increases. They imply that for a given low-loss material the wave speed, attenuation, and ellipticity of the particle motion for P and SI waves, and each of the energy characteristics for each wave type, can differ significantly from that for a corresponding homogeneous wave, provided the inhomogeneity of the wave increases to a value sufficiently close to its physical limit of $\pi/2$. The tilt η of the particle motion ellipse for inhomogeneous P and SI waves in low-loss media is approximately zero and shows no the dependence on the degree of inhomogeneity. Equations (3.7.2) through (3.7.43) provide explicit estimates to further evaluate the influence of inhomogeneity and the approximations introduced by the low-loss assumption on characteristics of P, SI, and SII waves.

Comparison of the wave speed, $v_{HS} = |\vec{v}_{HS}|$, and maximum attenuation, $A_{HS} \equiv |\vec{A}_{HS}|$, for homogeneous S waves with those predicted by the low-loss assumption for homogeneous S waves is provided explicitly from (3.5.5), (3.6.6), (3.7.3) and (3.7.5) by

$$\frac{v_{HS} - v_{HS_L}}{v_{HS}} = 1 - \frac{\sqrt{1+\chi_{HS}}}{\sqrt{2}\chi_{HS}} \tag{3.7.44}$$

and

$$\frac{A_{HS_L} - A_{HS}}{A_{HS}} = \frac{\chi_{HS}\sqrt{1+\chi_{HS}}}{\sqrt{2}} - 1. \tag{3.7.45}$$

Corresponding error-percentage curves computed as a function of the amount of intrinsic absorption are shown in Figure (3.7.46). The curves show that for solids with $Q_{HS}^{-1} \leq 0.1$ (Figure (3.7.46)a), the low-loss expressions for wave speed and absorption coefficient yield estimates that deviate from those predicted exactly for homogeneous S waves by amounts less than 0.35 and 0.65 percent, respectively. The corresponding curves for solids with moderate to large amounts of intrinsic

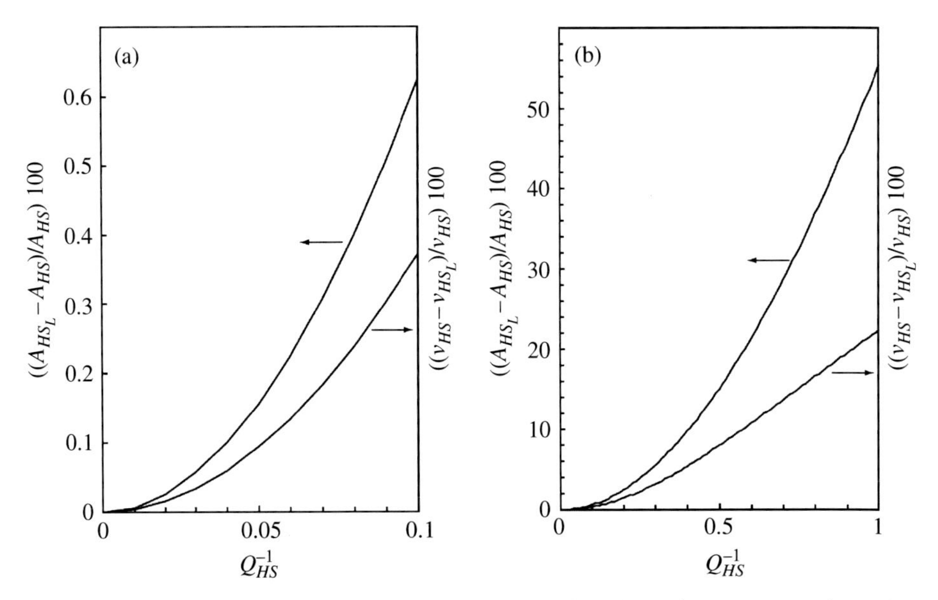

Figure (3.7.46). Error percentages for wave speed and maximum attenuation of homogeneous S waves as predicted by the low-loss approximation relative to that predicted exactly. The error percentages are plotted for materials with (a) small ($Q_{HS}^{-1} < 0.1$) and (b) moderate to large amounts of absorption ($Q_{HS}^{-1} \geq 0.1$). The curves are valid for P waves upon replacement of the S subscript.

absorption $Q_{HS}^{-1} > 0.1$ (Figure (3.7.46)b) show that the low-loss expressions yield estimates that deviate from those predicted exactly for homogeneous S waves by amounts up to 35 and 55 percent, respectively. Expressions (3.7.44) and (3.7.45) and the corresponding curves in Figure (3.7.46) are valid for P waves upon replacement of the "S" subscript.

Comparison of the wave speed, $v_S = |\vec{v}_S|$, and maximum attenuation, $A_S = |\vec{A}_S|$, for inhomogeneous SI and SII waves with those predicted by the low-loss approximation for homogeneous S waves is provided explicitly from (3.6.2) and (3.7.7) by

$$\frac{v_{HS_L} - v_S}{v_{HS_L}} = 1 - \frac{\sqrt{2}\chi_{HS}}{\sqrt{1+\chi_S}} = 1 - \frac{\sqrt{2}\sqrt{1+Q_{HS}^{-2}}}{\sqrt{1+\sqrt{1+Q_{HS}^{-2}\sec^2\gamma}}} \tag{3.7.47}$$

and

$$\frac{A_S - A_{HS_L}}{A_{HS_L}} = \frac{\sqrt{-1+\chi_S}}{Q_{HS}^{-1}\sqrt{1+\chi_{HS}}} - 1 = \frac{\sqrt{-1+\sqrt{1+Q_{HS}^{-2}\sec^2\gamma}}}{Q_{HS}^{-1}\sqrt{1+\sqrt{1+Q_{HS}^{-2}}}} - 1. \tag{3.7.48}$$

Corresponding error-percentage curves for the wave speed in Figure (3.7.49) show that for materials with large amounts of absorption, for example $Q_{HS}^{-1} = 1$, the wave speed of an inhomogeneous SI or SII wave deviates significantly for nearly all degrees of inhomogeneity from that predicted for a corresponding homogeneous S wave using the low-loss approximation. For materials with small amounts of intrinsic absorption, $Q_{HS}^{-1} \leq 0.1$, Figure (3.7.49) shows that the amount of the deviation for wave speed increases with increasing Q_{HS}^{-1} and the degree of inhomogeneity of the wave. Expressions (3.7.47), (3.7.48), and the corresponding curves in Figure (3.7.49) are valid for P waves upon replacement of the S subscript.

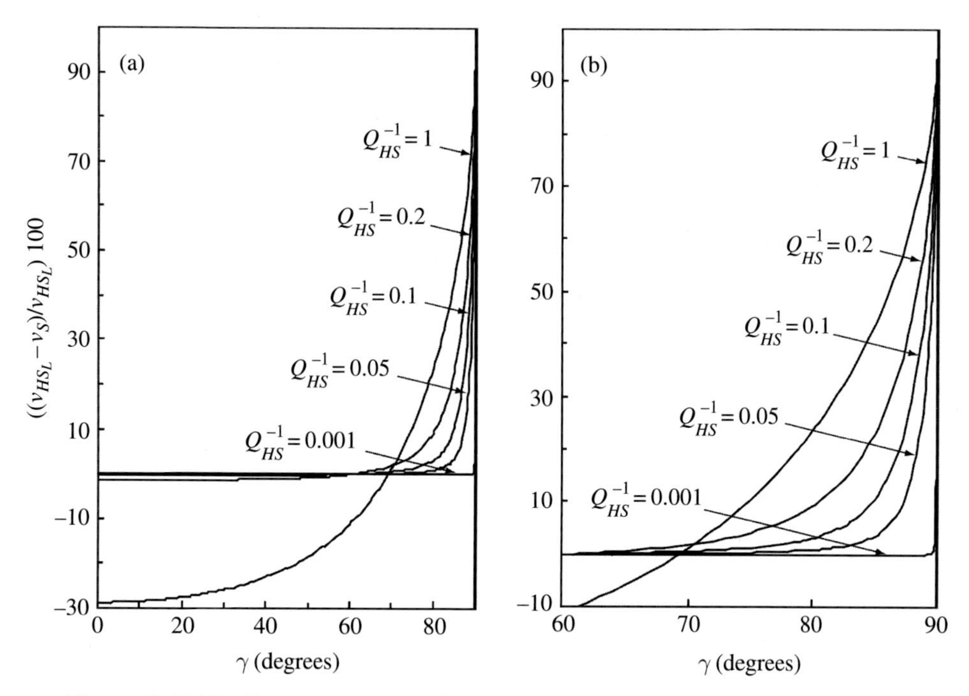

Figure (3.7.49). Error-percentages for wave speed of inhomogeneous S waves relative to that predicted by the low-loss approximation for homogeneous S waves. The error percentages are plotted on two scales as a function of the amount of inhomogeneity for materials ranging from those with a large amount of loss, $Q_{HS}^{-1} = 1$, to low-loss materials, $Q_{HS}^{-1} > 0.1$. The curves are valid for P waves upon replacement of the S subscript.

Error-percentage curves in Figure (3.7.50) show the attenuation in the direction of maximum attenuation of an inhomogeneous wave for relatively small amounts of inhomogeneity can deviate significantly from that predicted by the low-loss expressions for homogeneous waves. They indicate the deviations for degrees of inhomogeneity of greater than 20 degrees are significant even for low-loss materials. For materials with large amounts of absorption, e.g. $Q_{HS}^{-1} = 1$, the deviation is large for all degrees of inhomogeneity. For materials with small amounts of absorption the error percentage becomes infinite for sufficiently large values of inhomogeneity. Comparison of other characteristics of inhomogeneous waves with those of homogeneous waves in either

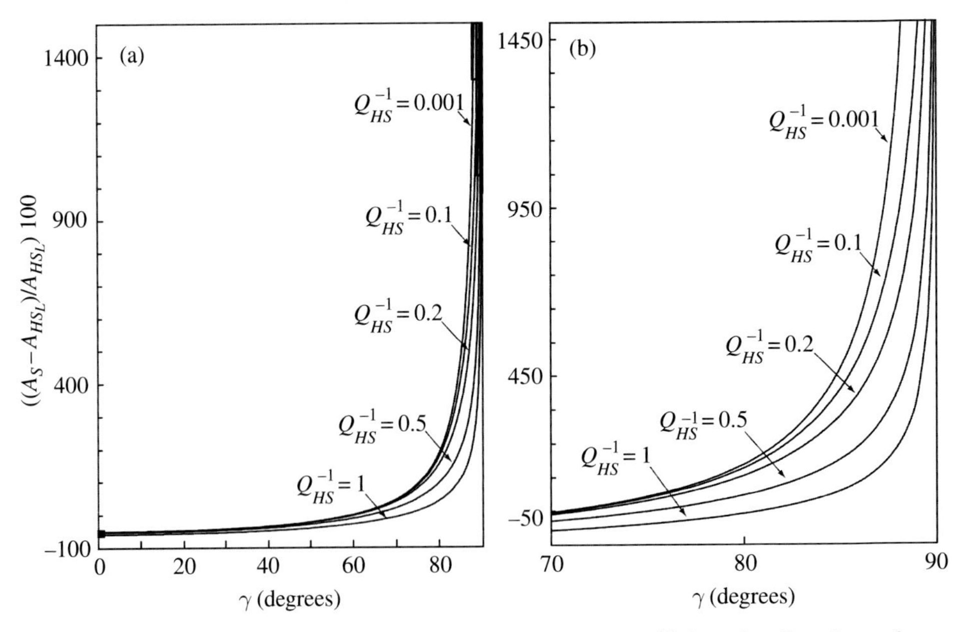

Figure (3.7.50). Error percentages for attenuation coefficient in direction of maximum attenuation for inhomogeneous S waves relative to that predicted by the low-loss approximation for homogeneous S waves. The error percentages are plotted at two scales as a function of the amount of inhomogeneity for materials ranging from those with a large amount of loss, $Q_{HS}^{-1} = 1$, to low-loss materials, $Q_{HS}^{-1} > 0.1$. The curves are valid for P waves upon replacement of the S subscript.

media with an arbitrary amount of intrinsic absorption or low-loss media can be readily derived in a similar fashion using (3.7.10) through (3.7.43).

3.8 P, SI, and SII Waves in Media with Equal Complex Lamé Parameters

An assumption sometimes made for elastic media is that of equal Lamé parameters, that is $\Lambda_R = M_R$. This assumption provides considerable simplification in some formulae and is a good first approximation for the elastic response of some Earth materials. An extension of the assumption for viscoelastic media is to assume that the imaginary parts of the complex Lamé parameter are equal as well, that is $\Lambda_I = M_I$.

Definitions (2.3.13) through (2.3.15) show that for the special class of solids with equal complex Lamé parameters the complex parameters of the material are related by the following expressions:

$$\Lambda = M, \tag{3.8.1}$$

$$K = \frac{5}{3} M, \tag{3.8.2}$$

$$E \equiv \frac{5}{2} M, \tag{3.8.3}$$

$$N = \frac{1}{4}. \tag{3.8.4}$$

Expressions (3.1.7) and (3.1.8) imply the complex velocities are related by

$$\frac{\beta^2}{\alpha^2} = \frac{1}{3}. \tag{3.8.5}$$

Expressions (3.5.1) through (3.5.9) and (3.6.5) and (3.6.6) imply the speed, maximum attenuation, and quality factor for homogeneous P and S waves in such solids are related by

$$v_{HP} = \sqrt{3}\, v_{HS}, \tag{3.8.6}$$

$$\left|\vec{A}_{HP}\right| = \left|\vec{A}_{HS}\right| / \sqrt{3} \tag{3.8.7}$$

and

$$Q_{HS}^{-1} = Q_{HP}^{-1}. \tag{3.8.8}$$

Expressions (3.6.12) through (3.6.23) and (3.6.58), (3.6.59), (3.6.78), and (3.6.80) imply for a given equal amount of inhomogeneity γ for inhomogeneous P, SI, and SII waves in HILV solids with equal Lamé parameters that the parameters of the inhomogeneous wave fields are related by

$$k_P^2 = \frac{1}{3} k_S^2, \tag{3.8.9}$$

$$|\vec{v}_P| = \sqrt{3}\, |\vec{v}_S|, \tag{3.8.10}$$

$$H_P = \frac{1}{3} H_S = \frac{1}{3} \frac{Q_{HS}^{-2}}{\chi_{HS}{}^2} \tan^2 \gamma, \tag{3.8.11}$$

$$Q_P^{-1} = \left(\frac{3 + 2H_S}{3 + H_S} \right) Q_{HS}^{-1} = \frac{3 + 2H_S}{3 + H_S} \frac{1 + H_S}{1 + 2H_S} Q_{SI}^{-1}, \tag{3.8.12}$$

$$Q_{SI}^{-1} = \left(\frac{1 + 2H_S}{1 + H_S} \right) Q_{HS}^{-1} = \frac{1 + 2H_S}{1 + H_S} \frac{1 + H_S + \sqrt{1 + H_S}}{2(1 + H_S)} Q_{SII}^{-1}, \tag{3.8.13}$$

and

$$Q_{SII}^{-1} = \left(\frac{2(1 + H_S)}{1 + H_S + \sqrt{1 + H_S}} \right) Q_{HS}^{-1}. \tag{3.8.14}$$

Expressions for other physical characteristics of general P, SI, and SII waves may be derived readily in a similar fashion for HILV solids with equal Lamé parameters.

This assumption will be shown to provide simplification of formulae used to describe a Rayleigh-Type surface wave on a HILV half space.

Equation (3.8.8) shows that the assumption of equal complex Lamé parameters implies that the fractional energy loss for homogeneous P waves equals that for homogeneous S waves. This assumption, valid for some elastic media, is used on occasion to describe anelastic media with very small amounts of absorption, but it is not a good approximation for many near-surface and crustal materials in the Earth.

3.9 P, SI, and SII Waves in a Standard Linear Solid

The characterization of linear viscoelastic behavior and the subsequent derivation of expressions describing the physical characteristics of P, SI, and SII waves in preceding chapters and sections are general and valid for any particular viscoelastic solid with a specific dependence of moduli on frequency. To illustrate the application of the general formulae consider a Standard Linear solid. The complex shear modulus for a Standard Linear solid is given by

$$M = M_r \frac{1 + i\omega\tau_e}{1 + i\omega\tau_p} \tag{3.9.1}$$

as a function of circular frequency ω, relaxation times τ_p and τ_e, and relaxed elastic modulus M_r (see Table (1.3.30).

The real and imaginary parts of (3.9.1) and general formulae (3.5.7), (3.5.5), and (3.5.2) immediately imply

$$v_{HSe} = \sqrt{\frac{M_R}{\rho}} = \sqrt{\frac{M_r}{\rho}} \sqrt{\frac{1 + \omega^2\tau_p\tau_e}{1 + \omega^2\tau_p^2}} \tag{3.9.2}$$

with the reciprocal quality factor and the speed of a homogeneous S wave given by

$$Q_{HS}^{-1} = \frac{M_I}{M_R} = \frac{\omega(\tau_e - \tau_p)}{1 + \omega^2\tau_p\tau_e} \tag{3.9.3}$$

and

$$v_{HS} = \sqrt{\frac{M_R}{\rho} \frac{2\left(1 + Q_{HS}^{-2}\right)}{1 + \sqrt{1 + Q_{HS}^{-2}}}}$$

$$= \sqrt{\frac{M_r}{\rho}} \sqrt{\frac{1 + \omega^2\tau_p\tau_e}{1 + \omega^2\tau_p^2}} \sqrt{2\left(1 + \left(\frac{\omega(\tau_e - \tau_p)}{1 + \omega^2\tau_p\tau_e}\right)^2\right) \Bigg/ \left(1 + \sqrt{1 + \left(\frac{\omega(\tau_e - \tau_p)}{1 + \omega^2\tau_p\tau_e}\right)^2}\right)}. \tag{3.9.4}$$

Substitution of (3.9.3) and (3.9.4) into (3.6.2) permits the dispersion relation for general inhomogeneous SI and SII waves in a Standard Linear solid to be readily specified as

$$v_S = \sqrt{\frac{M_r}{\rho}}\sqrt{\frac{1+\omega^2\tau_p\tau_e}{1+\omega^2\tau_p^2}}\sqrt{\frac{2\left(1+\left(\frac{\omega(\tau_e-\tau_p)}{1+\omega^2\tau_p\tau_e}\right)^2\right)}{1+\sqrt{1+\left(\frac{\omega(\tau_e-\tau_p)}{1+\omega^2\tau_p\tau_e}\right)^2}}}\sqrt{\frac{\left(1+\sqrt{1+\left(\frac{\omega(\tau_e-\tau_p)}{1+\omega^2\tau_p\tau_e}\right)^2}\right)}{1+\sqrt{1+\left(\frac{\omega(\tau_e-\tau_p)}{1+\omega^2\tau_p\tau_e}\right)^2\sec^2\gamma}}} \tag{3.9.5}$$

Substitution of (3.9.3) into (3.6.59) allows H_S to be specified as

$$H_S = H_{SI} = H_{SII} = \frac{Q_{HS}^{-2}}{\chi_{HS}^2}\tan^2\gamma$$
$$= \left(\left(\frac{\omega(\tau_e-\tau_p)}{1+\omega^2\tau_p\tau_e}\right)^2 \Big/ \left(1+\left(\frac{\omega(\tau_e-\tau_p)}{1+\omega^2\tau_p\tau_e}\right)^2\right)\right)\tan^2\gamma. \tag{3.9.6}$$

Hence, the frequency dependences of the reciprocal quality factors for inhomogeneous SI and SII waves in a Standard Linear solid are readily specified from (3.6.79) and (3.6.80) upon substitution of this expression for H_S into

$$Q_{SI}^{-1} = \left(\frac{1+2H_S}{1+H_S}\right)\frac{\omega(\tau_e-\tau_p)}{1+\omega^2\tau_p\tau_e} \tag{3.9.7}$$

for inhomogeneous SI waves and

$$Q_{SII}^{-1} = \left(\frac{2(1+H_S)}{1+H_S+\sqrt{1+H_S}}\right)\frac{\omega(\tau_e-\tau_p)}{1+\omega^2\tau_p\tau_e} \tag{3.9.8}$$

for inhomogeneous SII waves.

For Standard Linear viscoelastic solids with small amounts of absorption, the general low-loss expressions for wave speed (3.7.7) allow the dispersion relation for SI and SII waves to be readily written as

$$|\vec{v}_S| \approx |\vec{v}_S|_L = \sqrt{\frac{M_r}{\rho}}\sqrt{\frac{1+\omega^2\tau_p\tau_e}{1+\omega^2\tau_p^2}}\sqrt{\frac{2}{1+\sqrt{1+\left(\frac{\omega(\tau_e-\tau_p)}{1+\omega^2\tau_p\tau_e}\right)^2}}}. \quad (3.9.9)$$

Equation (3.7.22) implies that H_S is approximately given by

$$H_S = H_{SI} = H_{SII} \approx H_{S_L} = \left(\frac{\omega(\tau_e-\tau_p)}{1+\omega^2\tau_p\tau_e}\right)^2 \tan^2\gamma. \quad (3.9.10)$$

Hence, substitution of this low-loss approximation for H_S into (3.7.42) and (3.7.43) allows the frequency dependence for inhomogeneous SI and SII waves in a low-loss Standard Linear viscoelastic solid to be immediately specified.

The frequency dependence of other physical characteristics of P, SI, and SII waves in a Standard Linear solid or any other specific viscoelastic solid of choice may be written in a similar fashion using the general expression for the physical characteristic of interest derived in preceding sections for general viscoelastic media.

3.10 Displacement and Volumetric Strain

Measurements of seismic radiation fields using volumetric strain meters and three-component seismometers are of special interest in seismology. Corresponding expressions for volumetric strain and components of the displacement or velocity fields for general (homogeneous and inhomogeneous) P, SI, and SII waves in HILV media are useful for interpretation of near-surface measurements either in the field or in the laboratory (Borcherdt, 1988; Borcherdt *et al.*, 1989).

3.10.1 Displacement for General P and SI Waves

The physical displacement field for a general P wave may be written from (3.2.11) and identities (3.6.12) through (3.6.14) in terms of its propagation and attenuation vectors as

$$\vec{u}_R = |G_0 k_P| \exp\left[-\vec{A}_P \bullet \vec{r}\right]\left(\frac{\vec{P}_P}{|k_P|}\cos[\zeta_P(t)+\psi_P] + \frac{\vec{A}_P}{|k_P|}\sin[\zeta_P(t)+\psi_P]\right), \quad (3.10.1)$$

where

$$\zeta_P(t) \equiv \omega t - \vec{P}_P \bullet \vec{r} + \arg[G_0 k_P] - \pi/2 \quad (3.10.2)$$

and

$$\psi_P \equiv \tan^{-1}\left[\frac{Q_{HP}^{-1}}{1+\chi_{HP}}\right]. \quad (3.10.3)$$

Similarly, the physical displacement field for a general SI wave may be written from (3.3.23) in terms of its propagation vector, $\vec{P}_{SI}$, its attenuation vector, $\vec{A}_{SI}$, and its complex vector coefficient, $\vec{G}_0$, as

$$\vec{u}_R = \left|\vec{G}_0 k_S\right| \exp\left[-\vec{A}_{SI} \bullet \vec{r}\right] \left(\frac{\vec{P}_{SI} \times \hat{n}}{|k_S|} \cos[\zeta_{SI}(t) + \psi_S] + \frac{\vec{A}_{SI} \times \hat{n}}{|k_S|} \sin[\zeta_{SI}(t) + \psi_S] \right), \quad (3.10.4)$$

where

$$\zeta_S(t) \equiv \omega t - \vec{P}_{SI} \bullet \vec{r} + \arg[G_0 k_S] + \pi/2, \quad (3.10.5)$$

$$\psi_S \equiv \tan^{-1}\left[\frac{Q_{HS}^{-1}}{1+\chi_{HS}}\right], \quad (3.10.6)$$

and the "*SI*" subscript is introduced to avoid potential confusion with displacements for *SII* waves.

Components of the physical displacement field for P and SI waves, as might be inferred from seismometer recordings, can be derived upon specification of the directions of propagation and attenuation for the waves with respect to a rectangular coordinate system. Without loss of generality, the directions of propagation and attenuation shall be assumed to be in the x_1x_3 plane with propagation in the $+x_1$ direction. (For brevity, quantities requiring subscripts corresponding to components of the (x_1, x_2, x_3) coordinate system will be referenced using the notation (x, y, z).)

The angle of incidence or the direction that the propagation vector makes with the vertical shall be denoted by θ with $0 \le \theta \le \pi/2$ and the angle between the propagation and attenuation vector by γ, so that the direction the attenuation vector makes with respect to the vertical is given by $\theta - \gamma$. With this notation, the propagation and attenuation vectors for a general P wave may be written with identities (3.6.17) and (3.6.18) as

$$\vec{P}_P = \frac{\omega}{v_{HP}} \sqrt{\frac{1+\chi_P}{1+\chi_{HP}}} (\hat{x}_1 \sin\theta + \hat{x}_3 \cos\theta) \quad (3.10.7)$$

and

$$\vec{A}_P = \frac{\omega}{v_{HP}} \sqrt{\frac{-1+\chi_P}{1+\chi_{HP}}} (\hat{x}_1 \sin[\theta-\gamma] + \hat{x}_3 \cos[\theta-\gamma]), \quad (3.10.8)$$

where

$$\chi_P \equiv \sqrt{1 + Q_{HP}^{-2} \sec^2 \gamma}. \tag{3.10.9}$$

Substitution of these results into (3.10.1) allows the radial (x) and vertical (z) components of the physical displacement field for a general P wave to be written as

$$u_{Rx} = |G_0 k_P| \exp\left[-\vec{A}_P \bullet \vec{r}\right] F_P[\sin\theta] \cos[\zeta_P(t) + \psi_P - \Omega_P[\sin\theta]] \tag{3.10.10}$$

and

$$u_{Rz} = |G_0 k_P| \exp\left[-\vec{A}_P \bullet \vec{r}\right] F_P[\cos\theta] \cos[\zeta_P(t) + \psi_P - \Omega_P[\cos\theta]], \tag{3.10.11}$$

where

$$F_P[f[\theta]] \equiv \sqrt{\frac{(1+\chi_P) f^2[\theta] + (-1+\chi_P) f^2[\theta - \gamma]}{2\chi_{HP}}} \tag{3.10.12}$$

and

$$\Omega_P[f[\theta]] \equiv \tan^{-1}\left[\sqrt{\frac{-1+\chi_P}{1+\chi_P}} \frac{f[\theta-\gamma]}{f[\theta]}\right]. \tag{3.10.13}$$

If the P wave is homogeneous, then

$$F_P[f[\theta]] = f[\theta] \tag{3.10.14}$$

and

$$\Omega_P[f[\theta]] = \tan^{-1}\left[\frac{Q_{HP}^{-1}}{1+\chi_{HP}}\right] = \psi_P. \tag{3.10.15}$$

Hence, the expressions for the components of the physical displacement field for a homogeneous P wave simplify to

$$u_{Rx} = |G_0\, k_P| \exp\left[-\vec{A}_P \bullet \vec{r}\right] \sin[\theta] \cos[\zeta_P(t)] \tag{3.10.16}$$

and

$$u_{Rz} = |G_0\, k_P| \exp\left[-\vec{A}_P \bullet \vec{r}\right] \cos[\theta] \cos[\zeta_P(t)]. \tag{3.10.17}$$

For a general SI wave substitution of expressions analogous to (3.10.7) and (3.10.8) into (3.10.4) yields the components of the physical displacement field of a Type-I S wave as

$$u_{Rx} = \left|\vec{G}_0\, k_S\right| \exp\left[-\vec{A}_{SI} \bullet \vec{r}\right] F_{SI}[\cos\theta] \cos[\zeta_{SI}(t) + \psi_S - \Omega_{SI}[\cos\theta]] \tag{3.10.18}$$

and

$$u_{Rz} = \left|\vec{G}_0\ k_S\right| \exp\left[-\vec{A}_{SI} \bullet \vec{r}\right] F_{SI}[\sin\theta] \cos[\zeta_{SI}(t) + \psi_S - \Omega_{SI}[\sin\theta]], \quad (3.10.19)$$

where definitions for F_S and Ω_S are readily inferred from (3.10.12) and (3.10.13). The corresponding expressions for a homogeneous SI wave are

$$u_{Rx} = \left|\vec{G}_0\ k_S\right| \exp\left[-\vec{A}_{SI} \bullet \vec{r}\right] \cos[\theta] \cos[\zeta_{SI}(t)] \quad (3.10.20)$$

and

$$u_{Rz} = \left|\vec{G}_0\ k_S\right| \exp\left[-\vec{A}_{SI} \bullet \vec{r}\right] \sin[\theta] \cos[\zeta_{SI}(t)], \quad (3.10.21)$$

where

$$F_{SI}[f[\theta]] = f[\theta] \quad (3.10.22)$$

and

$$\Omega_{SI}[f[\theta]] = \tan^{-1}\left[\frac{Q_{HS}^{-1}}{1 + \chi_{HS}}\right] = \psi_S. \quad (3.10.23)$$

Expressions (3.10.18) through (3.10.23) for the displacement components of SI waves expressed in terms of θ, γ, and Q_{HS}^{-1} are similar to (3.10.10) through (3.10.17) for P waves upon replacement of Q_{HS}^{-1} with Q_{HP}^{-1}.

For homogeneous P and SI waves the radial and vertical amplitudes vary as the cosine or the sine of the angle of incidence, but the phase term is constant as a function of angle of incidence and depends only on the corresponding amount of intrinsic material absorption. For materials with small amounts of absorption ($Q_{HP}^{-1} \ll 1; Q_{HS}^{-1} \ll 1$), the phase terms may be approximated by

$$\Omega_P[f[\theta]] \approx \tan^{-1}\left[\frac{Q_{HP}^{-1}}{2}\right] \approx \frac{Q_{HP}^{-1}}{2} \quad (3.10.24)$$

and

$$\Omega_{SI}[f[\theta]] \approx \tan^{-1}\left[\frac{Q_{HS}^{-1}}{2}\right] \approx \frac{Q_{HS}^{-1}}{2}. \quad (3.10.25)$$

The expressions for the radial and vertical components of displacement for general P and SI waves, (3.10.10) through (3.10.23) show explicit dependences on the corresponding amplitude modulation factor F, and the phase modulation term Ω. Each of these terms depends on angle of incidence θ, wave inhomogeneity γ, and intrinsic material absorption Q_{HP}^{-1} or Q_{HS}^{-1} depending on the wave type being considered. To quantify these dependences for general P waves, numerical calculations for the amplitude modulation factor and the phase term are shown in Figures (3.10.26) and (3.10.27) for media with large ($Q_{HP}^{-1} = 0.2$),

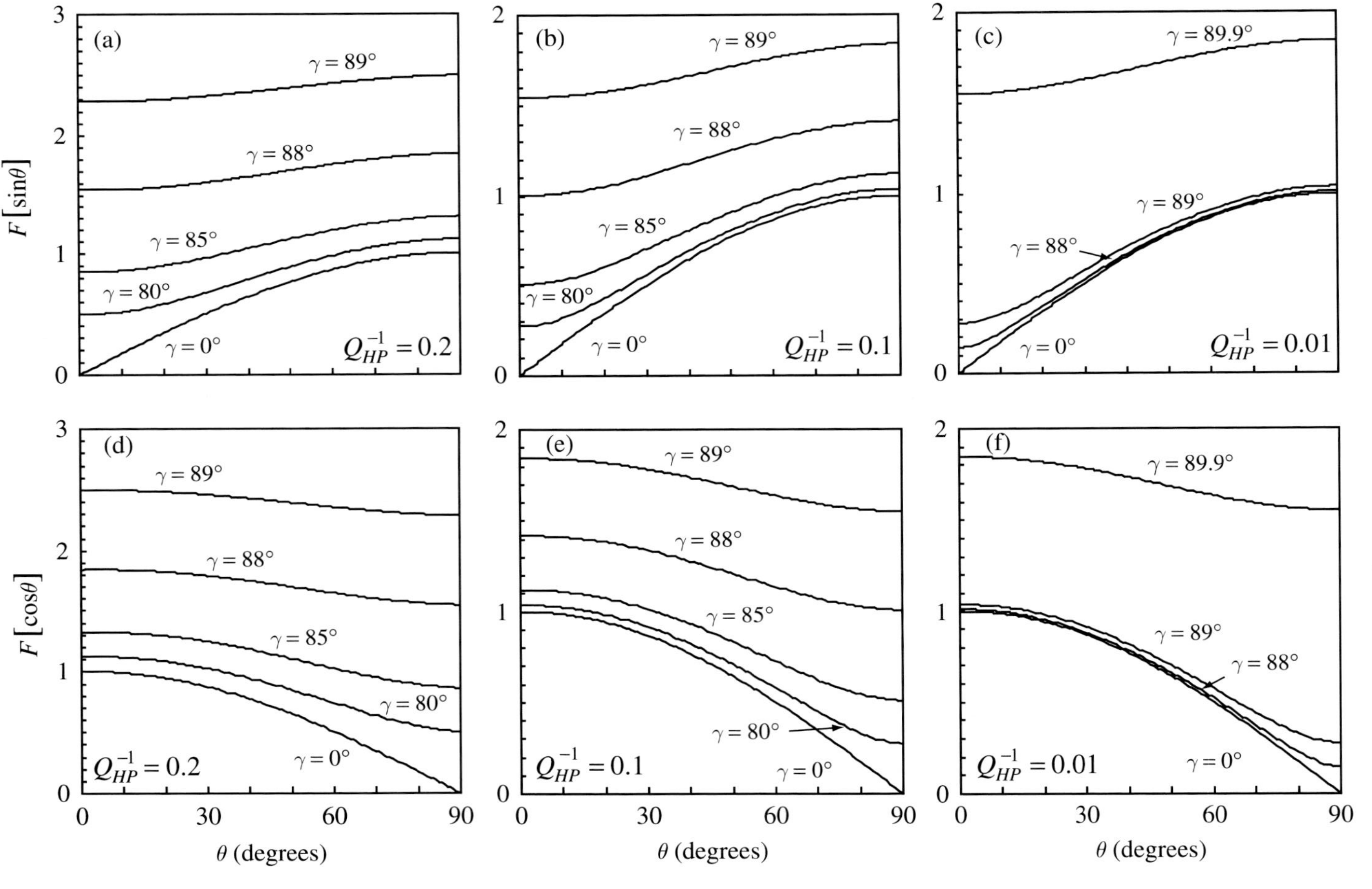

Figure (3.10.26). Amplitude modulation factors for the vertical and radial displacement components of inhomogeneous P waves computed as a function of angle of incidence θ and the degree of inhomogeneity γ for media with large, moderate, and small amounts of intrinsic absorption $\left(Q_{HP}^{-1} = 0.2, \text{a, d}; \; Q_{HP}^{-1} = 0.1, \text{b, e}; \; Q_{HP}^{-1} = 0.01, \text{c, f}\right)$. The curves are valid for general SI waves upon replacement of Q_{HP}^{-1} with Q_{HS}^{-1}.

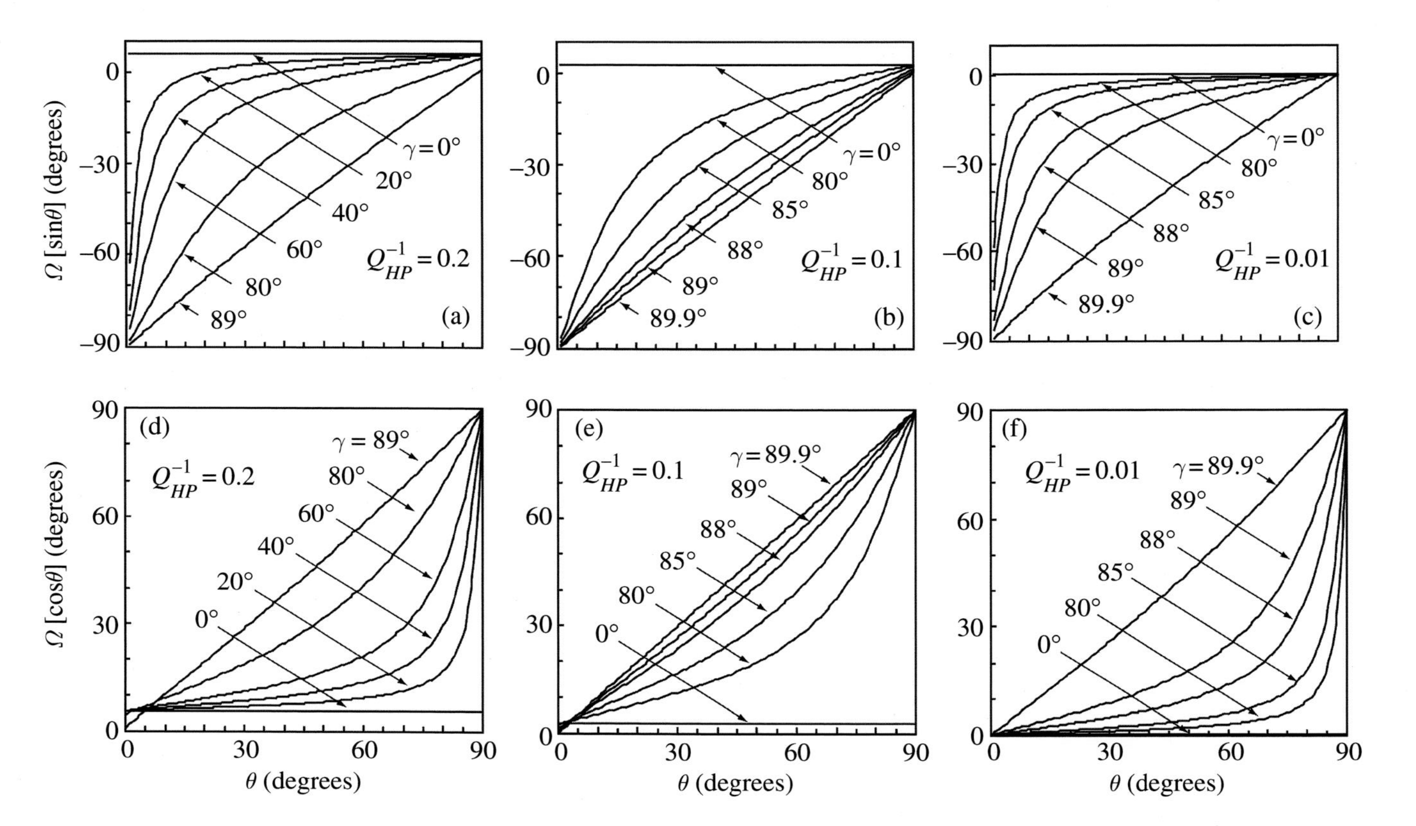

Figure (3.10.27). Phase terms for the vertical and radial displacement components for a general P wave, computed as a function of angle of incidence θ and wave-field inhomogeneity γ for media with large, moderate, and small amounts of intrinsic absorption $\left(Q_{HP}^{-1} = 0.2, \text{a, d}; \ Q_{HP}^{-1} = 0.1, \text{b, e}; \ Q_{HP}^{-1} = 0.01, \text{c, f}\right)$. The curves are valid for general SI waves upon replacement of Q_{HP}^{-1} with Q_{HS}^{-1}.

moderate ($Q_{HP}^{-1} = 0.1$), and small ($Q_{HP}^{-1} = 0.01$) amounts of intrinsic absorption. The curves as computed are valid for general SI waves upon replacement of Q_{HP}^{-1} with Q_{HS}^{-1}.

The calculations show that both the amplitude and the phase of the displacement components at each angle of incidence are influenced by the degree of inhomogeneity of the wave and the amount of intrinsic material absorption. The curves show that the dependence of the vertical and radial amplitudes on angle of incidence increasingly deviates from that of a cosinusoidal or sinusoidal dependence for corresponding homogeneous waves with increasing amounts of intrinsic absorption and wave-field inhomogeneity. For media with large amounts of absorption the dependences of the amplitude and phase modulation on angle of incidence differ significantly from those for a corresponding homogeneous wave ($\gamma = 0$) for degrees of inhomogeneity as small as 20°. For materials with small amounts of intrinsic absorption the amplitude and phase dependences on angle of incidence for inhomogeneous P and SI waves differ significantly from those for corresponding homogeneous waves only for large degrees of inhomogeneity, say $\gamma > \sim 85°$ and either large or small angles of incidence, say $\theta > \sim 75°$ or $\theta < \sim 15°$, depending on the component of interest (Figures (3.10.26) and (3.10.27)).

3.10.2 Volumetric Strain for a General P Wave

The volumetric strain associated with a general P wave as might be measured using a volumetric strain meter (Borcherdt, 1988) is given from (3.2.8) by

$$\Delta_R(t) = |G_0\, k_P| \exp\left[-\vec{A}_P \bullet \vec{r}\right] (k_{PI} \cos[\zeta_P(t)] + k_{PR} \sin[\zeta_P(t)]) \qquad (3.10.28)$$

with no such strain, of course, being associated with either type of S wave. Using identity (3.6.16) for a P wave and notation in the preceding section, this expression may be written as

$$\Delta_R(t) = |G_0\, k_P| \exp\left[-\vec{A}_P \bullet \vec{r}\right] |k_P| \cos[\zeta_P(t) - \psi_P - \pi/2]. \qquad (3.10.29)$$

Substituting the relations

$$\cos[\zeta_P(t)] = \frac{\vec{u}_R \bullet \vec{\xi}_{1P}}{D\left|\vec{\xi}_{1P}\right|^2} \qquad (3.10.30)$$

and

$$\sin[\zeta_P(t)] = \frac{\vec{u}_R \bullet \vec{\xi}_{2P}}{D\left|\vec{\xi}_{2P}\right|^2} \qquad (3.10.31)$$

into (3.10.28), where

$$D \equiv |G_0 \; k_P| \exp\left[-\vec{A}_P \bullet \vec{r}\right] \tag{3.10.32}$$

and $\vec{\xi}_{1P}$ and $\vec{\xi}_{2P}$ are defined by (3.2.12) and (3.2.13), allows the volumetric strain for an inhomogeneous P wave to be expressed in terms of the displacement field as

$$\Delta_R(t) = \frac{\omega}{v_{HP}}\left(\frac{\vec{u}_R \bullet \vec{\xi}_{2P}}{\left|\vec{\xi}_{2P}\right|^2} - \frac{Q_{HP}^{-1}}{1+\chi_{HP}}\frac{\vec{u}_R \bullet \vec{\xi}_{1P}}{\left|\vec{\xi}_{1P}\right|^2}\right) \tag{3.10.33}$$

and that for a homogeneous P wave by

$$\Delta_R(t) = \frac{\omega}{v_{HP}}\left(D\sqrt{1-\frac{\left(\vec{u}_R \bullet \vec{\xi}_{1P}\right)^2}{D^2}} - \frac{Q_{HP}^{-1}}{1+\chi_{HP}}\vec{u}_R \bullet \vec{\xi}_{1P}\right). \tag{3.10.34}$$

These expressions specify the volumetric strain for a general P wave as might be recorded on a dilatometer in a HILV whole space in terms of the corresponding displacement field for the wave as might be inferred from a three-component seismometer.

3.10.3 Simultaneous Measurement of Volumetric Strain and Displacement

Simultaneous measurement of volumetric strain and components of the displacement field as might be inferred from a dilatometer and a three-component seismometer allows information regarding the wave field and the material parameters to be inferred that cannot be inferred from either measurement alone. Expressions for elastic media relating displacement components and various components of the strain tensor have been derived by Benioff (1935), Benioff and Gutenberg (1952), Romney (1964), and Gupta (1966). Expressions derived here are valid for viscoelastic media (Borcherdt (1988).

Examples of additional information that can be derived from collocated simultaneous measurements for a general P wave propagating in a HILV whole space are implied by expressions (3.10.10) and (3.10.29). As a first example the general expression for the displacement field of a general P wave, (3.2.11), shows that the elliptical particle motion reaches its maximum along the major axis of the ellipse (i.e. $\left|\vec{\xi}_{1P}\right|$) and minimum along the minor axis of the ellipse (i.e. $\left|\vec{\xi}_{2P}\right|$) at times t_1 and t_2 when $\zeta_P(t_1)=0$ and $\zeta_P(t_2)=\pi/2$. Determination of these times from particle motion plots as might be inferred from seismometer measurements and substitution of these times into (3.10.29) shows that the ratio of corresponding values for volumetric strain is given by

$$\frac{\Delta_R(t_1)}{\Delta_R(t_2)} = -\frac{Q_{HP}^{-1}}{1+\chi_{HP}}. \tag{3.10.35}$$

Hence, measurements of volumetric strain amplitudes during the passage of an inhomogeneous P wave at times of maximum and minimum particle motion as inferred from collocated seismometer measurements when substituted into (3.10.35) permit an independent estimate of the intrinsic material absorption Q_{HP}^{-1}.

As a second example, involving inhomogeneous wave fields, consider the ratio of the maximum amplitude of the radial component of displacement as implied by (3.10.10) to the maximum volumetric strain as implied by (3.10.29), namely

$$\frac{\max[u_{Rx}]}{\max[\Delta_R]} = \frac{v_{HP}}{\omega}\sqrt{\frac{1+\chi_{HP}}{2\chi_{HP}}}F_P[\sin\theta]. \tag{3.10.36}$$

This expression shows that if the maximum amplitude of the radial component of displacement and the volumetric strain can be determined from measurements, then if independent estimates of the material parameters, v_{HP} and Q_{HP}^{-1}, are available then the function $F_P[\sin\theta]$ which depends on the angle of incidence and inhomogeneity of the P wave can be estimated. In addition the phase lag between the radial component of displacement and the volumetric strain is given by

$$2\psi_P - \Omega_P[\sin\theta] + \pi/2. \tag{3.10.37}$$

Hence measurement of this phase lag together with the estimates of the material parameters yields an estimate of $\Omega_P[\sin\theta]$. This estimate together with one for $F_P[\sin\theta]$ provides two expressions that can be solved to yield estimates of the angle of incidence and degree of inhomogeneity for the P wave.

If the wave field is known to be homogeneous from particle motion plots then the phase lag between the radial component of displacement and volumetric strain is

$$\psi_P + \frac{\pi}{2} = \tan^{-1}\left[\frac{Q_{HP}^{-1}}{1+\chi_{HP}}\right] + \frac{\pi}{2}. \tag{3.10.38}$$

Hence, measurement of the phase lag between radial displacement and volumetric strain permits an independent estimate of the intrinsic material absorption Q_{HP}^{-1}.

As another example, if the wave field is known to be homogeneous, the ratio of the maximum amplitude of one of the displacement components, say the vertical, from (3.10.11), and the maximum volumetric strain is given by

$$\frac{\max[u_{Rz}]}{\max[\Delta_R]} = \frac{v_{HP}}{\omega}\cos\theta\sqrt{\frac{1+\chi_{HP}}{2\chi_{HP}}}. \tag{3.10.39}$$

This expression suggests that if the material parameters are known, then this ratio of maximum amplitudes allows the angle of incidence to be estimated. Conversely, if the angle of incidence is known, as might be estimated from

$$\theta = \tan^{-1}[u_{Rx}/u_{Rz}], \tag{3.10.40}$$

then the ratio of maximum displacement and maximum volumetric strain might be used to estimate one or the other of the material parameters v_{HP} or Q_{HP}^{-1}.

Determination of the times that the volumetric strain signal vanishes and reaches its maximum value, namely $\zeta_P(t_1) = \psi_P + \pi$ and $\zeta_P(t_2) = \psi_P + \pi/2$, and corresponding determination of the horizontal or vertical displacement at these times yields

$$\frac{u_{Rx}(t_2)}{u_{Rx}(t_1)} = \frac{u_{Rz}(t_2)}{u_{Rz}(t_1)} = \frac{Q_{HP}^{-1}}{1 + \chi_{HP}}. \tag{3.10.41}$$

Hence, these measurements provide another means of estimating the intrinsic material absorption Q_{HP}^{-1} of the media.

For materials with small amounts of intrinsic absorption $Q_{HP}^{-1} \ll 1$ and $\chi_{\mathrm{HP}} \approx 1$, so (3.10.39) implies the ratio of maximum displacement of the vertical component to the maximum volumetric strain for a homogeneous P wave is given approximately by

$$\frac{\max[u_{Rz}]}{\max[\Delta_R]} \approx \frac{v_{HP}}{\omega} \cos\theta \tag{3.10.42}$$

and the phase lag in radians between the radial component of displacement and volumetric strain is, from (3.10.37), approximately

$$\psi_P + \frac{\pi}{2} \approx \tan^{-1}\left[\frac{Q_{HP}^{-1}}{2}\right] + \frac{\pi}{2} \approx \frac{Q_{HP}^{-1}}{2} + \frac{\pi}{2}. \tag{3.10.43}$$

Consequently, for a homogeneous P wave in low-loss anelastic media, (3.10.42) shows that the ratio of maximum amplitudes of vertical displacement to maximum volumetric strain provides an approximate estimate of one of the local constitutive properties, namely homogeneous P velocity, as scaled by the cosine of the angle of incidence and the inverse of circular frequency. Equation (3.10.43) shows that measurement of the phase lag provides an approximate estimate of another of the local constitutive properties, namely Q_{HP}^{-1}.

The measurability of the phase shift ψ_P due to intrinsic absorption depends on both the amount of intrinsic absorption and the period of the wave. For low-loss materials with $Q_{HP}^{-1} = 0.01$, the phase shift ψ_P for a one-second wave is about 0.8 millisecond (ms) and that for a 20 second wave about 16 ms. Hence, the phase shift for materials with $Q_{HP}^{-1} = 0.01$ should be detectable in the time domain using 20-second signals sampled at rates greater than about 100 sps. For shorter-period waves (e.g. one-second waves) in materials with $Q_{HP}^{-1} = 0.01$ the phase shift could be ignored at sampling rates of 100 sps and the volumetric strain signal could be regarded as an amplitude-scaled displacement signal for a homogeneous P wave in low-loss media. For elastic media, $Q_{HP}^{-1} = 0$, so definition (3.10.3) implies phase shift vanishes, that is $\psi_P = 0$.

3.11 Problems

(1) Compare characteristics of an inhomogeneous wave in linear anelastic media with those of an inhomogeneous wave in elastic media.

(2) Show that an inhomogeneous plane P wave with the direction of phase propagation not perpendicular to the direction of maximum attenuation may propagate in linear anelastic media but cannot in elastic media.

(3) Use the method of separation of variables to derive a solution to the scalar Helmholtz equation for a harmonic plane S wave as specified in Appendix 3, Chapter 11.3. Show that the complex amplitude of the solution is independent of the spatial coordinates. (*Hint*: see the procedure used in Appendix 3 to derive solutions for P waves.)

(4) Describe basic differences in the physical characteristics of homogeneous and inhomogeneous waves. Show that the expressions for wave speed, particle motion, energy flux, kinetic energy density, potential energy density, total energy density, mean rate of energy dissipation, energy velocity, and reciprocal quality factor depend on the degree of inhomogeneity.

(5) Describe basic differences in the physical characteristics of inhomogeneous Type-I and Type-II S waves. Show that the physical characteristics of particle motion, energy flux, kinetic energy density, potential energy density, total energy density, mean rate of energy dissipation, energy velocity, and reciprocal quality factor are different for the two types of S waves.

(6) Use expressions for the physical characteristics of P, SI, and SII waves in low-loss anelastic media to show that each of the characteristics, as identified in Problems 4 and 5, depends on the degree of inhomogeneity of the waves with the only exception being tilt of the particle motion ellipse for P and Type-I S waves.

(7) Derive expressions showing the dependences of wave speed and the reciprocal quality factor on frequency for inhomogeneous P, SI, and SII waves in a Maxwell solid.

(8) Derive expressions for the time-averaged energy flux for Type-I and Type-II S waves, namely

$$\langle \vec{\mathcal{J}}_{SI} \rangle = \left| \vec{G}_0 \right|^2 \exp[-2\vec{A}_S \cdot \vec{r}] \frac{\omega}{2} \left(\rho\omega^2 \vec{P}_S + 4\left(\vec{P}_S \times \vec{A}_S \right) \times \left(M_I \vec{P}_S - M_R \vec{A}_S \right) \right)$$

and

$$\langle \vec{\mathcal{J}}_{SII} \rangle = \frac{|D_{SII}|^2}{h_S} \exp[-2\vec{A}_S \bullet \vec{r}] \frac{\omega}{2} \left(\rho\omega^2 \vec{P}_S + 2\left(\vec{P}_S \times \vec{A}_S \right) \right.$$
$$\left. \times \left(M_I \vec{P}_S - M_R \vec{A}_S \right) \right),$$

from (2.5.7) or (2.5.11). Show that the mean energy fluxes for the two types of S waves are equal if and only if the waves are homogeneous.

(9) Derive expressions (3.5.8) and (3.5.9) for complex wave numbers k_P and k_S in terms of the wave speed and Q^{-1} for a corresponding homogeneous wave using (3.5.1) through (3.5.5). Explain why wave speed (v_{HS}, v_{HP}) and Q^{-1} $\left(Q_{HS}^{-1}, Q_{HP}^{-1}\right)$ for homogeneous S and P waves may be used together with density to characterize the response of a viscoelastic material for a given frequency.

(10) Use definitions and equations in Sections 3.1, 3.5, and 3.6 to derive identities (3.6.12) through (3.6.24) involving the wave speed (v_{HS}, v_{HP}) and Q^{-1} $\left(Q_{HS}^{-1}, Q_{HP}^{-1}\right)$ for homogeneous S and P waves.

(11) Use expressions derived for components of the displacement field and the volumetric strain to show that simultaneous collocated measurements of each can be used to infer constitutive properties of the media.

4

Framework for Single-Boundary Reflection–Refraction and Surface-Wave Problems

Analytic closed-form solutions for problems of body- and surface-wave propagation in layered viscoelastic media will show that the waves in anelastic media are predominantly inhomogeneous with the degree of inhomogeneity dependent on angle of incidence and intrinsic absorption. Hence, it is reasonable to expect from results in the previous chapter that the physical characteristics of refracted waves in layered anelastic media also will vary with the angle of incidence and be dependent on the previous travel path of the wave. These concepts are not encountered for waves in elastic media, because the waves traveling through a stack of layers are homogeneous with their physical characteristics such as phase speed not dependent on the angle of incidence. This chapter will provide the framework and solutions for each of the waves needed to derive analytic solutions for various reflection–refraction and surface-wave problems in subsequent chapters.

4.1 Specification of Boundary

To set up the mathematical framework for considering reflection–refraction problems at a single viscoelastic boundary and surface-wave problems, consider two infinite HILV media denoted by V and V' with a common plane boundary in welded contact (Figure (4.1.3)). For reference, the locations of the media are described by a rectangular coordinate system specified by coordinates (x_1, x_2, x_3) or (x, y, z) with the space occupied by V described by $x_3 > 0$, the space occupied by V' by $x_3 > 0$, and the plane boundary by $x_3 = 0$. For problems involving a viscoelastic half space V' will be considered a vacuum.

Wave propagation in multiple layers of viscoelastic media will be considered in subsequent chapters by extending the notation used here for single-boundary problems. The notation will be extended by eliminating the prime superscript and introducing an additional subscript to distinguish parameters for the mth layer. This additional subscript is not introduced for single-boundary problems for purposes of simplicity.

The response of a HILV medium V of density ρ shall be characterized by complex shear and bulk moduli or real parameters corresponding to the wave speed and reciprocal quality factor for homogeneous S and P waves in V, namely v_{HS}, v_{HP}, Q_{HS}^{-1}, and Q_{HP}^{-1} as indicated by (3.5.1) through (3.5.5). The corresponding complex wave numbers for homogeneous S and P waves may also be used to characterize the material as the following expressions from (3.6.12) indicate:

$$k_S = \frac{\omega}{v_{HS}}\left(1 - i\frac{Q_{HS}^{-1}}{1+\chi_{HS}}\right), \qquad k_P = \frac{\omega}{v_{HP}}\left(1 - i\frac{Q_{HP}^{-1}}{1+\chi_{HP}}\right), \tag{4.1.1}$$

where $\chi_{HS} \equiv \sqrt{1+Q_{HS}^{-2}}$ and $\chi_{HP} \equiv \sqrt{1+Q_{HP}^{-2}}$.

Similarly, the response of HILV medium V' of density ρ' shall be characterized by K' and M' or real parameters corresponding to the wave speed and reciprocal quality factor for homogeneous S and P waves in V', namely v'_{HS}, v'_{HP}, $Q_{HS}^{\prime -1}$, and $Q_{HP}^{\prime -1}$, or complex wave numbers for homogeneous S and P waves, namely

$$k'_S = \frac{\omega}{v'_{HS}}\left(1 - i\frac{Q_{HS}^{\prime -1}}{1+\chi'_{HS}}\right), \qquad k'_P = \frac{\omega}{v'_{HP}}\left(1 - i\frac{Q_{HP}^{\prime -1}}{1+\chi'_{HP}}\right). \tag{4.1.2}$$

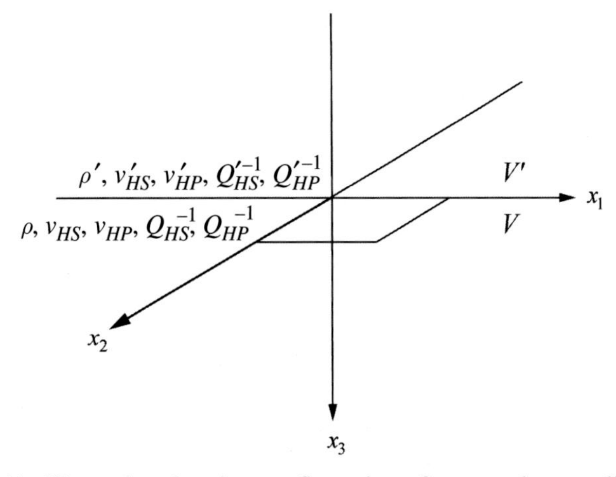

Figure (4.1.3). Illustration showing configuration of rectangular coordinate system and welded plane boundary at $x_3 = 0$ between two infinite HILV media V and V'.

4.2 Specification of Waves

Harmonic motions of media V and V' are governed by the equation of motion (2.3.17). Solutions of the equations of motion corresponding to Helmholtz equations (3.1.3) and (3.1.4) which satisfy the divergenceless gauge (3.1.2) in each medium describe harmonic motions of the media. To consider two-dimensional surface-wave and reflection–refraction problems, the solutions shall be restricted without loss of generality to those that have components of propagation and attenuation only in a plane perpendicular to the planar boundary, namely in the x_1x_3 plane. In addition, only solutions propagating in the $+x_1$ direction need be considered with incident waves assumed to originate in medium V. Upon consideration of a specific reflection–refraction problem or a particular surface wave problem, application of the boundary conditions and (3.1.2) implies that the complex wave number k associated with each solution is the same. Hence, without loss of generality the complex wave number k for each of the solutions in media V and V' are assumed to be equal.

For medium V, a set of general displacement potential solutions that satisfy the preceding restrictions for two-dimensional surface-wave and reflection–refraction problems (Borcherdt, 1971, 1973a, 1973b, 1977, 1982) as inferred from (3.1.12) is

$$\phi = \phi_1 + \phi_2 = \sum_{j=1}^{2} B_j \exp\left[i\left(\omega t - \vec{K}_{\phi_j} \bullet \vec{r}\right)\right] \tag{4.2.1}$$

and

$$\vec{\psi} = \vec{\psi}_1 + \vec{\psi}_2 = \sum_{j=1}^{2} \vec{C}_j \exp\left[i\left(\omega t - \vec{K}_{\psi_j} \bullet \vec{r}\right)\right], \tag{4.2.2}$$

where the complex wave vectors are given by

$$\vec{K}_{\phi_j} = \vec{P}_{\phi_j} - i\vec{A}_{\phi_j} = k\hat{x}_1 + (-1)^j d_\alpha \hat{x}_3, \tag{4.2.3}$$

$$\vec{K}_{\psi_j} = \vec{P}_{\psi_j} - i\vec{A}_{\psi_j} = k\hat{x}_1 + (-1)^j d_\beta \hat{x}_3, \tag{4.2.4}$$

and the corresponding propagation and attenuation vectors are given by

$$\vec{P}_{\phi_j} = k_R\hat{x}_1 + (-1)^j d_{\alpha_R} \hat{x}_3, \tag{4.2.5}$$

$$\vec{P}_{\psi_j} = k_R\hat{x}_1 + (-1)^j d_{\beta_R} \hat{x}_3, \tag{4.2.6}$$

and

$$\vec{A}_{\phi_j} = -k_I\hat{x}_1 + (-1)^{j+1} d_{\alpha_I} \hat{x}_3, \tag{4.2.7}$$

$$\vec{A}_{\psi_j} = -k_I\hat{x}_1 + (-1)^{j+1} d_{\beta_I} \hat{x}_3, \tag{4.2.8}$$

for $j = 1, 2$ with

$$d_\alpha \equiv \sqrt{k_P^2 - k^2}, \tag{4.2.9}$$

$$d_\beta \equiv \sqrt{k_S^2 - k^2}, \tag{4.2.10}$$

where "$\sqrt{\ }$" is understood to indicate the principal value of the square root of a complex number $z=x+iy=z_R+iz_I$ (Kreysig, 1967, p. 535) defined in terms of its argument as

$$\sqrt{z} \equiv \text{principal value}\left[(z)^{1/2}\right] \equiv \sqrt{z}\exp\left[i\frac{\arg z}{2}\right] \tag{4.2.11}$$

with

$$-\pi < \arg z \leq \pi \tag{4.2.12}$$

or in terms of the positive square root of real numbers by

$$\sqrt{z} = \sqrt{\frac{|z| + z_R}{2}} + i\ \text{sign}[z_I]\sqrt{\frac{|z| - z_R}{2}}, \tag{4.2.13}$$

where

$$\text{sign}[z_I] \equiv \left\{ \begin{array}{rl} 1 & \text{if} \quad z_I \geq 0 \\ -1 & \text{if} \quad z_I < 0 \end{array} \right\}. \tag{4.2.14}$$

Use of the principal value of the square root implies $d_{\alpha_R} \geq 0$ and $d_{\beta_R} \geq 0$. This selection of a root ensures a known direction of propagation for each of the specified solutions with respect to the boundary. A non-negative value for k_R ensures propagation in the positive x_1 direction. For purposes of generality $\vec{C}_j\,(j = 1,2)$ is interpreted as an arbitrary complex vector defined by

$$\vec{C}_j = C_{j1}\hat{x}_1 + C_{j2}\hat{x}_2 + C_{j3}\hat{x}_3 \tag{4.2.15}$$

such that $\nabla \bullet \vec{\psi}_j = 0$ and C_{ji} (j = 1, 2; i = 1, 2, 3) are arbitrary but fixed complex numbers. This interpretation permits (4.2.2) to describe both Type-I and Type-II S waves, depending on the characteristics of $\vec{C}_j\,(j = 1,2)$, where the complex amplitudes B_j and $\vec{C}_j\,(j = 1,2)$ are independent of the spatial coordinates (Appendix 3).

A set of solutions in medium V' that correspond to waves propagating away from and toward the boundary with circular frequency ω are specified symbolically by attaching primes to each of the wave parameters in (4.2.1) through (4.2.10). The appropriate solutions for V' are

$$\phi' = \phi'_1 + \phi'_2 = \sum_{j=1}^{2} B'_j \exp\left[i\left(\omega t - \vec{K}'_{\phi_j} \bullet \vec{r}\right)\right] \tag{4.2.16}$$

and

$$\vec{\psi}' = \vec{\psi}'_1 + \vec{\psi}'_2 = \sum_{j=1}^{2} \vec{C}'_j \exp\left[i\left(\omega t - \vec{K}'_{\psi_j} \bullet \vec{r}\right)\right] \tag{4.2.17}$$

where

$$\vec{K}'_{\phi_j} = \vec{P}'_{\phi_j} - i\vec{A}'_{\phi_j} = k\hat{x}_1 + (-1)^j d'_\alpha \hat{x}_3, \tag{4.2.18}$$

$$\vec{K}'_{\psi_j} = \vec{P}'_{\psi_j} - i\vec{A}'_{\psi_j} = k\hat{x}_1 + (-1)^j d'_\beta \hat{x}_3, \tag{4.2.19}$$

$$\vec{P}'_{\phi_j} = k_R\hat{x}_1 + (-1)^j d'_{\alpha_R} \hat{x}_3, \tag{4.2.20}$$

$$\vec{P}'_{\psi_j} = k_R\hat{x}_1 + (-1)^j d'_{\beta_R} \hat{x}_3, \tag{4.2.21}$$

$$\vec{A}'_{\phi_j} = -k_I\hat{x}_1 + (-1)^{j+1} d'_{\alpha_I} \hat{x}_3, \tag{4.2.22}$$

$$\vec{A}'_{\psi_j} = -k_I\hat{x}_1 + (-1)^{j+1} d'_{\beta_I} \hat{x}_3, \tag{4.2.23}$$

for $j = 1, 2$ with

$$d'_\alpha = \sqrt{k'^2_P - k^2} \tag{4.2.24}$$

and

$$d'_\beta = \sqrt{k'^2_S - k^2}. \tag{4.2.25}$$

For reflection–refraction problems involving a single boundary the amplitudes of the waves in V' that propagate toward the boundary are set to zero, that is $B'_2 = 0$ and $\vec{C}'_2 = 0$.

The nature of the complex vectors $\vec{C}_j$ and $\vec{C}'_1$ determines whether the displacement potential solutions represent Type-I S or Type-II S solutions. If the complex vectors are of the simple form $\vec{C}_j = C_j\,\hat{n}$, where C_j is an arbitrary, but fixed, complex number and $\hat{n}$ is a real unit vector, then as shown in (3.3.18) and in (3.3.19) $\hat{n}$ must be perpendicular to the plane (x_1, x_3) containing the propagation and attenuation vectors and, hence, parallel to $\hat{x}_2$. So, for SI solutions only the $\hat{x}_2$ component of the vector displacement potential $\vec{\psi}_j$ is needed, that is $\vec{C}_j = C_{j2}\hat{x}_2 \ (j = 1, 2)$ and $\vec{C}'_1 = C'_{12}\,\hat{x}_2$. Hence, problems involving P and SI solutions are more easily specified using the solutions of the equations of motion involving the displacement potentials ϕ_j and the $\hat{x}_2$ component of $\vec{\psi}_j$ as opposed to solutions of equation (3.3.16) involving the vector displacement field $\vec{u}$, which would require two components of the displacement field vector to describe P and SI solutions.

If the complex vectors $\vec{C}_j$ and $\vec{C}'_1$ are not of simple form, then the special case of interest for two-dimensional problems under consideration is that in which the $\hat{x}_2$ component of the vectors is zero. For this case the vector displacement potential solution describes a Type-II S wave as described by equations (3.3.34) through (3.4.13). The particle motion for a Type-II S wave is unidirectional and parallel to $\hat{x}_2$. Hence, problems involving SII solutions are more easily described using the solutions of equation (3.3.16) for the vector displacement field, as opposed to using two components of the vector displacement potential $\vec{\psi}$ to describe Type-II S solutions.

Solutions of the equation of motion involving the displacement field (3.3.16) for Type-II S waves in media V and V' are easily specified by

$$\vec{u} = \vec{u}_1 + \vec{u}_2 = \sum_{j=1}^{2} D_j \exp\left[i\left(\omega t - \vec{K}_{u_j} \bullet \vec{r}\right)\right] \hat{x}_2, \tag{4.2.26}$$

and

$$\vec{u}' = \vec{u}_1' + \vec{u}_2' = \sum_{j=1}^{2} D_j' \exp\left[i\left(\omega t - \vec{K}'_{u_j} \bullet \vec{r}\right)\right] \hat{x}_2, \tag{4.2.27}$$

where

$$\vec{K}_{u_j} = \vec{P}_{u_j} - i\vec{A}_{u_j} = k\hat{x}_1 + (-1)^j d_\beta \hat{x}_3, \tag{4.2.28}$$

$$\vec{K}'_{u_j} = \vec{P}'_{u_j} - i\vec{A}'_{u_j} = k\hat{x}_1 + (-1)^j d'_\beta \hat{x}_3, \tag{4.2.29}$$

$$\vec{P}_{u_j} = k_R\hat{x}_1 + (-1)^j d_{\beta_R} \hat{x}_3, \tag{4.2.30}$$

$$\vec{P}'_{u_j} = k_R\hat{x}_1 + (-1)^j d'_{\beta_R} \hat{x}_3, \tag{4.2.31}$$

$$\vec{A}_{u_j} = -k_I\hat{x}_1 + (-1)^{j+1} d_{\beta_I} \hat{x}_3, \tag{4.2.32}$$

$$\vec{A}'_{u_j} = -k_I\hat{x}_1 + (-1)^{j+1} d'_{\beta_I} \hat{x}_3, \tag{4.2.33}$$

for $j = 1, 2$ with d_β and d'_β defined by (4.2.10) and (4.2.25). For reflection–refraction problems involving a single boundary the amplitudes of the waves in V' that propagate toward the boundary are set to zero, that is $D'_2 = 0$.

For ease of physical interpretation of reflected and refracted waves specified with respect to the boundary, it is expedient to introduce notation designating the angles θ that the propagation vectors make with respect to the vertical and the angles γ between the corresponding attenuation and propagation vectors. The angles θ are termed the angles of incidence, reflection, or refraction depending on the corresponding wave. The angles γ specify the degree of the inhomogeneity of the corresponding wave and in turn together with θ specify the angles that the corresponding attenuation vectors make with respect to the vertical. An illustration of this notation is given in Figure (5.2.18) for the reflection–refraction problem of an incident general SI wave.

In terms of these angles, the propagation vectors in medium V may be written as

$$\vec{P}_{\phi_j} = k_R\hat{x}_1 + (-1)^j d_{\alpha_R}\hat{x}_3 = |\vec{P}_{\phi_j}|\left(\sin\left[\theta_{\phi_j}\right]\hat{x}_1 + (-1)^{j+1}\cos\left[\theta_{\phi_j}\right]\hat{x}_3\right), \quad (4.2.34)$$

$$\vec{P}_{\psi_j} = k_R\hat{x}_1 + (-1)^j d_{\beta_R}\hat{x}_3 = |\vec{P}_{\psi_j}|\left(\sin\left[\theta_{\psi_j}\right]\hat{x}_1 + (-1)^{j+1}\cos\left[\theta_{\psi_j}\right]\hat{x}_3\right), \quad (4.2.35)$$

$$\vec{P}_{u_j} = k_R\hat{x}_1 + (-1)^j d_{\beta_R}\hat{x}_3 = |\vec{P}_{u_j}|\left(\sin\left[\theta_{u_j}\right]\hat{x}_1 + (-1)^{j+1}\cos\left[\theta_{u_j}\right]\hat{x}_3\right), \quad (4.2.36)$$

and the attenuation vectors as

$$\begin{aligned}\vec{A}_{\phi_j} &= -k_I\hat{x}_1 + (-1)^{j+1} d_{\alpha_I}\hat{x}_3 \\ &= |\vec{A}_{\phi_j}|\left(\sin\left[\theta_{\phi_j} - \gamma_{\phi_j}\right]\hat{x}_1 + (-1)^{j+1}\cos\left[\theta_{\phi_j} - \gamma_{\phi_j}\right]\hat{x}_3\right),\end{aligned} \quad (4.2.37)$$

$$\begin{aligned}\vec{A}_{\psi_j} &= -k_I\hat{x}_1 + (-1)^{j+1} d_{\beta_I}\hat{x}_3 \\ &= |\vec{A}_{\psi_j}|\left(\sin\left[\theta_{\psi_j} - \gamma_{\psi_j}\right]\hat{x}_1 + (-1)^{j+1}\cos\left[\theta_{\psi_j} - \gamma_{\psi_j}\right]\hat{x}_3\right),\end{aligned} \quad (4.2.38)$$

$$\begin{aligned}\vec{A}_{u_j} &= -k_I\hat{x}_1 + (-1)^{j+1} d_{\beta_I}\hat{x}_3 \\ &= |\vec{A}_{u_j}|\left(\sin\left[\theta_{u_j} - \gamma_{u_j}\right]\hat{x}_1 + (-1)^{j+1}\cos\left[\theta_{u_j} - \gamma_{u_j}\right]\hat{x}_3\right).\end{aligned} \quad (4.2.39)$$

The corresponding propagation vectors in medium V' are

$$\vec{P}'_{\phi_j} = k_R\hat{x}_1 + (-1)^j d'_{\alpha_R}\hat{x}_3 = |\vec{P}'_{\phi_j}|\left(\sin\left[\theta'_{\phi_j}\right]\hat{x}_1 + (-1)^{j+1}\cos\left[\theta'_{\phi_j}\right]\hat{x}_3\right), \quad (4.2.40)$$

$$\vec{P}'_{\psi_j} = k_R\hat{x}_1 + (-1)^j d'_{\beta_R}\hat{x}_3 = |\vec{P}'_{\psi_j}|\left(\sin\left[\theta'_{\psi_j}\right]\hat{x}_1 + (-1)^{j+1}\cos\left[\theta'_{\psi_j}\right]\hat{x}_3\right), \quad (4.2.41)$$

$$\vec{P}'_{u_j} = k_R\hat{x}_1 + (-1)^j d'_{\beta_R}\hat{x}_3 = |\vec{P}'_{u_j}|\left(\sin\left[\theta'_{u_j}\right]\hat{x}_1 + (-1)^{j+1}\cos\left[\theta'_{u_j}\right]\hat{x}_3\right), \quad (4.2.42)$$

and the corresponding attenuation vectors in medium V' are

$$\begin{aligned}\vec{A}'_{\phi_j} &= -k_I\hat{x}_1 + (-1)^{j+1} d'_{\alpha_I}\hat{x}_3 \\ &= |\vec{A}'_{\phi_j}|\left(\sin\left[\theta'_{\phi_j} - \gamma'_{\phi_j}\right]\hat{x}_1 + (-1)^{j+1}\cos\left[\theta'_{\phi_j} - \gamma'_{\phi_j}\right]\hat{x}_3\right),\end{aligned} \quad (4.2.43)$$

$$\begin{aligned}\vec{A}'_{\psi_j} &= -k_I\hat{x}_1 + (-1)^{j+1} d'_{\beta_I}\hat{x}_3 \\ &= |\vec{A}'_{\psi_j}|\left(\sin\left[\theta'_{\psi_j} - \gamma'_{\psi_j}\right]\hat{x}_1 + (-1)^{j+1}\cos\left[\theta'_{\psi_j} - \gamma'_{\psi_j}\right]\hat{x}_3\right),\end{aligned} \quad (4.2.44)$$

$$\begin{aligned}\vec{A}'_{u_j} &= -k_I\hat{x}_1 + (-1)^{j+1} d'_{\beta_I}\hat{x}_3 \\ &= |\vec{A}'_{u_j}|\left(\sin\left[\theta'_{u_j} - \gamma'_{u_j}\right]\hat{x}_1 + (-1)^{j+1}\cos\left[\theta'_{u_j} - \gamma'_{u_j}\right]\hat{x}_3\right).\end{aligned} \quad (4.2.45)$$

For later reference, the complex wave number k for each of the solutions may be written explicitly in terms of the directions of propagation and attenuation for each solution and the parameters of the material using (3.6.21), (3.6.22), and (3.6.23) as

$$
\begin{aligned}
k_{\psi_j} &= \frac{\omega}{v_{HS}}\left(\sqrt{\frac{1+\chi_{S\psi_j}}{1+\chi_{HS}}}\sin\theta_{\psi_j} - i\sqrt{\frac{-1+\chi_{S\psi_j}}{1+\chi_{HS}}}\sin\left[\theta_{\psi_j}-\gamma_{\psi_j}\right]\right),\\
k_{u_j} &= \frac{\omega}{v_{HS}}\left(\sqrt{\frac{1+\chi_{Su_j}}{1+\chi_{HS}}}\sin\theta_{u_j} - i\sqrt{\frac{-1+\chi_{Su_j}}{1+\chi_{HS}}}\sin\left[\theta_{u_j}-\gamma_{u_j}\right]\right),\\
k_{\phi_j} &= \frac{\omega}{v_{HP}}\left(\sqrt{\frac{1+\chi_{P\phi_j}}{1+\chi_{HP}}}\sin\theta_{\phi_j} - i\sqrt{\frac{-1+\chi_{P\phi_j}}{1+\chi_{HP}}}\sin\left[\theta_{\phi_j}-\gamma_{\phi_j}\right]\right),\\
k'_{\psi_j} &= \frac{\omega}{v'_{HS}}\left(\sqrt{\frac{1+\chi'_{S\psi_j}}{1+\chi'_{HS'}}}\sin\theta'_{\psi_j} - i\sqrt{\frac{-1+\chi'_{S\psi_j}}{1+\chi'_{HS}}}\sin\left[\theta'_{\psi_j}-\gamma'_{\psi_j}\right]\right),\\
k'_{u_j} &= \frac{\omega}{v'_{HS}}\left(\frac{\sqrt{1+\chi'_{Su_j}}}{\sqrt{1+\chi'_{HS}}}\sin\theta'_{u_j} - i\sqrt{\frac{-1+\chi'_{Su_j}}{1+\chi'_{HS}}}\sin\left[\theta'_{u_j}-\gamma'_{u_j}\right]\right),\\
k'_{\phi_j} &= \frac{\omega}{v'_{HP}}\left(\sqrt{\frac{1+\chi'_{P\phi_j}}{1+\chi'_{HP}}}\sin\theta'_{\phi_j} - i\sqrt{\frac{-1+\chi'_{P\phi_j}}{1+\chi'_{HP}}}\sin\left[\theta'_{\phi_j}-\gamma'_{\phi_j}\right]\right),
\end{aligned}
\tag{4.2.46}
$$

where

$$
\begin{aligned}
&\chi_{S\psi_j} \equiv \sqrt{1+Q_{HS}^{-2}\sec^2\gamma_{\psi_j}}, \quad \chi_{Su_j} \equiv \sqrt{1+Q_{HS}^{-2}\sec^2\gamma_{u_j}}, \quad \chi_{P\phi_j} \equiv \sqrt{1+Q_{HP}^{-2}\sec^2\gamma_{\phi_j}},\\
&\chi'_{S\psi_j} \equiv \sqrt{1+Q_{HS}'^{-2}\sec^2\gamma_{\psi_j}}, \quad \chi'_{Su_j} \equiv \sqrt{1+Q_{HS}'^{-2}\sec^2\gamma'_{u_j}}, \quad \chi'_{P_{\phi_j}} \equiv \sqrt{1+Q_{HP}'^{-2}\sec^2\gamma'_{\phi_j}}
\end{aligned}
\tag{4.2.47}
$$

and

$$
\begin{aligned}
\chi_{HS} &\equiv \sqrt{1+Q_{HS}^{-2}}, & \chi_{HP} &\equiv \sqrt{1+Q_{HP}^{-2}},\\
\chi'_{HS} &\equiv \sqrt{1+Q_{HS}'^{-2}}, & \chi'_{HP} &\equiv \sqrt{1+Q_{HP}'^{-2}}
\end{aligned}
\tag{4.2.48}
$$

for $j=1,2$. Application of the boundary conditions will imply the complex wave number for each of the solutions is equal. Hence, without loss of generality, each wave number is designated by k with $k=k_{\psi_j}=k_{u_j}=k_{\phi_j}=k'_{\psi_j}=k'_{\psi_j}=k'_{u_j}=k'_{\phi_j}$ for $j=1,2$.

4.3 Problems

(1) Express the complex wave number k as specified in the solutions for an incident P wave in (4.2.1), (4.2.3), and (4.2.9) in terms of the directions of phase propagation θ_{ϕ_1} and maximum attenuation $(\theta_{\phi_1} - \gamma_{\phi_1})$ for the incident wave, material parameters, and circular frequency as

$$k = \frac{\omega}{v_{HP}} \left(\sqrt{\frac{1 + \chi_{P\phi_1}}{1 + \chi_{HP}}} \sin \theta_{\phi_1} - i \sqrt{\frac{-1 + \chi_{P\phi_1}}{1 + \chi_{HP}}} \sin \left[\theta_{\phi_1} - \gamma_{\phi_1} \right] \right).$$

(2) Derive the expression for the wave number for an incident homogeneous P wave using the expression derived in Problem 1 for
   (a) an arbitrary viscoelastic solid,
   (b) an elastic solid, and
   (c) a Voight solid.

(3) Derive an expression for the wave speed and absorption coefficient along the viscoelastic interface of an incident P wave using the expression for the complex wave number derived in Problem 1. For a fixed angle of incidence how do the apparent wave speed and absorption coefficient along the interface change as
   (a) the amount of intrinsic absorption in the incident wave increases, that is as $Q_{HP}^{-1} \to \infty$, and
   (b) the degree of inhomogeneity of the incident wave approaches its physical limit, that is $\gamma_{\phi_1} \to \pi/2$.

5

General P, SI, and SII Waves Incident on a Viscoelastic Boundary

A theoretical closed-form solution for the problem of a general P, SI, or SII wave incident on a plane welded boundary between HILV media, V and V', is one for which the characteristics of the reflected and refracted waves are expressed in terms of the assumed characteristics of the incident wave. Application of the boundary conditions at the boundary allows the amplitude and phase for the reflected and refracted waves to be expressed in terms of the properties of the media and those given for the incident wave. The directions of the propagation and attenuation vectors for the reflected and refracted waves are determined in terms of those of the incident wave by showing that the complex wave number for each solution must be the same. For problems involving incident P and SI waves, the boundary conditions are most readily applied using the solutions involving displacement potentials, namely, (4.2.1), (4.2.2), (4.2.16), and (4.2.17). For problems involving incident SII waves, the boundary conditions can be applied most easily using solutions involving only one component of the displacement field, namely (4.2.26) and (4.2.27).

5.1 Boundary-Condition Equations for General Waves

The welded boundary between media V and V' is specified mathematically by requiring that the stress and displacement are continuous across the boundary. For purposes of brevity, application of these boundary conditions to the general solutions specifying each type of wave as incident, reflected, or refracted allows a general set of equations to be derived from which a particular problem of interest can be solved by choosing the incident wave of interest.

Substitution of solutions (4.2.1), (4.2.2), (4.2.16), and (4.2.17) into the expressions relating the displacement field and the displacement potentials, namely (3.1.1) and (3.1.2), applying identities in Appendix 2, and requiring that the displacements on each side of the boundary when evaluated at $x_3 = 0$ are equal yields a set of equations describing the displacement fields for the wave fields on the two sides of the boundary. Substituting these expressions for the displacement fields into the expressions relating the components of the displacement fields to components of the strain tensor (2.3.5) and substituting these in turn into the expression relating stress and strain (2.3.12) yields a set of equations for the components of the stress tensor on both sides of the boundary acting on the plane $x_3 = 0$. The resulting continuity of displacement and stress equations with k taken as an arbitrary complex number yet to be specified in terms of the propagation and attenuation characteristics of the incident wave are

$$u_1 = u_1';$$
$$k(B_1 + B_2) + d_\beta(C_{12} - C_{22}) = k\left(B_1' + B_2'\right) + d_\beta'\left(C_{12}' - C_{22}'\right), \quad (5.1.1)$$

$$u_2 = u_2';$$
$$d_\beta(C_{11} - C_{21}) + k(C_{13} - C_{23}) = d_\beta'\left(C_{11}' - C_{21}'\right) + k\left(C_{13}' - C_{23}'\right), \quad (5.1.2)$$

$$u_3 = u_3';$$
$$d_\alpha(B_1 - B_2) - k(C_{12} + C_{22}) = d_\alpha'\left(B_1' - B_2'\right) - k\left(C_{12}' + C_{22}'\right), \quad (5.1.3)$$

$$p_{31} = p_{31}';$$
$$M\left[2kd_\alpha(B_1 - B_2) + \left(d_\beta^2 - k^2\right)(C_{12} + C_{22})\right]$$
$$= M'\left[2kd_\alpha'\left(B_1' - B_2'\right) + \left(d_\beta'^2 - k^2\right)\left(C_{12}' + C_{22}'\right)\right], \quad (5.1.4)$$

$$p_{32} = p_{32}';$$
$$Md_\beta\left[d_\beta(C_{11} + C_{21}) + k(C_{13} - C_{23})\right]$$
$$= M'd_\beta'\left[d_\beta'\left(C_{11}' + C_{21}'\right) + k\left(C_{13}' - C_{23}'\right)\right], \quad (5.1.5)$$

$$p_{33} = p_{33}';$$
$$M\left[-\left(d_\beta^2 - k^2\right)(B_1 + B_2) + 2d_\beta k(C_{12} - C_{22})\right]$$
$$= M'\left[-\left(d_\beta'^2 - k^2\right)\left(B_1' + B_2'\right) + 2d_\beta' k\left(C_{12}' - C_{22}'\right)\right]. \quad (5.1.6)$$

Choice of solutions for a single incident wave and choice of solutions for reflected and refracted waves with phase propagation away from the boundary will allow these equations to be solved for the complex amplitudes of the reflected and refracted waves in terms of the complex amplitude and complex wave number for the specified incident wave.

5.2 Incident General SI Wave

The physical problem of a general SI wave incident on a viscoelastic boundary is specified by choosing solutions (4.2.1), (4.2.2), (4.2.16), and (4.2.17) to represent the incident SI and reflected and refracted SI and P waves, where the amplitudes of the incident P solution and the incident P and SI solutions in V' with phase propagation toward the boundary are set to zero, that is

$$B_1 = B_2' = C_{21}' = C_{22}' = C_{23}' = 0. \qquad (5.2.1)$$

The boundary conditions imply that no SII waves are generated at the boundary, hence without loss of generality the complex vector amplitude of the vector displacement potential $\vec{\psi}$ is of simple form for incident, reflected, and refracted SI waves and the $\hat{x}_1$ and $\hat{x}_3$ components for each of the waves vanish, namely

$$C_{11} = C_{13} = C_{21} = C_{23} = C_{11}' = C_{13}' = 0. \qquad (5.2.2)$$

5.2.1 Specification of Incident General SI Wave

The expression for the assumed or specified incident general SI wave as given by (4.2.2) through (4.2.38) with $j = 1$ is

$$\vec{\psi}_1 = C_{12}\,\hat{x}_2 \exp\left[i\left(\omega t - \vec{K}_{\psi_1} \bullet \vec{r}\right)\right], \qquad (5.2.3)$$

where

$$\vec{K}_{\psi_1} = \vec{P}_{\psi_1} - i\vec{A}_{\psi_1} = k\hat{x}_1 + (-1)^j d_\beta \hat{x}_3, \qquad (5.2.4)$$

$$\vec{P}_{\psi_1} = k_R\hat{x}_1 - d_{\beta_R}\hat{x}_3 = \left|\vec{P}_{\psi_1}\right|\left(\sin\left[\theta_{\psi_1}\right]\hat{x}_1 + \cos\left[\theta_{\psi_1}\right]\hat{x}_3\right), \qquad (5.2.5)$$

$$\vec{A}_{\psi_1} = -k_I\hat{x}_1 + d_{\beta_I}\hat{x}_3 = \left|\vec{A}_{\psi_1}\right|\left(\sin\left[\theta_{\psi_1} - \gamma_{\psi_1}\right]\hat{x}_1 + \cos\left[\theta_{\psi_1} - \gamma_{\psi_1}\right]\hat{x}_3\right), \qquad (5.2.6)$$

and the complex wave number k, written in terms of the given parameters of the incident general SI wave, is given by

$$k = \left|\vec{P}_{\psi_1}\right| \sin\theta_{\psi_1} - i\left|\vec{A}_{\psi_1}\right| \sin\left[\theta_{\psi_1} - \gamma_{\psi_1}\right], \qquad (5.2.7)$$

where, to ensure propagation of the incident wave in V, the magnitudes of the propagation and attenuation vectors for the incident wave are specified from (3.1.20) and (3.1.21) in terms of the given parameters of the material, $k_S = \omega/\beta = \omega/\sqrt{M/\rho}$, and the given degree of inhomogeneity γ_{ψ_1} by

$$\left|\vec{P}_{\psi_1}\right| = \sqrt{\frac{1}{2}\left(\text{Re}\left[k_S^2\right] + \sqrt{\left(\text{Re}\left[k_S^2\right]\right)^2 + \left(\text{Im}\left[k_S^2\right]\right)^2 \sec^2\gamma_{\psi_1}}\right)}$$
$$= \frac{\omega}{v_{HS}}\sqrt{\frac{1+\chi_{S\psi_1}}{1+\chi_{HS}}} \tag{5.2.8}$$

and

$$\left|\vec{A}_{\psi_1}\right| = \sqrt{\frac{1}{2}\left(-\text{Re}\left[k_S^2\right] + \sqrt{\left(\text{Re}\left[k_S^2\right]\right)^2 + \left(\text{Im}\left[k_S^2\right]\right)^2 \sec^2\gamma_{\psi_1}}\right)}$$
$$= \frac{\omega}{v_{HS}}\sqrt{\frac{-1+\chi_{S\psi_1}}{1+\chi_{HS}}}, \tag{5.2.9}$$

with γ_{ψ_1} chosen such that $\gamma_{\psi_1} = 0$ for elastic media and $\gamma_{\psi_1} < 90°$, for anelastic media with $\chi_{S\psi_1} = \sqrt{1 + Q_{HS}^{-2}\sec^2\gamma_{\psi_1}}$ and $\chi_{HS} = \sqrt{1 + Q_{HS}^{-2}}$, where the expressions in terms of $\chi_{S\psi_1}$ follow from (3.6.17) and (3.6.18).

The complex wave number k for the incident general SI wave (5.2.7) as given by (4.2.46) shows an explicit dependence on an assumed degree of inhomogeneity and angle of incidence in anelastic media, namely

$$k = \frac{\omega}{v_{HS}}\left(\sqrt{\frac{1+\chi_{S_{\psi_1}}}{1+\chi_{HS}}}\sin\theta_{\psi_1} - i\sqrt{\frac{-1+\chi_{S_{\psi_1}}}{1+\chi_{HS_{\psi_1}}}}\sin\left[\theta_{\psi_1} - \gamma_{\psi_1}\right]\right). \tag{5.2.10}$$

In elastic media k shows only a dependence on angle of incidence as indicated by

$$k = \frac{\omega}{v_{HS}}\sin\theta_{\psi_1}. \tag{5.2.11}$$

For the special case that the incident SI wave is assumed to be homogeneous, (5.2.7) and (5.2.10) for k simplify for HILV media to

$$k = \left(\left|\vec{P}_{\psi_1}\right| - i\left|\vec{A}_{\psi_1}\right|\right)\sin\theta_{\psi_1} = k_S\sin\theta_{\psi_1} \tag{5.2.12}$$

and

$$k = \frac{\omega}{v_{HS}}\left(1 - i\frac{Q_{HS}^{-1}}{1+\chi_{HS}}\right)\sin\theta_{\psi_1}. \tag{5.2.13}$$

Expressions (5.2.5) through (5.2.13) show explicitly that the complex wave number k and the magnitude of the propagation and attenuation vectors for the incident general SI wave are completely determined upon specification of (1) the angle of incidence θ_{ψ_1} for the propagation vector, (2) the degree of inhomogeneity γ_{ψ_1} or the angle of incidence for the attenuation vector $\theta_{\psi_1} - \gamma_{\psi_1}$ in anelastic media, (3) the material parameters as specified by ρ, k_S or v_{HS} and Q_{HS}^{-1}, and (4) the circular frequency ω.

The expression for the assumed or specified general incident SI wave as given by (5.2.3) through (5.2.13) is a general formulation needed to consider two-dimensional wave-propagation problems in layered viscoelastic media. The general formulation as initially presented (Borcherdt, 1971, 1982) allows the incident SI wave to be chosen as either a homogeneous or an inhomogeneous wave for problems in which the incident medium V is anelastic and as only a homogeneous wave if V is elastic.

The general formulation is needed in order to consider wave propagation in anelastic media with more than one boundary. It will be shown in subsequent sections that plane waves refracted at boundaries between anelastic media with different amounts of intrinsic absorption are in general inhomogeneous. Hence, in anelastic media with more than one boundary the refracted wave will encounter the second layer as an inhomogeneous wave. It is not necessary to consider such problems for elastic media, because in those situations in which the refracted elastic wave is inhomogeneous the directions of the propagation and attenuation vectors are parallel and perpendicular to the boundary, respectively. The problem of an incident homogeneous wave as considered by Lockett (1962), Cooper and Reiss (1966), and Cooper (1967) is a special case of the more general formulation presented here.

5.2.2 Propagation and Attenuation Vectors; Generalized Snell's Law

Expressions for the solutions describing the reflected and refracted general P and SI waves as specified by (4.2.1), (4.2.2), (4.2.16), and (4.2.17) are in medium V

$$\phi_2 = B_2 \exp\left[i\left(\omega t - \vec{K}_{\phi_2} \bullet \vec{r}\right)\right] \tag{5.2.14}$$

and

$$\vec{\psi}_2 = C_{22} \exp\left[i\left(\omega t - \vec{K}_{\psi_2} \bullet \vec{r}\right)\right] \hat{x}_2, \tag{5.2.15}$$

and in medium V'

$$\phi'_1 = B'_1 \exp\left[i\left(\omega t - \vec{K}'_{\phi_1} \bullet \vec{r}\right)\right] \tag{5.2.16}$$

and

$$\vec{\psi}'_1 = C'_{12} \exp\left[i\left(\omega t - \vec{K}'_{\psi_1} \bullet r\right)\right] \hat{x}_2, \tag{5.2.17}$$

where $\vec{K}_{\phi_2}$, $\vec{K}_{\psi_2}$, $\vec{K}'_{\phi_1}$, $\vec{K}'_{\psi_1}$ and the other parameters of the reflected and transmitted waves are defined in (4.2.3) through (4.2.44). Parameters of the incident, reflected and transmitted waves for the problem of an incident general SI wave are illustrated in Figure (5.2.18).

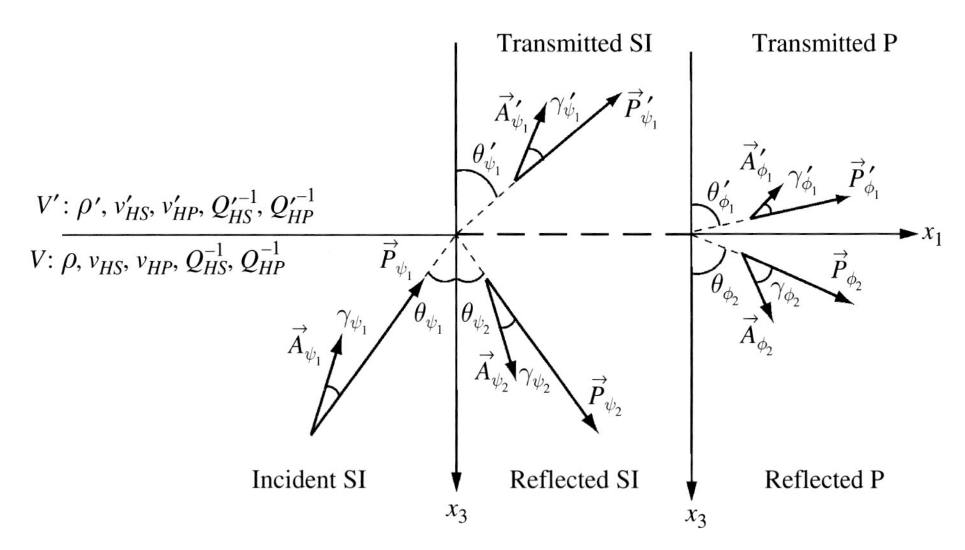

Figure (5.2.18). Diagram illustrating notation for direction and magnitude of propagation and attenuation vectors for the reflected and refracted waves due to a general SI wave incident on a welded boundary between HILV media.

Specification of the complex wave number k using (5.2.10) in terms of the given directions of propagation and attenuation $(\theta_{\psi 1}, \theta_{\psi_1} - \gamma_{\psi_1})$ for the incident SI wave and the parameters of the media ρ and k_S or v_{HS} and Q^{-1}_{HS} implies that $d_\beta = \sqrt{k_S^2 - k^2}$, $d_\alpha = \sqrt{k_P^2 - k^2}$, $d'_\beta = \sqrt{k'^2_S - k^2}$, and $d'_\alpha = \sqrt{k'^2_P - k^2}$ are determined. Hence, the magnitude and directions of the propagation and attenuation vectors for the reflected and refracted general SI and P waves as defined by (4.2.5) through (4.2.8) and (4.2.20) through (4.2.23) are determined in terms of the given directions for the incident general SI wave.

Equality of the complex wave number k assumed without loss of generality in each of the displacement-potential solutions for the incident, reflected, and refracted waves as evident in expressions (4.2.34) through (4.2.44) yields a generalized form of Snell's Law. Equality of the real part of k is familiar from optics and elasticity theory. It implies that

$$\begin{aligned} k_R = \left|\vec{P}_{\psi_1}\right| \sin\theta_{\psi_1} = \left|\vec{P}_{\psi 2}\right| \sin\theta_{\psi 2} = \left|\vec{P}_{\phi 2}\right| \sin\theta_{\phi 2} = \left|\vec{P}'_{\psi_2}\right| \sin\theta'_{\psi 2} \\ = \left|\vec{P}'_{\phi_2}\right| \sin\theta'_{\phi 2}, \end{aligned} \tag{5.2.19}$$

which may be rewritten in terms of the phase speeds of the waves as

$$\frac{k_R}{\omega} = \frac{\sin\theta_{\psi 1}}{\left|\vec{v}_{\psi 1}\right|} = \frac{\sin\theta_{\psi 2}}{\left|\vec{v}_{\psi 2}\right|} = \frac{\sin\theta_{\phi 2}}{\left|\vec{v}_{\phi 2}\right|} = \frac{\sin\theta'_{\psi 2}}{\left|\vec{v}'_{\psi 2}\right|} = \frac{\sin\theta'_{\phi 2}}{\left|\vec{v}'_{\phi 2}\right|}. \tag{5.2.20}$$

Equality of the imaginary part of k implies

$$\begin{aligned} -k_I &= \left|\vec{A}_{\psi_1}\right| \sin\left[\theta_{\psi_1} - \gamma_{\psi_1}\right] = \left|\vec{A}_{\psi 2}\right| \sin\left[\theta_{\psi 2} - \gamma_{\psi 2}\right] = \left|\vec{A}_{\phi 2}\right| \sin\left[\theta_{\phi_2} - \gamma_{\phi 2}\right] \\ &= \left|\vec{A}'_{\psi 2}\right| \sin\left[\theta'_{\psi_2} - \gamma'_{\psi_2}\right] = \left|\vec{A}'_{\phi_2}\right| \sin\left[\theta'_{\phi_2} - \gamma'_{\phi_2}\right]. \end{aligned} \tag{5.2.21}$$

Equality of the real part of the complex wave number k as indicated in (5.2.19) and (5.2.20) shows that the component of the propagation vector along the boundary for the reflected and refracted general P and SI waves must be equal to that of the incident general SI wave. When rewritten as (5.2.20), it shows that the reciprocal of the apparent phase velocity along the boundary of the general reflected and refracted waves must equal that of the general incident wave. It also shows that the $\hat{x}_1$ component of the complex wave vector $\vec{K}$ for each of the reflected and refracted waves is equal to that of the assumed incident wave.

Equality of the imaginary part of the complex wave number k as indicated in (5.2.21) implies that the component of the attenuation vector along the boundary of the reflected and refracted waves must be equal to that of the incident wave.

The results as stated specifically by (5.2.20) and (5.2.21) may be restated as an extension or generalization of Snell's Law as

Theorem (5.2.22). Generalized Snell's Law – For the problem of a general SI wave incident on a welded viscoelastic boundary in a plane perpendicular to the boundary,

(1) the reciprocal of the apparent phase velocity along the boundary of the general reflected and refracted waves is equal to that of the given general incident wave,

and

(2) the apparent attenuation along the boundary of the general reflected and refracted waves is equal to that of the given general incident wave.

The *Generalized Snell's Law* as stated in Theorem (5.2.22) and described explicitly by (5.2.19) through (5.2.21) together with the conditions to ensure propagation in V and V', namely (3.1.20) and (3.1.21), provide the laws of reflection and refraction for plane waves incident on parallel plane, welded viscoelastic boundaries in a plane perpendicular to the boundaries. Specifically, they permit the directions of phase propagation and maximum attenuation for each of the reflected and refracted waves to be determined in terms of the corresponding given directions for the incident wave and the given material parameters.

Generalized Snell's Law has important implications for wave propagation in layered viscoelastic media. It implies that in order to describe the propagation of waves through layered anelastic media two types of rays are needed, rays perpendicular to surfaces of constant amplitude and rays perpendicular to surfaces of constant phase, where the angle between the rays at each boundary is determined by the angles of incidence of the respective rays and the contrast in material properties at the boundary.

Theorem (5.2.22) includes the results for elastic media as a special case. If the media are elastic and the incident wave is homogeneous, then the component of attenuation for the incident wave along the boundary vanishes. Hence, the components of attenuation for the reflected and refracted waves must vanish or be perpendicular to the boundary. The result of phase propagation parallel to the boundary and attenuation perpendicular to the boundary for elastic media is of course a classical result known to occur for waves refracted at angles of incidence beyond a so-called elastic critical angle. A similar result cannot occur in anelastic media, because inhomogeneous waves for which the degree of inhomogeneity is 90° cannot propagate in an anelastic viscoelastic medium (Theorem (3.1.18)). Hence, the problems of reflection and refraction of incident inhomogeneous waves in viscoelastic media have no counterpart in elastic media.

At first appearance the first part of Generalized Snell's Law concerning equality of apparent phase velocity along the boundary is a result similar to the result for elastic media. However, the result as stated for viscoelastic media is considerably more general, in that the phase velocities of the incident, reflected, and refracted waves as indicated in (5.2.20) are those of general waves which may be inhomogeneous. As a result the phase velocities, maximum attenuations, and other physical characteristics of the reflected and refracted waves are dependent on the assumed degree of inhomogeneity of the incident wave, the angle of incidence, the contrast in viscoelastic material parameters at the boundary, and frequency. Generalized Snell's Law implies the magnitude of the velocity ratios and the refraction indices as implied by (5.2.20) are not constant as a function of angle of incidence as they are for elastic media. Instead, for anelastic viscoelastic media, they vary with angle of incidence as do other physical characteristics of the refracted waves, such as wave speed and absorption coefficient. Hence, the implications of Generalized Snell's Law for characteristics of the wave fields propagated in layered anelastic media are more profound than they are for elastic media.

5.2.3 Amplitude and Phase

Selection of a single incident SI wave as indicated in (5.2.1) and (5.2.2) reduces the boundary condition equations (5.1.1) through (5.1.6) to four equations in four unknowns. The resulting equations to be solved for the complex amplitude of the reflected and refracted waves in terms of that for the incident SI wave (C_{12}) are

$$kB_2 + d_\beta(C_{12} - C_{22}) = kB_1' + d_\beta' C_{12}', \tag{5.2.23}$$

$$-d_\alpha B_2 - k(C_{12} + C_{22}) = d'_\alpha B'_1 - kC'_{12}, \tag{5.2.24}$$

$$M\left[-2kd_\alpha B_2 + \left(d_\beta^2 - k^2\right)(C_{12} + C_{22})\right] = M'\left[2kd'_\alpha B'_1 + \left(d'^2_\beta - k^2\right)C'_{12}\right], \tag{5.2.25}$$

$$M\left[-\left(d_\beta^2 - k^2\right)B_2 + 2d_\beta k(C_{12} - C_{22})\right] = M'\left[-\left(d'^2_\beta - k^2\right)B'_1 + 2d'_\beta kC'_{12}\right]. \tag{5.2.26}$$

Written in matrix form the equations to be solved are

$$\begin{bmatrix} -k & d_\beta & k & d'_\beta \\ -d_\alpha & -k & -d'_\alpha & k \\ -2Mk\,d_\alpha & -M\left(d_\beta^2 - k^2\right) & 2M'k\,d'_\alpha & M'\left(d'^2_\beta - k^2\right) \\ M\left(d_\beta^2 - k^2\right) & 2k\,d_\beta & -M'\left(d'^2_\beta - k^2\right) & 2M'k\,d'_\beta \end{bmatrix} \begin{bmatrix} B_2 \\ C_{22} \\ B'_1 \\ C'_{12} \end{bmatrix} = C_{12}\begin{bmatrix} d_\beta \\ k \\ M\left(d_\beta^2 - k^2\right) \\ 2Mk\,d_\beta \end{bmatrix}. \tag{5.2.27}$$

Equation (5.2.27) implies that the amplitude and phase of the displacement potentials for the reflected and refracted general P and SI waves are determined in terms of those for the given incident general SI wave and the given wave number k for the incident wave as specified by (5.2.10). The existence of a solution to (5.2.27) shows that a solution to the problem can be found if the complex wave number k is the same for each of the assumed wave-field solutions. Hence, with k given by (5.2.10) for the incident general SI wave, equality of k for each of the appropriate assumed solutions as specified by (4.2.34) through (4.2.44) is established. Or equivalently, Generalized Snell's Law as stated in Theorem (5.2.22) for the problem of an incident general SI wave is established. Hence, from a mathematical point of view the desired solution for the directions and magnitudes of propagation and attenuation and the amplitude and phase of the reflected and refracted general SI and P waves in terms of those for the general incident SI wave is established.

5.2.4 Conditions for Homogeneity and Inhomogeneity

Considering first the case of the reflected SI wave, the expressions for the propagation and attenuation vectors in terms of k and d_β, namely (4.2.35) and (4.2.38), immediately imply that

$$|\vec{P}_{\psi_1}| = |\vec{P}_{\psi_2}| \tag{5.2.28}$$

and

$$|\vec{A}_{\psi_1}| = |\vec{A}_{\psi_2}|, \tag{5.2.29}$$

which indicates that the phase velocity and maximum attenuation of the reflected SI wave equals that of the incident wave. Hence, (5.2.19) and (5.2.21) imply that the angles of reflection and incidence for the propagation vectors of the reflected and incident SI waves are equal, that is

$$\theta_{\psi_1} = \theta_{\psi_2}, \tag{5.2.30}$$

and that the degrees of inhomogeneity for the reflected and incident SI waves are equal, that is

$$\gamma_{\psi_1} = \gamma_{\psi_2}. \tag{5.2.31}$$

Hence, the angles of reflection and incidence for the attenuation vectors of the reflected and incident SI waves are equal, that is

$$\theta_{\psi_2} - \gamma_{\psi_2} = \theta_{\psi_1} - \gamma_{\psi_1}. \tag{5.2.32}$$

Conditions of homogeneity and inhomogeneity for the reflected SI wave may be restated formally from (5.2.31) as

Theorem (5.2.33). For the problem of a general SI wave incident on a welded viscoelastic boundary, the reflected SI wave is homogeneous if and only if the incident SI wave is homogeneous.

The contrapositive of Theorem (5.2.33) implies that the reflected SI wave is inhomogeneous if and only if the incident wave is inhomogeneous.

The conditions for inhomogeneity of the reflected P wave and the transmitted P and SI waves are described by the following results.

Theorem (5.2.34). For the problem of a general SI wave incident on a welded viscoelastic boundary, if the incident SI wave is homogeneous $(\gamma_{\psi_1} = 0)$ *and not normally incident* $(0 < \theta_{\psi_1} \le \pi/2)$, *then*

(1) the reflected P wave is homogeneous if and only if

$$Q_{HS}^{-1} = Q_{HP}^{-1} \quad \text{and} \quad \sin^2\theta_{\psi_1} \le \frac{k_P^2}{k_S^2} = \frac{M_R}{K_R + \frac{4}{3}M_R} = \frac{v_{HS}^2}{v_{HP}^2},$$

(2) the transmitted SI wave is homogeneous if and only if

$$Q_{HS}^{-1} = Q_{HS}^{\prime -1} \qquad \text{and} \qquad \sin^2 \theta_{\psi_1} \leq \frac{k_S^{\prime 2}}{k_S^2} = \frac{\rho' M_R}{\rho M_R'} = \frac{v_{HS}^2}{v_{HS}^{\prime 2}},$$

(3) the transmitted P wave is homogeneous if and only if

$$Q_{HS}^{-1} = Q_{HP}^{\prime -1} \qquad \text{and} \qquad \sin^2 \theta_{\psi_1} \leq \frac{k_P^{\prime 2}}{k_S^2} = \frac{\rho' M_R}{\rho \left(K_R' + \frac{4}{3} M_R'\right)} = \frac{v_{HS}^2}{v_{HP}^{\prime 2}}.$$

A lemma useful in proving Theorem (5.2.34) is

Lemma (5.2.35). k_P^2/k_S^2 is a real number if and only if $Q_{HS}^{-1} = Q_{HP}^{-1}$, in which case $k_P^2/k_S^2 = M_R/\left(K_R + \frac{4}{3} M_R\right) = v_{HS}^2/v_{HP}^2$.

The proof of this lemma follows from the following result implied by equations (3.5.8) and (3.5.9), namely,

$$\frac{k_P^2}{k_S^2} = \frac{v_{HS}^2}{v_{HP}^2} \frac{1 - i\, Q_{HP}^{-1}}{1 - i\, Q_{HS}^{-1}} \frac{1 + \sqrt{1 + Q_{HS}^{-2}}}{1 + \sqrt{1 + Q_{HP}^{-2}}}. \tag{5.2.36}$$

To prove the "only if" part of (5.2.34) assume the reflected P wave is homogeneous (i.e. $\gamma_{\phi_2} = 0$), then (5.2.19), (5.2.21), (3.1.13), (3.1.14), and (4.2.9) imply

$$k^2 = k_P^2 \sin^2 \theta_{\phi_2} \tag{5.2.37}$$

and

$$d_\alpha^2 = k_P^2 - k^2 = k_P^2 \cos^2 \theta_{\phi_2}. \tag{5.2.38}$$

These relations, together with (5.2.12), imply k_P^2/k_S^2 is a real number for $\theta_{\psi_1} \neq 0$ and

$$\frac{d_\alpha^2}{k^2} = \frac{k_P^2}{k_S^2 \sin^2 \theta_{\psi_1}} - 1 \geq 0. \tag{5.2.39}$$

Hence, Lemma (5.2.40) and (5.2.39) yield the desired conclusion that

$$Q_{HS}^{-1} = Q_{HP}^{-1} \qquad \text{and} \qquad \sin^2 \theta_{\psi_1} \leq \frac{k_P^2}{k_S^2} \leq \frac{M_R}{K_R + \frac{4}{3} M_R} = \frac{v_{HS}^2}{v_{HP}^2}. \tag{5.2.41}$$

Conversely, if (5.2.41) is valid, then k_P^2/k_S^2 is a real number from which it follows that d_α^2/k^2 is a non-negative real number. Therefore, d_α/k is a real number, say c, which implies

$$d_{\alpha_R} = c\, k_R \qquad \text{and} \qquad d_{\alpha_I} = c\, k_I. \tag{5.2.42}$$

Substitution of (5.2.42) into (4.2.34) and (4.2.37) with $j=2$ shows that the propagation and attenuation vectors for the reflected P wave are parallel, which yields the desired conclusion, namely the reflected P wave is homogeneous. The results for the transmitted SI and P waves in Theorem (5.2.34) may be proved in a similar fashion.

An implication of a part of the contrapositive of Theorem (5.2.34) indicates that the reflected P and transmitted P and SI waves are in general inhomogeneous as stated efficiently in the next theorem.

Theorem (5.2.43). For the problem of a general SI wave incident on a welded viscoelastic boundary, if the incident SI wave is homogeneous $(\gamma_{\psi_1} = 0)$ *and not normally incident* $(0 < \theta_{\psi_1} \leq \pi/2)$, *and*

(1) *if* $Q_{HS}^{-1} \neq Q_{HP}^{-1}$, *then the reflected P wave is inhomogeneous,*

(2) *if* $Q_{HS}^{-1} \neq Q_{HP}'^{-1}$, *then the transmitted P wave is inhomogeneous,*

(3) *if* $Q_{HS}^{-1} \neq Q_{HS}'^{-1}$, *then the transmitted SI wave is inhomogeneous.*

If the incident SI wave is both homogeneous and normally incident, then (5.2.19) and (5.2.21) of Generalized Snell's Law imply the following result:

Theorem (5.2.44). For the problem of a general SI wave incident on a welded viscoelastic boundary, if the incident P wave is homogeneous $(\gamma_{\phi_1} = 0)$ *and normally incident* $(\theta_{\phi_1} = 0)$, *then the reflected and refracted SI and P waves are homogeneous.*

If the media are elastic, then Theorems (5.2.34) and (5.2.44) with $Q_{HS}^{-1} = Q_{HP}^{-1} = Q_{HS}'^{-1} = Q_{HP}'^{-1} = 0$ yield the well-known conditions for homogeneity and inhomogeneity of the reflected and refracted waves.

For realistic anelastic materials the fractional energy loss for homogeneous P waves is not equal to that of corresponding S waves. In addition, a boundary between anelastic media implies, in general, that the intrinsic absorption for corresponding wave types on the two sides of the boundary is different. Consequently, the preceding results establish the important result for anelastic media that with the exception of normal incidence, the reflected P and transmitted P and SI waves are in general inhomogeneous even if the incident SI wave is homogeneous. In contrast, for elastic media the preceding results show that the reflected P wave and transmitted P and SI waves are homogeneous for all angles of incidence less than a possible critical angle.

As the intrinsic attenuations across anelastic boundaries in the Earth are, in general, not equal (i.e. $Q_{HS}^{-1} \neq Q_{HP}^{-1}, Q_{HS}^{-1} \neq Q_{HS}'^{-1}, Q_{HS}^{-1} \neq Q_{HP}'^{-1}$) the preceding results establish for the problem under consideration and suggest, in general, the important result that P and S body waves in a layered anelastic Earth are in general inhomogeneous for all angles of incidence except normal even if the incident wave is homogeneous. This important result implies that the theoretical characteristics of body waves in a layered anelastic Earth in addition to attenuation are theoretically distinct from those for homogeneous waves in a layered elastic Earth. Specifically, the characteristics of inhomogeneous waves as derived in Chapter 3 imply that their velocity is less, their maximum attenuation is greater, their particle motions are elliptical, their directions of maximum energy flux are different from that of phase propagation, their mean kinetic and potential energy densities are not equal, and their mean rate of energy dissipation is not in the direction of maximum attenuation in comparison with the corresponding characteristics of homogeneous waves. In addition, for viscoelastic media two types of inhomogeneous S waves need to be considered, one with elliptical particle motions and the other with linear particle motion, each attenuating at a different rate.

An important theoretical implication of Generalized Snell's Law is that the distinguishing physical characteristics of inhomogeneous waves vary as the angle of incidence of the incident wave varies. Consequently, the characteristics of a body wave propagating through a stack of anelastic layers will depend on its previous travel path through the stack. For example, if it entered the stack at normal incidence, it will travel faster and attenuate less in each layer than it will if it entered the stack at a non-normal angle of incidence. In contrast, for layered elastic models for waves propagating as body waves at angles less than critical, characteristics of the waves at any point in the stack are uniquely determined by elastic material parameters and hence they are not dependent on the previous travel path.

Considering the results derived for the problem of an incident general SI wave, it is of interest to reconsider the correspondence principle as stated by Bland (1960) and discussed in Chapter 2. Boundary condition equation (5.2.27) is valid for elastic media with material parameters ($M, M', k_S, k_P, k_S', k_P'$) and the wave number k being a real number. For elastic media the incident SI wave is assumed homogeneous, hence (5.2.11) implies the wave number for the incident wave in elastic media is given by $k = k_S \sin \theta_{\psi_1}$. However, conversely, if the real material parameters are replaced with corresponding complex material parameters for the solution to the elastic problem, then the complex wave number, as given by $k = k_S \sin \theta_{\psi_1}$ for the incident wave, describes only the problem of an incident homogeneous wave in viscoelastic media. Similarly, if wave numbers used to describe the reflected and refracted waves are similar to those used to describe body waves in elastic media, except that the material parameters are replaced with complex numbers, then the expressions for the wave numbers of the reflected and refracted waves, namely $k = k_S \sin \theta_{\psi_2}$, $k = k_P \sin \theta_{\phi_2}$, $k = k_S' \sin \theta'_{\psi_2}$, and $k = k_P' \sin \theta'_{\phi_2}$, are each those of homogeneous waves. Theorems (5.2.34) and (5.2.43) show that in general the reflected and refracted waves are inhomogeneous waves, hence the expressions for the wave numbers obtained by replacement of the real material parameters with complex ones yield expressions for the wave numbers of the reflected and refracted waves in viscoelastic media that do not correspond to those needed to describe an inhomogeneous wave. Expressions for the complex wave numbers needed to describe reflected and refracted waves are given by (4.2.46).

The preceding observations indicate that the reflection–refraction problem for viscoelastic media is not *identical* in the sense used by Bland (1960) to that for elastic media with the only exception being that elastic is replaced by viscoelastic. As additional evidence that the problems are not "identical", the concept of an incident inhomogeneous wave in elastic media is not well defined as it is for anelastic media as shown following Theorem (3.1.19). Hence, one of the criteria for application of the correspondence principle, as stated by Bland (1960, p. 67), is not satisfied, indicating that the principle cannot be applied to solve reflection–refraction problems for viscoelastic media.

The boundary condition equation (5.2.27) derived for viscoelastic media is similar to that derived for elastic media except that the material parameters are in general complex numbers and the wave number k is that corresponding to a general SI wave with arbitrary, but fixed, angles of inhomogeneity and phase incidence. Hence, the boundary condition equation for viscoelastic media can be written using that derived for elastic media; however, its solution requires that the given complex wave number k be separated into its real and imaginary parts in order to interpret the physical characteristics of incident inhomogeneous waves for which there are no corresponding counterparts in elastic media.

5.2.5 Conditions for Critical Angles

Critical angles as predicted by elasticity have been used in a variety of ways in seismology. They have been used to calculate the travel time for the classic refraction arrival (Ewing *et al.*, 1957, p. 93). They have been used to facilitate interpretation of "wide-angle" reflection data. As a consequence, it is of interest to investigate the concept for layered viscoelastic media.

A critical angle for an elastic reflected wave is generally thought of as the minimum angle of incidence for which total internal reflection occurs. For a transmitted wave it is generally regarded as the minimum angle of incidence for which the corresponding transmitted solution represents a wave with phase propagation parallel to the boundary and maximum attenuation perpendicular to the boundary. Angles of incidence greater than the critical angle are often termed super-critical or simply angles of incidence beyond the critical angle. For simplicity here, all angles of incidence for which the corresponding propagation vector is parallel to the interface will be referred to as critical angles for the corresponding wave with the smallest of these referred to as the minimum critical angle.

The familiar conditions for elastic media that give rise to phase propagation parallel to the boundary or elastic critical angles for the problem of an incident SI wave follow immediately from Theorem (5.2.34) with $Q_{HS}^{-1} = Q_{HP}^{-1} = Q_{HS}'^{-1} = Q_{HP}'^{-1} = 0$ and the definitions of d_α, d_β, d'_α, and d'_β given by (4.2.9), (4.2.10), (4.2.24), and (4.2.25).

Theorem (5.2.45). For the problem of a homogeneous SI (SV) wave incident on a plane boundary between elastic media V and V', conditions for phase propagation parallel to the boundary or critical angles are as follows:

(1) each angle of incidence satisfying $\sin\theta_{\psi_1} \geq v_{HS}/v_{HP}$ *is a critical angle for the reflected P wave,*

(2) if $v'_{HP} < v_{HS}$, *then no critical angles exist for the transmitted waves,*

(3) if $v'_{HS} < v_{HS} < v'_{HP}$, *then the angles of incidence satisfying* $\sin\theta_{\psi_1} \geq v_{HS}/v'_{HP}$ *are critical angles for the transmitted P wave and no critical angles exist for the transmitted SI wave,*

(4) if $v_{HS} < v'_{HS}$, *then the angles of incidence satisfying* $\sin\theta_{\psi_1} \geq v_{HS}/v'_{HP}$ *are critical angles for the transmitted P wave and the angles of incidence satisfying* $\sin\theta_{\psi_1} \geq v_{HS}/v'_{HS}$ *are critical angles for the transmitted SI (SV) wave.*

These results for elastic media show that critical angles or phase propagation parallel to the boundary exist whenever the phase speeds of the reflected or transmitted homogeneous waves exceed those of the incident wave. For such situations for a particular reflected or transmitted wave all angles of incidence greater than the minimum critical angle yield phase propagation parallel to the boundary. The phase speed of these inhomogeneous waves varies as the apparent phase speed of the incident SI wave varies along the boundary. This example illustrates that the phase speed of the only type of inhomogeneous wave that can exist in elastic media (see Theorem (3.1.17)) is not unique for a given degree of inhomogeneity, namely 90°, as it is for inhomogeneous waves in anelastic media.

If the incident medium V is elastic and the refraction medium V' is anelastic, then the conditions for critical angles change. The definition of d_α (4.2.9) and Theorems (5.2.45) and (3.1.18) immediately imply the following result.

Theorem (5.2.46). For the problem of a homogeneous SI wave incident in an elastic medium V on a plane boundary with anelastic medium V',

(1) each angle of incidence satisfying $\sin\theta_{\psi_1} \geq v_{HS}/v_{HP}$ *is a critical angle for the reflected P wave,*

and

(2) no critical angles exist for either transmitted wave.

This theorem shows that as a result of the finite relaxation times associated with an anelastic medium phase propagation for the resultant transmitted waves is never exactly parallel to the boundary, but away from the boundary for all angles of incidence, in contrast to the situation described in Theorem (5.2.45) for elastic media.

If medium V is anelastic, then the conditions for the existence of critical angles are specified by the following results.

Theorem (5.2.47). For the problem of a general (homogeneous or inhomogeneous) SI wave incident on a plane boundary between an anelastic medium V and a viscoelastic medium V', if the angle of incidence θ_{ψ_1} $(\theta_{\psi_1} \neq \pi/2)$ is a critical angle for

(1) the reflected P wave, then

$$\tan \gamma_{\psi_1} = \left(\sin^2 \theta_{\psi_1} - \frac{k_{P_R} k_{P_I}}{k_{S_R} k_{S_I}} \right) \Big/ \left(\sin \theta_{\psi_1} \cos \theta_{\psi_1} \right),$$

(2) the transmitted SI wave, then

$$\tan \gamma_{\psi_1} = \left(\sin^2 \theta_{\psi_1} - \frac{k'_{S_R} k'_{S_I}}{k_{S_R} k_{S_I}} \right) \Big/ \left(\sin \theta_{\psi_1} \cos \theta_{\psi_1} \right),$$

and

(3) the transmitted P wave , then

$$\tan \gamma_{\psi_1} = \left(\sin^2 \theta_{\psi_1} - \frac{k'_{P_R} k'_{P_I}}{k_{S_R} k_{S_I}} \right) \Big/ \left(\sin \theta_{\psi_1} \cos \theta_{\psi_1} \right).$$

To prove the first part of the theorem assume θ_{ψ_1} is a critical angle for the reflected P wave, that is $\vec{P}_{\phi_2}$ is parallel to the boundary, so (4.2.34) implies $d_{\alpha R}=0$ and hence

$$\operatorname{Im}\left[d_\alpha^2\right] = 2 d_{\alpha_R} d_{\alpha_I} = 2 k_{P_R} k_{P_I} - 2 k_R k_I = 0. \tag{5.2.48}$$

Equations (5.2.7) and (3.1.14) imply

$$\begin{aligned} \operatorname{Im}\left[k^2\right] &= -2 \left|\vec{P}_{\psi_1}\right| \left|\vec{A}_{\psi_1}\right| \sin \theta_{\psi_1} \sin\left[\theta_{\psi_1} - \gamma_{\psi_1}\right] \\ &= -2 k_{S_R} k_{S_I} \frac{\sin \theta_{\psi_1} \sin\left[\theta_{\psi_1} - \gamma_{\psi_1}\right]}{\cos \gamma_{\psi_1}}. \end{aligned} \tag{5.2.49}$$

Substitution of (5.2.49) into (5.2.48) yields

$$\sin \theta_{\psi_1} \sin\left[\theta_{\psi_1} - \gamma_{\psi_1}\right] = \cos \gamma_{\psi_1} \left(k_{P_R} k_{P_I}\right) / \left(k_{S_R} k_{S_I}\right),$$

which simplifies with trigonometric identities for $\theta_{\psi_1} \neq \pi/2$ to the desired result. Proofs of parts (2) and (3) of the theorem are similar.

An immediate corollary of Theorem (5.2.47) is that if medium V′ is elastic, then for a given value of γ_{ψ_1} there exists at most one angle of incidence, namely $\theta_{\psi_1} = \gamma_{\psi_1}$ such that θ_{ψ_1} is a critical angle. Theorem (5.2.47) implies that for an anelastic–viscoelastic boundary, if phase propagation occurs exactly parallel to the boundary, then it occurs for at most one angle of incidence. Hence, Theorem (5.2.47) indicates that phase propagation exactly parallel to the boundary rarely occurs for anelastic boundaries, while Theorem (5.2.45) shows that for elastic boundaries it occurs for a range in angles of incidence for problems in which the speeds of the reflected or refracted waves are greater than those of the incident waves. As a consequence the concept of a critical angle as defined for elastic boundaries with phase propagation parallel to the boundary is not applicable in the same way to problems involving anelastic boundaries.

5.3 Incident General P Wave

Derivation of the solution for the problem of a general P wave incident on a welded viscoelastic boundary is analogous to the preceding derivation for the problem of an incident SI wave. Key steps of the derivation are briefly provided here. They illustrate the analogy and yield the desired solution for the problem of an incident general P wave (Borcherdt, 1971, 1982).

5.3.1 Specification of Incident General P Wave

The wave solutions for the physical problem of a general (homogeneous or inhomogeneous) P wave incident on a welded viscoelastic boundary at an arbitrary angle of incidence and polarized in a plane perpendicular to the boundary are specified by solutions (4.2.1), (4.2.2), (4.2.16), and (4.2.17) where the amplitudes of the incident SI wave in V and the P and SI waves in V' propagating toward the boundary are set equal to zero, namely

$$C_{11} = C_{12} = C_{13} = C'_{21} = C'_{22} = C'_{23} = B'_2 = 0. \tag{5.3.1}$$

The boundary conditions imply that no SII waves are generated at the boundary, hence without loss of generality

$$C_{21} = C_{23} = C'_{11} = C'_{13} = 0. \tag{5.3.2}$$

The assumed or specified incident P wave is given by (4.2.1) through (4.2.37) with $j = 1$. These expressions for the assumed incident general P wave are:

$$\phi_1 = B_1 \exp\left[i\left(\omega t - \vec{K}_{\phi_1} \bullet \vec{r}\right)\right], \tag{5.3.3}$$

where the complex wave vector is

$$\vec{K}_{\phi_1} = \vec{P}_{\phi_1} - i\vec{A}_{\phi_1}, \tag{5.3.4}$$

the propagation and attenuation vectors are

$$\vec{P}_{\phi_1} = k_R\hat{x}_1 - d_{\alpha_R}\hat{x}_3 = |\vec{P}_{\phi_1}|\left(\sin\left[\theta_{\phi_1}\right]\hat{x}_1 + \cos\left[\theta_{\phi_1}\right]\hat{x}_3\right), \tag{5.3.5}$$

$$\vec{A}_{\phi_1} = -k_I\hat{x}_1 + d_{\alpha_I}\hat{x}_3 = |\vec{A}_{\phi_1}|\left(\sin\left[\theta_{\phi_1} - \gamma_{\phi_1}\right]\hat{x}_1 + \cos\left[\theta_{\phi_1} - \gamma_{\phi_1}\right]\hat{x}_3\right), \tag{5.3.6}$$

the complex wave number for the assumed general P wave is

$$k = |\vec{P}_{\phi_1}|\sin\theta_{\phi_1} - i|\vec{A}_{\phi_1}|\sin\left[\theta_{\phi_1} - \gamma_{\phi_1}\right], \tag{5.3.7}$$

where, to ensure propagation of the incident general P wave in V, the magnitudes of the propagation and attenuation vectors for the incident wave are specified from (3.1.20) and (3.1.21) in terms of the given parameters of the material, $k_P = \omega/\alpha = \omega\Big/\sqrt{\left(K + \frac{4}{3}M\right)/\rho}$ or v_{HP} and Q_{HP}^{-1}, and the given degree of inhomogeneity γ_{ϕ_1} by

$$|\vec{P}_{\phi_1}| = \sqrt{\frac{1}{2}\left(\mathrm{Re}[k_P^2] + \sqrt{\left(\mathrm{Re}[k_P^2]\right)^2 + \left(\mathrm{Im}[k_P^2]\right)^2\sec^2\gamma_{\phi_1}}\right)} = \frac{\omega}{v_{HP}}\sqrt{\frac{1+\chi_{P\phi_1}}{1+\chi_{HP}}} \tag{5.3.8}$$

and

$$|\vec{A}_{\phi_1}| = \sqrt{\frac{1}{2}\left(-\mathrm{Re}[k_P^2] + \sqrt{\left(\mathrm{Re}[k_P^2]\right)^2 + \left(\mathrm{Im}[k_P^2]\right)^2\sec^2\gamma_{\phi_1}}\right)}$$
$$= \frac{\omega}{v_{HP}}\sqrt{\frac{-1+\chi_{P\phi_1}}{1+\chi_{HP}}}, \tag{5.3.9}$$

where γ_{ϕ_1} is chosen such that $\gamma_{\phi_1} = 0$ for elastic media and $\gamma_{\phi_1} < 90°$ for anelastic media with $\chi_{P\phi_1} = \sqrt{1 + Q_{HP}^{-2}\sec^2\gamma_{\phi_1}}$ and $\chi_{HP} = \sqrt{1 + Q_{HP}^{-2}}$. k may be written explicitly in terms of the given angle of incidence θ_{ϕ_1}, degree of inhomogeneity γ_{ϕ_1} of the incident wave, and material parameters as

$$k = \frac{\omega}{v_{HP}}\left(\sqrt{\frac{1+\chi_{P\phi_1}}{1+\chi_{HP}}}\sin\theta_{\phi_1} - i\sqrt{\frac{-1+\chi_{P\phi_1}}{1+\chi_{HP}}}\sin\left[\theta_{\phi_1} - \gamma_{\phi_1}\right]\right) \tag{5.3.10}$$

with the corresponding expression for k in elastic media being

$$k = \frac{\omega}{v_{HP}} \sin \theta_{\phi_1}. \tag{5.3.11}$$

Hence, (5.3.5) through (5.3.11) show explicitly that the complex wave number k and the propagation and attenuation vectors for the incident general P wave are completely determined upon specification of the angle of incidence θ_{ϕ_1}, degree of inhomogeneity γ_{ϕ_1}, circular frequency , and material parameters ρ and k_P or v_{HP} and Q_{HP}^{-1}.

5.3.2 Propagation and Attenuation Vectors; Generalized Snell's Law

Expressions for the solutions describing the reflected and refracted general P and SI waves are specified by (4.2.1), (4.2.2), (4.2.16), and (4.2.17) with other parameters of the reflected and refracted waves defined in (4.2.3) through (4.2.44). Parameters for the problem of a general P wave incident on a welded boundary are illustrated in Figure (5.3.12).

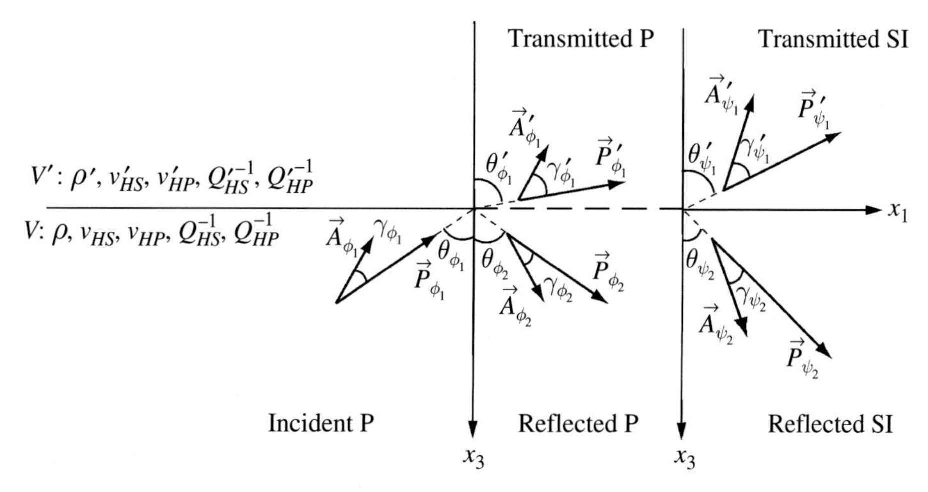

Figure (5.3.12). Diagram illustrating notation for directions and magnitude of propagation and attenuation vectors for the reflected and refracted waves due to a general P wave incident on a welded boundary between HILV media.

Specification of the complex wave number k in terms of the given directions of propagation and attenuation (θ_{ϕ_1}, $\theta_{\phi_1} - \gamma_{\phi_1}$) for the incident P wave and the parameters of the media (k_S, k_P) implies that the magnitude and directions of the propagation and attenuation vectors for the reflected and refracted general SI and P waves as defined by (4.2.5) through (4.2.8) and (4.2.20) through (4.2.23) are determined in terms of the given directions for the incident general P wave. These expressions and equality of the complex wave number k for each of the solutions yields the components of Generalized Snell's Law, namely

$$k_R = \left|\vec{P}_{\phi_1}\right| \sin\theta_{\phi_1} = \left|\vec{P}_{\psi_2}\right| \sin\theta_{\psi_2} = \left|\vec{P}_{\phi_2}\right| \sin\theta_{\phi_2} = \left|\vec{P}'_{\psi_2}\right| \sin\theta'_{\psi_2} = \left|\vec{P}'_{\phi_2}\right| \sin\theta'_{\phi_2} \tag{5.3.13}$$

or

$$\frac{k_R}{\omega} = \frac{\sin\theta_{\phi_1}}{\left|\vec{v}_{\phi_1}\right|} = \frac{\sin\theta_{\psi_2}}{\left|\vec{v}_{\psi_2}\right|} = \frac{\sin\theta_{\phi_2}}{\left|\vec{v}_{\phi_2}\right|} = \frac{\sin\theta'_{\psi_2}}{\left|\vec{v}'_{\psi_2}\right|} = \frac{\sin\theta'_{\phi_2}}{\left|\vec{v}'_{\phi_2}\right|} \tag{5.3.14}$$

and

$$-k_I = \left|\vec{A}_{\phi_1}\right| \sin\left[\theta_{\phi_1} - \gamma_{\phi_1}\right] = \left|\vec{A}_{\psi_2}\right| \sin\left[\theta_{\psi_2} - \gamma_{\psi_2}\right] = \left|\vec{A}_{\phi_2}\right| \sin\left[\theta_{\phi_2} - \gamma_{\phi_2}\right] = \left|\vec{A}'_{\psi_2}\right| \sin\left[\theta'_{\psi_2} - \gamma'_{\psi_2}\right] = \left|\vec{A}'_{\phi_2}\right| \sin\left[\theta'_{\phi_2} - \gamma'_{\phi_2}\right], \tag{5.3.15}$$

which in turn may be restated as

Theorem (5.3.16). Generalized Snell's Law – For the problem of a general P wave incident on a welded viscoelastic boundary in a plane perpendicular to the boundary,

(1) the reciprocal of the apparent phase velocity along the boundary of the general reflected and refracted waves is equal to that of the given general incident wave,

and

(2) the apparent attenuation along the boundary of the general reflected and refracted waves is equal to that of the given general incident wave.

A discussion of the implications of Generalized Snell's Law is entirely analogous to that provided for the problem of an incident SI wave. For brevity a similar discussion is omitted here.

5.3.3 Amplitude and Phase

Specifications (5.3.1) and (5.3.2) imply that the boundary-condition equations (5.1.1) through (5.1.6) reduce to four equations in four unknowns. The resulting equations to be solved are

$$k(B_1 + B_2) - d_\beta C_{22} = k B'_1 + d'_\beta C'_{12}, \tag{5.3.17}$$

$$d_\alpha(B_1 - B_2) - k C_{22} = d'_\alpha B'_1 - k C'_{12}, \tag{5.3.18}$$

$$M\left[2k d_\alpha(B_1 - B_2) + \left(d_\beta^2 - k^2\right) C_{22}\right] = M'\left[2k d'_\alpha B'_1 + \left(d'^2_\beta - k^2\right) C'_{12}\right], \tag{5.3.19}$$

$$M\left[-\left(d_\beta^2 - k^2\right)(B_1 + B_2) - 2\,d_\beta k\,C_{22}\right] = M'\left[-\left(d_\beta'^2 - k^2\right)B_1' + 2\,d_\beta' k\,C_{12}'\right]. \tag{5.3.20}$$

Written in matrix notation the equations to be solved for the complex amplitude of the reflected and refracted general P and SI waves in terms of that for the incident general P wave (B_1) are

$$\begin{bmatrix} -k & d_\beta & k & d_\beta' \\ d_\alpha & k & d_\alpha' & -k \\ 2Mkd_\alpha & -M\left(d_\beta^2 - k^2\right) & 2M'kd_\alpha' & M'\left(d_\beta'^2 - k^2\right) \\ -M\left(d_\beta^2 - k^2\right) & -2Mkd_\beta & M'\left(d_\beta'^2 - k^2\right) & -2M'k\,d_\beta' \end{bmatrix} \begin{bmatrix} B_2 \\ C_{22} \\ B_1' \\ C_{12}' \end{bmatrix}$$
$$= B_1 \begin{bmatrix} k \\ d_\alpha \\ 2\,M\,k\,d_\alpha \\ M\left(d_\beta^2 - k^2\right) \end{bmatrix}. \tag{5.3.21}$$

As for the incident SI problem, the existence of the solution to (5.3.21) confirms that the complex wave number, k, is the same in each of the solutions for the incident, reflected, and transmitted waves. Hence, with k given by (5.3.10), Generalized Snell's Law (5.3.16) is established for the problem of a general (homogeneous or inhomogeneous) P wave incident on a plane boundary between viscoelastic media. Equation (5.3.21) implies that the amplitude and phase of the displacement potentials for the reflected and refracted general P and SI waves are determined in terms of those for the given incident general P wave and the given wave number k for the incident wave. Hence, from a mathematical point of view the desired solution for the directions and magnitudes of propagation and attenuation and the amplitude and phase of the reflected and refracted general SI and P waves in terms of those for the general P wave incident on a plane boundary between specified HILV media is established.

5.3.4 Conditions for Homogeneity and Inhomogeneity

The proofs for conditions for homogeneity and inhomogeneity of the reflected and refracted waves generated by an incident general P wave are analogous to those for the incident general SI wave problem. For brevity, only the results are given here.

Theorem (5.3.22). For the problem of a general P wave incident on a welded viscoelastic boundary, the reflected P wave is homogeneous if and only if the incident P wave is homogeneous.

The conditions for the reflected SI and transmitted P and SI waves are as follows.

Theorem (5.3.23). For the problem of a general P wave incident on a welded viscoelastic boundary, if the incident P wave is homogeneous $(\gamma_{\phi_1} = 0)$ *and not normally incident* $(0 < \theta_{\phi_1} \leq \pi/2)$, *then*

(1) *the reflected SI wave is homogeneous if and only if* $Q_{HS}^{-1} = Q_{HP}^{-1}$,

(2) *the transmitted P wave is homogeneous if and only if*

$$Q_{HP}^{-1} = Q_{HP}'^{-1} \qquad \textit{and} \qquad \sin^2\theta_{\phi_1} \leq \frac{k_P'^2}{k_P^2} = \frac{\rho'\left(K_R + \frac{4}{3}M_R\right)}{\rho\left(K_R' + \frac{4}{3}M_R'\right)} = \frac{v_{HP}^2}{v_{HP}'^2},$$

(3) *the transmitted SI wave is homogeneous if and only if*

$$Q_{HP}^{-1} = Q_{HS}'^{-1} \qquad \textit{and} \qquad \sin^2\theta_{\phi_1} \leq \frac{k_S'^2}{k_P^2} = \frac{\rho'\left(K_R + \frac{4}{3}M_R\right)}{\rho M_R'} = \frac{v_{HP}^2}{v_{HS}'^2}.$$

Theorem (5.3.24). For the problem of a general P wave incident on a welded viscoelastic boundary, if the incident P wave is homogeneous $(\gamma_{\phi_1} = 0)$ *and not normally incident* $(0 < \theta_{\phi_1} \leq \pi/2)$, *and*

(1) *if* $Q_{HP}^{-1} \neq Q_{HS}^{-1}$, *then the reflected SI wave is inhomogeneous,*

(2) *if* $Q_{HP}^{-1} \neq Q_{HP}'^{-1}$, *then the transmitted P wave is inhomogeneous,*

(3) *if* $Q_{HP}^{-1} \neq Q_{HS}'^{-1}$, *then the transmitted P wave is inhomogeneous.*

Theorem (5.3.25). For the problem of a general P wave incident on a welded viscoelastic boundary, if the incident P wave is homogeneous $(\gamma_{\phi_1} = 0)$ *and normally incident* $(\theta_{\phi_1} = 0)$, *then the reflected and refracted P and SI waves are homogeneous.*

If the media are elastic, then Theorems (5.3.23) and (5.3.25) with $Q_{HS}^{-1} = Q_{HP}^{-1} = Q_{HS}'^{-1} = Q_{HP}'^{-1} = 0$ yield the well-known conditions for homogeneity and inhomogeneity of reflected and refracted elastic waves.

These results are completely analogous to those for the incident SI wave problem. They imply that for layered anelastic media the reflected SI wave and the transmitted waves will, in general, be inhomogeneous. A detailed discussion of the implications of this result is provided following Theorem (5.2.44).

5.3.5 Conditions for Critical Angles

The results for critical angles or angles of incidence for which phase propagation occurs parallel to the boundary also are analogous to those derived for the incident general SI wave problem. They are restated here for completeness.

Results for elastic boundaries implied by Theorem (5.3.23) with $Q_{HS}^{-1} = Q_{HP}^{-1} = Q_{HS}'^{-1} = Q_{HP}'^{-1} = 0$ are given by the following theorem.

Theorem (5.3.26). For the problem of a homogeneous P wave incident on a plane boundary between elastic media V and V', conditions for phase propagation parallel to the boundary or critical angles are as follows:

(1) no critical angles exist for either reflected wave,

(2) if $v_{HP} > v'_{HP}$, then no critical angles exist for the transmitted waves,

(3) if $v'_{HS} < v_{HP} < v'_{HP}$, then the angles of incidence satisfying $\sin\theta_{\phi_1} \geq v_{HP}/v'_{HP}$ are critical angles for the transmitted P wave and no critical angles exist for the transmitted SI wave,

(4) if $v_{HP} < v'_{HS}$ then the angles of incidence satisfying $\sin\theta_{\phi_1} \geq v_{HS}/v'_{HP}$ are critical angles for the transmitted P wave and angles of incidence satisfying $\sin\theta_{\phi_1} \geq v_{HP}/v'_{HS}$ are critical angles for the transmitted SI (SV) wave.

Results for an elastic–anelastic boundary restated for the incident P wave problem are as follows.

Theorem (5.3.27). For the problem of a homogeneous P wave incident in an elastic medium V on a plane boundary with anelastic medium V', no critical angles exist for either reflected or transmitted waves.

Results for an anelastic–viscoelastic boundary are as follows.

Theorem (5.3.28). For the problem of a general (homogeneous or inhomogeneous) P wave incident on a plane boundary between an anelastic medium V and viscoelastic medium V', if the angle of incidence θ_{ϕ_1} $(\theta_{\phi_1} \neq \pi/2)$ is a critical angle for

(1) the transmitted SI wave, then

$$\tan\gamma_{\phi_1} = \left(\sin^2\theta_{\phi_1} - \frac{k'_{S_R}k'_{S_I}}{k_{P_R}k_{P_I}}\right) \Big/ \left(\sin\theta_{\phi_1}\cos\theta_{\phi_1}\right),$$

(2) the transmitted P wave, then

$$\tan\gamma_{\phi_1} = \left(\sin^2\theta_{\phi_1} - \frac{k'_{P_R}k'_{P_I}}{k_{P_R}k_{P_I}}\right) \Big/ \left(\sin\theta_{\phi_1}\cos\theta_{\phi_1}\right).$$

As for the incident SI problem, Theorem (5.3.26) for elastic boundaries shows that critical angles or phase propagation parallel to the boundary exists whenever the wave speeds of the transmitted waves exceed those of the incident wave. For elastic–anelastic and anelastic–viscoelastic boundaries Theorems (5.3.27) and (5.3.28) show phase propagation exactly parallel to the boundary or a critical angle in general does not occur, but if it does then it occurs for at most one angle of incidence.

5.4 Incident General SII Wave

The problem of a general Type-II S wave incident on a welded boundary with particle motion parallel to the boundary and perpendicular to the plane of incidence is less cumbersome than incident SI and P wave problems, because only general SII waves are reflected and refracted at the boundary. Results for the problem of an incident general SII wave problem, as initially derived by Borcherdt (1977), are briefly stated for completeness as a rigorous basis to establish the nature of energy flow at a viscoelastic boundary.

5.4.1. Specification of Incident General SII Wave

Solutions for the displacement field to represent incident, reflected, and refracted SII waves are given by (4.2.26) and (4.2.27). To consider the problem of an incident general SII wave, the amplitude of the SII wave in V' with phase propagation toward the boundary is set to zero, that is

$$D_2' = 0. \tag{5.4.1}$$

The expression for the incident SII problem rewritten here for convenience is

$$\vec{u}_1 = D_1 \exp\left[i\left(\omega t - \vec{K}_{u_1} \bullet \vec{r}\right)\right]\hat{x}_2, \tag{5.4.2}$$

where the complex wave vector is

$$\vec{K}_{u_1} = \vec{P}_{u_1} - i\vec{A}_{u_1}, \tag{5.4.3}$$

the propagation and attenuation vectors are

$$\vec{P}_{u_1} = k_R\hat{x}_1 - d_{\beta_R}\hat{x}_3 = \left|\vec{P}_{u_1}\right|\left(\sin\left[\theta_{u_1}\right]\hat{x}_1 + \cos\left[\theta_{u_1}\right]\hat{x}_3\right) \tag{5.4.4}$$

and

$$\vec{A}_{u_1} = -k_I\hat{x}_1 + d_{\beta_I}\hat{x}_3 = \left|\vec{A}_{u_1}\right|\left(\sin\left[\theta_{u_1} - \gamma_{u_1}\right]\hat{x}_1 + \cos\left[\theta_{u_1} - \gamma_{u_1}\right]\hat{x}_3\right), \tag{5.4.5}$$

the complex wave number k for the assumed general SII wave is

$$k = \left|\vec{P}_{u_1}\right|\sin\theta_{u_1} - i\left|\vec{A}_{u_1}\right|\sin\left[\theta_{u_1} - \gamma_{u_1}\right] \tag{5.4.6}$$

with the propagation and attenuation vectors specified in terms of the given material parameters, k_S, and the given degree of inhomogeneity, γ_{u_1}, to ensure propagation by

$$\left|\vec{P}_{u_1}\right| = \sqrt{\frac{1}{2}\left(\mathrm{Re}\left[k_S^2\right] + \sqrt{\left(\mathrm{Re}\left[k_S^2\right]\right)^2 + \sec^2\gamma_{u_1}\left(\mathrm{Im}\left[k_S^2\right]\right)^2}\right)}$$
$$= \frac{\omega}{v_{HS}}\sqrt{\frac{1+\chi_{Su_1}}{1+\chi_{HS}}} \tag{5.4.7}$$

and

$$\left|\vec{A}_{u_1}\right| = \sqrt{\frac{1}{2}\left(-\mathrm{Re}\left[k_S^2\right] + \sqrt{\left(\mathrm{Re}\left[k_S^2\right]\right)^2 + \sec^2\gamma_{u_1}\left(\mathrm{Im}\left[k_S^2\right]\right)^2}\right)}$$
$$= \frac{\omega}{v_{HS}}\sqrt{\frac{-1+\chi_{Su_1}}{1+\chi_{HS}}}, \tag{5.4.8}$$

where θ_{u_1} and γ_{u_1} are arbitrary, but fixed, given parameters of the incident wave satisfying $0 \leq \theta_{u_1} \leq \pi/2$ and $0 \leq \gamma_{u_1} < \pi/2$. k expressed explicitly in terms of the angle of incidence, the degree of inhomogeneity of the SII wave, and parameters of the medium is given by

$$k = \frac{\omega}{v_{HS}}\left(\sqrt{\frac{1+\chi_{Su_1}}{1+\chi_{HS}}}\sin\theta_{u_1} - i\sqrt{\frac{-1+\chi_{Su_1}}{1+\chi_{HS}}}\sin\left[\theta_{u_1} - \gamma_{u_1}\right]\right), \tag{5.4.9}$$

which simplifies for elastic media to

$$k = \frac{\omega}{v_{HS}}\sin\theta_{u_1}. \tag{5.4.10}$$

Hence, the complex wave number k and the magnitudes of the propagation and attenuation vectors for the incident SII are completely determined by (5.4.4) through (5.4.10) upon specification of the angle of incidence θ_{u_1}, the degree of inhomogeneity γ_{u_1}, the circular frequency ω, and the material parameters for V as specified by ρ and k_S or v_{HS} and Q_{HS}^{-1}.

5.4.2 Propagation and Attenuation Vectors; Generalized Snell's Law

Expressions for the solutions describing the reflected and refracted general SII waves are specified by (4.2.26) and (4.2.27) with other parameters of the reflected and refracted waves defined in (4.2.28) through (4.2.45). Parameters for the problem of a general SII wave incident on a welded boundary are illustrated in Figure (5.4.11).

Specification of the complex wave number k in terms of the given directions of propagation and attenuation $(\theta_{\psi_1}, \theta_{\psi_1} - \gamma_{\psi_1})$ for the incident SII wave and the

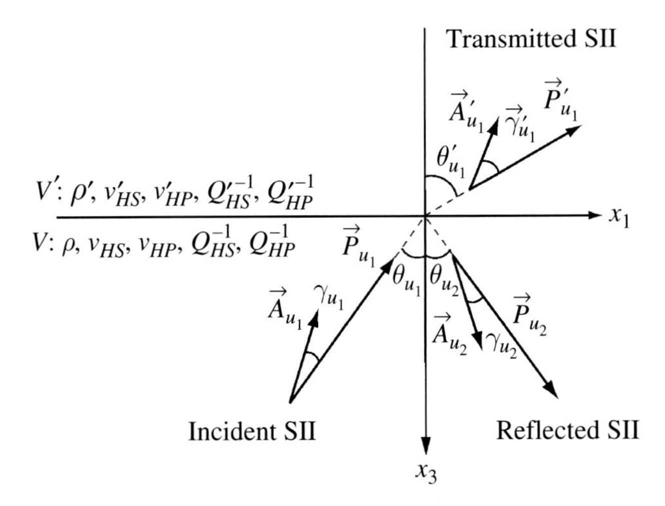

Figure (5.4.11). Diagram illustrating notation for directions and magnitude of propagation and attenuation vectors for the reflected and refracted waves due to a general Type-II S wave incident on a welded boundary between HILV media.

parameters of the media ρ, k_S, or v_{HS} and Q_{HS}^{-1} implies that the magnitude and directions of the propagation and attenuation vectors for the reflected and refracted general SII waves as defined by (4.2.30) through (4.2.33) are determined in terms of the given directions for the incident general SII wave. These expressions and equality of the complex wave number k for each of the solutions yields the components of the Generalized Snell's Law for the problem of an incident general SII wave, namely

$$k_R = \left|\vec{P}_{u_1}\right| \sin\theta_{u_1} = \left|\vec{P}_{u_2}\right| \sin\theta_{u_2} = \left|\vec{P}'_{u_2}\right| \sin\theta'_{u_2} \tag{5.4.12}$$

or

$$\frac{k_R}{\omega} = \frac{\sin\theta_{u_1}}{\left|\vec{v}_{u_1}\right|} = \frac{\sin\theta_{u_2}}{\left|\vec{v}_{u_2}\right|} = \frac{\sin\theta'_{u_2}}{\left|\vec{v}'_{u_2}\right|} \tag{5.4.13}$$

and

$$-k_I = \left|\vec{A}_{u_1}\right| \sin\left[\theta_{u_1} - \gamma_{u_1}\right] = \left|\vec{A}_{u_2}\right| \sin\left[\theta_{u_2} - \gamma_{u_2}\right] = \left|\vec{A}'_{u_2}\right| \sin\left[\theta'_{u_2} - \gamma'_{u_2}\right], \tag{5.4.14}$$

which in turn may be restated as

Theorem (5.4.15). Generalized Snell's Law – For the problem of a general SII wave incident on a welded viscoelastic boundary in a plane perpendicular to the boundary,

(1) the reciprocal of the apparent phase velocity along the boundary of the general reflected and refracted waves is equal to that of the given general incident wave,

and

(2) the apparent attenuation along the boundary of the general reflected and refracted waves is equal to that of the given general incident wave.

The discussion of the implications of Generalized Snell's Law is entirely analogous to that provided for the problem of an incident SI wave. A similar discussion is omitted here.

5.4.3 Amplitude and Phase

The welded contact between media V and V' is specified mathematically by requiring that the displacement and stress are continuous across the boundary. Substitution of solutions (4.2.26) and (4.2.27) for the displacement fields into the expressions relating the components of the displacement fields to components of the strain tensor, (2.3.5), and substituting these in turn into the expression relating stress and strain, (2.3.12), yields a set of equations for the components of the stress on both sides of the boundary acting on the plane $x_3 = 0$. The resulting continuity of displacement and stress equations for the problem of a Type-II S wave incident on a welded boundary are

$$D_1 + D_2 = D_1' \tag{5.4.16}$$

and

$$M\, d_\beta (D_1 - D_2) = M' d_\beta' D_1'. \tag{5.4.17}$$

These equations readily imply the desired solutions for the complex amplitude of the reflected and refracted general SII waves in terms of that given for the incident general SII wave, namely

$$D_2 = D_1 \left(M d_\beta - M' d_\beta' \right) \Big/ \left(M d_\beta + M' d_\beta' \right) \tag{5.4.18}$$

$$D_1' = D_1 \left(2 M d_\beta \right) \Big/ \left(M d_\beta + M' d_\beta' \right), \tag{5.4.19}$$

for $M\, d_\beta \neq M'\, d_\beta'$. These solutions establish that the boundary conditions for the specified problem are satisfied with the assumed form of the solutions (4.2.26) and (4.2.27). Hence, the amplitude and phase of the displacement for the reflected and refracted general SII waves may be expressed in terms of that for the given incident general SII wave. The results show that a general Type-II S wave incident on a plane welded viscoelastic boundary generates only general Type-II S waves.

Solutions (5.4.18) and (5.4.19) confirm that a complex wave number, k, is the same in each of the solutions for the incident, reflected, and transmitted waves. Hence, Generalized Snell's Law (5.4.15) is established for the problem of a general Type-II S wave incident on a plane boundary between viscoelastic media. With k given by (5.4.9) or (5.4.10), expressions (5.4.18) and (5.4.19) yield solutions for the amplitude and phase of the reflected and refracted general SII waves in terms of those for the incident general SII wave. Hence, from a mathematical point of view the solution for the problem of a general SII wave incident on a viscoelastic boundary is established.

5.4.4 Conditions for Homogeneity and Inhomogeneity

The proofs for conditions for homogeneity and inhomogeneity of the reflected and refracted waves generated by a general Type-II S wave are entirely analogous to those for the incident general SI wave problem. For brevity, only the results for the incident SII problem are given here (see Borcherdt (1977) for details).

Theorem (5.4.20). For the problem of a general SII wave incident on a welded viscoelastic boundary, the reflected SII wave is homogeneous if and only if the incident SII wave is homogeneous.

The condition for the transmitted SII wave is as follows.

Theorem (5.4.21). For the problem of a general SII wave incident on a welded viscoelastic boundary, if the incident SII wave is homogeneous $(\gamma_{u_1} = 0)$ *and not normally incident* $(0 < \theta_{u_1} \leq \pi/2)$*, then the transmitted SII wave is homogeneous if and only if*

$$Q_{HS}^{-1} = Q_{HS}'^{-1} \qquad \text{and} \qquad \sin^2 \theta_{u_1} \leq \frac{k_S'^2}{k_S^2} = \frac{\rho' M_R}{\rho M_R'} = \frac{v_{HS}^2}{v_{HS}'^2}.$$

Theorem (5.4.22). For the problem of a general SII wave incident on a welded viscoelastic boundary, if the incident SII wave is homogeneous $(\gamma_{u_1} = 0)$ *and not normally incident* $\left(0 < \theta_{u_1} \leq \pi/2\right)$ *and if* $(Q_{HS}^{-1} \neq Q_{HS}'^{-1})$*, then the transmitted SII wave is inhomogeneous.*

Theorem (5.4.23). For the problem of a general SII wave incident on a welded viscoelastic boundary, if the incident SII wave is homogeneous $(\gamma_{\phi_1} = 0)$ *and normally incident* $\left(\theta_{\phi_1} = 0\right)$*, then the reflected and refracted SII waves are homogeneous.*

These results show that SII waves transmitted in a layered anelastic Earth will, in general, be inhomogeneous with the physical characteristics dependent on the angle of incidence, because for most anelastic Earth boundaries $Q_{HS}^{-1} \neq Q_{HS}'^{-1}$.

5.4.5 Conditions for Critical Angles

The results for critical angles or angles of incidence for which phase propagation occurs parallel to the boundary are analogous to those derived for the incident SI wave problem. They are stated here for completeness.

Theorem (5.4.24). For the problem of a homogeneous SII wave incident on a plane boundary between elastic media V and V′, conditions for phase propagation parallel to the boundary or critical angles are as follows:

(1) no critical angles exist for the reflected wave,

(2) if $v_{HS} > v'_{HS}$, *then no critical angles exist for the transmitted waves,*

(3) if $v_{HS} < v'_{HS}$ *then the angles of incidence satisfying* $\sin\theta_{u_1} \geq v_{HS}/v'_{HS}$ *are critical angles for the transmitted SII wave.*

Theorem (5.4.25). For the problem of a homogeneous SII wave incident on a plane boundary between an elastic medium V and an anelastic medium V′, no critical angles exist for either reflected or transmitted waves.

Theorem (5.4.26). For the problem of a general (homogeneous or inhomogeneous) SII wave incident in an anelastic medium V on a plane boundary with a viscoelastic medium V′, if the angle of incidence θ_{u_1} $\left(\theta_{u_1} \neq \pi/2\right)$ *is a critical angle for the transmitted SII wave, then*

$$\tan\gamma_{u_1} = \left(\sin^2\theta_{u_1} - \frac{k'_{S_R}k'_{S_I}}{k_{S_R}k_{S_I}}\right) \Big/ \left(\sin\theta_{u_1}\cos\theta_{u_1}\right).$$

5.4.6 Energy Flux and Energy Flow Due to Wave Field Interactions

The simplicity of the incident general SII problem facilitates a detailed treatment of the energy flow and energy dissipation associated with the incident, reflected, and refracted waves at viscoelastic boundaries. A detailed treatment for the incident general SII wave problem provides a framework from which to consider energy flow for the more cumbersome incident SI and P wave problems. The development presented here follows that initially derived by Borcherdt (1977).

A general result which describes the intuitively evident situation for energy flow across the boundary is

Theorem (5.4.27). For steady-state radiation fields described by (4.2.26) and (4.2.27), the normal component of mean energy flux at a plane welded boundary between viscoelastic media is continuous.

This result follows immediately from the fact that for the prescribed radiation fields the mean energy flux as given by (3.4.17) may be rewritten in terms of the components of the physical stress tensor from (3.4.20) in medium V as

$$\langle\vec{\mathcal{J}}\rangle = -\left(\langle\dot{u}_R\, p_{R12}\rangle\hat{x}_1 + \langle\dot{u}_R\, p_{R32}\rangle\hat{x}_3\right) \tag{5.4.28}$$

and in medium V' as

$$\langle \vec{\mathcal{J}}' \rangle = -(\langle \dot{u}'_R \, p'_{R12} \rangle \hat{x}_1 + \langle \dot{u}'_R \, p'_{R32} \rangle \hat{x}_3). \tag{5.4.29}$$

Consequently, continuity of displacement and stress at the boundary together with the preceding equations implies the desired conclusion, namely,

$$\langle \vec{\mathcal{J}} \rangle \bullet \hat{x}_3 = \langle \vec{\mathcal{J}}' \rangle \bullet \hat{x}_3, \tag{5.4.30}$$

which is the desired result that the normal component of mean energy flux across the boundary is continuous.

Theorem (5.4.31). For the problem of a SII wave incident on a welded viscoelastic boundary, the total mean (time averaged) energy flux is given by

$$\langle \vec{\mathcal{J}} \rangle = \langle \vec{\mathcal{J}}_1 \rangle + \langle \vec{\mathcal{J}}_2 \rangle + \langle \vec{\mathcal{J}}_{12} \rangle + \langle \vec{\mathcal{J}}_{21} \rangle, \tag{5.4.32}$$

where

(1) $\langle \vec{\mathcal{J}}_j \rangle$ *is the mean energy flux associated with the propagation of the incident* $(j = 1)$ *and the reflected* $(j = 2)$ *SII waves as specified by*

$$\langle \vec{\mathcal{J}}_j \rangle = -(\langle \dot{u}_{R_j} ((p_{R12})_j \hat{x}_1 + (p_{R32})_j \hat{x}_3) \rangle) \qquad \text{for } j = 1, 2, \tag{5.4.33}$$

and

(2) $\langle \vec{\mathcal{J}}_{ij} \rangle$ *is the mean energy flux due to the velocity field of the incident,* $(j = 1)$ *wave interacting with the stress field of the reflected* $(j = 2)$ *wave and vice versa, as specified by*

$$\langle \vec{\mathcal{J}}_{ij} \rangle = -(\langle \dot{u}_{R_i} ((p_{R12})_j \, \hat{x}_1 + (p_{R32})_j \, \hat{x}_3) \rangle) \text{ for } i, j = 1, 2 \text{ and } i \neq j. \tag{5.4.34}$$

This result follows immediately from the fact that the total displacement field and the total stress field in medium V are the sums of those associated with the incident and reflected waves, that is,

$$\vec{u}_R \hat{x}_2 = u_{R_1} \hat{x}_2 + u_{R_2} \hat{x}_2 \tag{5.4.35}$$

and

$$(p_{Ri2}) = (p_{Ri2})_1 + (p_{Ri2})_2 \qquad \text{for } i = 1, 3, \tag{5.4.36}$$

where $(p_{Ri2})_j$ denotes the real component of the stress acting on the i plane in the $\hat{x}_2$ direction with $j=1$ for the incident wave and $j=2$ for the reflected wave. Hence, substitution of (5.4.35) and (5.4.36) into (5.4.28) yields

$$\begin{aligned} \langle \vec{\mathcal{J}} \rangle = & - \left(\left\langle \dot{u}_{R_1} \left((p_{R12})_1 \hat{x}_1 + (p_{R32})_1 \hat{x}_3 \right) \right\rangle + \left\langle \dot{u}_{R_2} \left((p_{R12})_2 \hat{x}_1 + (p_{R32})_2 \hat{x}_3 \right) \right\rangle \right) \\ & - \left(\left\langle \dot{u}_{R_1} (p_{R12})_2 \hat{x}_1 + (p_{R32})_2 \hat{x}_3 \right\rangle + \left\langle \dot{u}_{R_2} (p_{R12})_1 \hat{x}_1 + (p_{R32})_1 \hat{x}_3 \right\rangle \right), \end{aligned} \tag{5.4.37}$$

which establishes desired results (5.4.32) through (5.4.34).

Theorem (5.4.31) shows that in general the total mean energy flux associated with the superimposed incident and reflected waves is the sum of the mean energy fluxes associated with the propagation of the two individual wave fields plus the mean energy flux due to the stress field of the incident wave interacting with the velocity field of the reflected wave plus the mean energy flux due to the stress field of the reflected wave interacting with the velocity field of the incident wave.

The mean energy flux associated with the propagation of the individual incident and reflected SII waves is given in terms of the wave parameters from (3.4.6) by

$$\left\langle \vec{\mathcal{J}}_j \right\rangle = |D_j|^2 \exp\left[-2\vec{A}_{u_j} \bullet \vec{r} \right] \frac{\omega}{2} \left(M_R \vec{P}_{u_j} + M_I \vec{A}_{u_j} \right) \tag{5.4.38}$$

for $j=1, 2$.

Substitution of the physical displacement field as specified by (3.3.38) and the expressions for the components of the physical stress tensor as given by (3.4.21) and (3.4.22) into (5.4.37) shows that the mean energy flux for the individual waves in medium V is given by (5.4.38) and the wave–field interaction terms are given by

$$\begin{aligned} \left\langle \vec{\mathcal{J}}_{ij} \right\rangle = & \left(|D_1||D_2| \exp\left[-\left(\vec{A}_{u_1} + \vec{A}_{u_2} \right) \bullet \vec{r} \right] \frac{\omega}{2} |M| \right) \\ & \left(\cos\left[\Omega + (-1)^j \delta \right] \vec{P}_{u_j} + (-1)^j \sin\left[\Omega + (-1)^j \delta \right] \vec{A}_{u_j} \right) \end{aligned} \tag{5.4.39}$$

for $i, j=1, 2$ and $i \neq j$, where

$$\Omega \equiv \left(\vec{P}_{u_1} - \vec{P}_{u_2} \right) \bullet \vec{r} + \arg[D_2/D_1] \tag{5.4.40}$$

and

$$\delta \equiv \tan^{-1}[M_I/M_R] = \tan^{-1}\left[Q_{HS}^{-1} \right]. \tag{5.4.41}$$

Hence, the normal component of the mean energy flux due to the interaction of the incident and reflected SII waves is, from (5.4.39), given by

$$\vec{x}_3 \cdot \left(\langle \vec{\mathcal{J}}_{12} \rangle + \langle \vec{\mathcal{J}}_{21} \rangle\right) = \left(|D_1||D_2| \exp\left[-\left(\vec{A}_{u_1} + \vec{A}_{u_2} \right) \cdot \vec{r} \right] \omega \sin \Omega \right) \left(M_I \vec{P}_{u_1} - M_R \vec{A}_{u_1} \right) \cdot \hat{x}_3. \tag{5.4.42}$$

This expression immediately implies the familiar result that if medium V is elastic, in which case $M_I = 0$ and $\vec{A}_{u_1} = 0$, then the normal component of energy flux due to interaction of the waves is zero. However, if medium V is anelastic, then (5.4.42) shows that there is a net flow of energy normal to the boundary due to the interaction of the velocity and stress fields of the incident wave with those of the reflected wave. This expression also shows that this normal component of energy flow varies sinusoidally with distance from the boundary. It shows that at some distances there is a mean flow of energy toward the boundary and at other distances the mean flow of energy due to interaction is away from the boundary.

The condition of continuity of the normal component of mean energy flux across the boundary as stated in Theorem (5.4.27) may be rewritten using (5.4.32) as

$$\left(\langle \vec{\mathcal{J}}_1 \rangle + \langle \vec{\mathcal{J}}_2 \rangle + \langle \vec{\mathcal{J}}_{12} \rangle + \langle \vec{\mathcal{J}}_{21} \rangle\right) \cdot \hat{x}_3 = \langle \vec{\mathcal{J}}'_1 \rangle \cdot \hat{x}_3. \tag{5.4.43}$$

Upon defining energy reflection (R) and transmission (T) coefficients as the normal component of mean energy flux normalized by that carried by the incident wave as

$$R \equiv -\left(\langle \vec{\mathcal{J}}_2 \rangle \cdot \hat{x}_3\right) / \left(\langle \vec{\mathcal{J}}_1 \rangle \cdot \hat{x}_3\right) \tag{5.4.44}$$

and

$$T \equiv \left(\langle \vec{\mathcal{J}}'_1 \rangle \cdot \hat{x}_3\right) / \left(\langle \vec{\mathcal{J}}_1 \rangle \cdot \hat{x}_3\right) \tag{5.4.45}$$

and interaction coefficients (IC_{ij}) as the normal component of the energy flow due to interaction normalized by that carried by the incident wave as

$$IC_{ij} \equiv -\left(\langle \vec{\mathcal{J}}_{ij} \rangle \cdot \hat{x}_3\right) / \left(\langle \vec{\mathcal{J}}_1 \rangle \cdot \hat{x}_3\right) \qquad \text{for } i, j = 1, 2 \text{ and } i \neq j, \tag{5.4.46}$$

the condition of conservation of energy at the boundary may be written as

$$R + T + IC_{12} + IC_{21} = 1. \tag{5.4.47}$$

Substitution of (5.4.38) and (5.4.39) into the expressions for reflection, transmission, and interaction coefficients allows them to be written in terms of the parameters of the waves as

$$R = |D_2/D_1|^2, \tag{5.4.48}$$

$$T = \left|D_1'/D_1\right|^2 \left(\mathrm{Re}\left[M' d_\beta'\right] \Big/ \mathrm{Re}\left[M d_\beta\right]\right), \tag{5.4.49}$$

$$IC_{ij} = (-1)^j \ \mathrm{Re}[D_2/D_1] - \left(\mathrm{Im}[D_2/D_1]\mathrm{Im}\left[M d_\beta\right]/\mathrm{Re}\left[M d_\beta\right]\right) \quad \text{for } i,\, j = 1, 2 \text{ and } i \neq j. \tag{5.4.50}$$

Simplification of these expressions with D_2/D_1 and D_1'/D_1 as given by the boundary condition equations (5.4.18) and (5.4.19) confirms equation (5.4.47) for conservation of energy at the boundary. For elastic media, $\mathrm{Im}[M d_\beta] = 0$, hence the interaction coefficients are equal and opposite in sign so that $IC_{12} + IC_{21} = 0$ and the familiar result for elastic media follows, namely $R + T = 1$.

Energy partition curves showing the relative amounts of energy reflected and transmitted as various wave types as computed for elastic media (Ewing, Jardetsky, and Press, 1957, pp. 88–89) have been used frequently in seismology to interpret various phases. Equation (5.4.47) shows that such curves for anelastic media must account for the energy flow across the boundary due to interaction of the various waves.

For anelastic viscoelastic media the preceding results show that energy flows as a result of interaction of superimposed wave fields. Consequently, for such media, it is reasonable to expect that energy also would be dissipated due to interaction of superimposed wave fields. The mean rate of energy dissipation per unit volume for a steady-state radiation field is given from (3.4.53) by

$$\langle \mathscr{D} \rangle = -\nabla \cdot \langle \vec{\mathscr{I}} \rangle. \tag{5.4.51}$$

Equation (5.4.32) shows that in medium V, the total mean rate of energy dissipation per unit volume is given by

$$\nabla \cdot \langle \vec{\mathscr{I}} \rangle = \nabla \cdot \langle \vec{\mathscr{I}}_1 \rangle + \nabla \cdot \langle \vec{\mathscr{I}}_2 \rangle + \nabla \cdot \langle \vec{\mathscr{I}}_{12} \rangle + \nabla \cdot \langle \vec{\mathscr{I}}_{21} \rangle. \tag{5.4.52}$$

Hence the total mean rate of energy dissipation partitions into that associated with the incident wave ($\langle \mathscr{D}_1 \rangle \equiv \nabla \cdot \langle \vec{\mathscr{I}}_1 \rangle$), the reflected wave ($\langle \mathscr{D}_2 \rangle \equiv \nabla \cdot \langle \vec{\mathscr{I}}_2 \rangle$), and that associated with their interaction, ($\langle \mathscr{D}_{12} \rangle \equiv \nabla \cdot \langle \vec{\mathscr{I}}_{12} \rangle$) and ($\langle \mathscr{D}_{21} \rangle \equiv \nabla \cdot \langle \vec{\mathscr{I}}_{21} \rangle$), that is

$$\langle \mathscr{D} \rangle = \langle \mathscr{D}_1 \rangle + \langle \mathscr{D}_2 \rangle + \langle \mathscr{D}_{12} \rangle + \langle \mathscr{D}_{21} \rangle. \tag{5.4.53}$$

The mean rate of energy dissipation for the incident and reflected waves is given in terms of the wave parameters from (3.4.52) by

$$\langle \mathscr{D}_j \rangle = \frac{\left|\vec{D}_j\right|^2}{h_S} \exp\left[-2\vec{A}_{u_j} \cdot \vec{r}\,\right] \omega \left(\rho\omega^2 \vec{P}_{u_j} \cdot \vec{A}_{u_j} + 2M_I \left|\vec{P}_{u_j} \times \vec{A}_{u_j}\right|^2\right) \quad \text{for } j = 1, 2. \tag{5.4.54}$$

The vector identity (11.2.9), rewritten here for a general vector solution of the form (3.1.12) as

$$\nabla \bullet \vec{G} = -i\left(\vec{P} - i\vec{A}\right) \bullet \vec{G} \tag{5.4.55}$$

and applied to (5.4.39), yields expressions for the mean rates of energy dissipation due to interaction of the incident and reflected waves,

$$\begin{aligned}\langle \mathscr{D}_{ij} \rangle = |D_1||D_2| \exp\left[-\left(\vec{A}_{u_1} + \vec{A}_{u_2}\right) \bullet \vec{r}\right] |M|\omega \Big(&\cos\left[\Omega + (-1)^j \delta\right] \left(\vec{P}_{u_1} \bullet \vec{A}_{u_1}\right) \\ +(-1)^j \sin\left[\Omega + (-1)^j \delta\right] &\left(\left(\vec{A}_{u_1} \bullet \hat{x}_1\right)^2 - \left(\vec{P}_{u_1} \bullet \hat{x}_3\right)^2\right)\Big)\end{aligned} \tag{5.4.56}$$

for $i, j = 1, 2$ and $i \neq j$, with the sum of the interaction coefficients given by

$$\begin{aligned}\langle \mathscr{D}_{12} \rangle + \langle \mathscr{D}_{21} \rangle = &|D_1||D_2| \exp\left[-\left(\vec{A}_{u_1} + \vec{A}_{u_2}\right) \bullet \vec{r}\right] 2\omega \cos[\Omega] \\ &\left(M_R \vec{P}_{u_1} \bullet \vec{A}_{u_1} + M_I \left(\left(\vec{A}_{u_1} \bullet \hat{x}_1\right)^2 - \left(\vec{P}_{u_1} \bullet \hat{x}_3\right)^2\right)\right).\end{aligned} \tag{5.4.57}$$

Equation (5.4.57) shows the expected result that if medium V is anelastic, then the mean rate of energy dissipation per unit volume due to interaction of the incident and reflected waves is not zero. Hence, a finite amount of energy is dissipated due to the interaction of the two superimposed wave fields, a phenomena not encountered when considering elastic reflection–refraction problems.

The preceding results derived to account for the energy flow and dissipation at a viscoelastic boundary are extendable in a straightforward way to problems of incident P and SI waves. Proofs for the problems of incident P and SI waves are similar to those presented here. However, they are more cumbersome, because the number of interaction coefficients increases to six in medium V and two in medium V'.

For later reference, notation used for the incident SII problem is extended here for the problems of incident P or SI waves. The notation is extended as follows: (1) reflection and transmission coefficients for P and SI waves are distinguished by addition of subscripts corresponding to wave type, (2) interaction coefficients corresponding to interaction of the velocity and stress fields of the incident wave with those of the reflected P and SI waves are distinguished by addition of subscripts corresponding to wave type, (3) interaction coefficients corresponding to interaction of the velocity and stress fields of the reflected P and SI waves or the transmitted P and SI waves are distinguished by addition of an ordered pair of subscripts corresponding to wave type. Using these notational extensions for the problems of either an incident P or an incident SI wave, the condition for conservation of energy at the boundary (5.4.47) can be written as

$$R_P + R_{SI} + T_P + T_{SI} + IC_{12_P} + IC_{21_P} + IC_{12_{SI}} \\ + IC_{21_{SI}} + IC_{SIP} + IC_{PSI} + IC'_{SIP} + IC'_{PSI} = 1. \tag{5.4.58}$$

Similarly, the condition of energy dissipation in the incident medium V (5.4.53) may be written as

$$\langle \mathcal{D} \rangle = \langle \mathcal{D}_1 \rangle + \langle \mathcal{D}_{2_P} \rangle + \langle \mathcal{D}_{2_{SI}} \rangle + \langle \mathcal{D}_{12_P} \rangle + \langle \mathcal{D}_{21_P} \rangle \\ + \langle \mathcal{D}_{12_{SI}} \rangle + \langle \mathcal{D}_{21_{SI}} \rangle + \langle \mathcal{D}_{SIP} \rangle + \langle \mathcal{D}_{PSI} \rangle. \tag{5.4.59}$$

As an additional simplification for later reference corresponding interaction coefficients may be combined by defining interaction coefficients for

interaction of reflected P and SI waves with incident wave,

$$IC_P \equiv IC_{12_P} + IC_{21_P}, \qquad IC_{SI} \equiv IC_{12_{SI}} + IC_{21_{SI}}, \tag{5.4.60}$$

interaction of reflected P and SI,

$$IC_r \equiv IC_{SIP} + IC_{PSI}, \tag{5.4.61}$$

interaction of transmitted P and SI,

$$IC_t \equiv IC'_{SIP} + IC'_{PSI}, \tag{5.4.62}$$

and total normalized mean energy flux due to interaction,

$$IC_{total} \equiv IC_P + IC_{SI} + IC_r + IC_t, \tag{5.4.63}$$

where the wave-type subscript is understood to indicate interaction with the incident wave and the subscripts "r" and "t" are introduced to designate the total interaction between the two reflected waves and between the two transmitted waves, respectively. Proofs of these expressions for the problems of incident P and SI problems are similar to those for the incident SII problem. For brevity, they are left to the reader.

5.5 Problems

(1) Describe why it is necessary to consider the problem of incident inhomogeneous waves in order to consider wave propagation in layered linear anelastic media.

(2) Describe how Snell's Law for a plane wave incident on a plane elastic boundary differs from the Generalized Snell's Law for a plane wave incident on a plane viscoelastic boundary.

(3) Use Generalized Snell's Law for an incident homogeneous SI wave to show that the wave speed and reciprocal quality factor for the transmitted SI wave vary with angle of incidence, unless there is no contrast in intrinsic absorption for homogeneous S waves across the boundary.

(4) Prove that for the problem of a general P wave incident on a welded viscoelastic boundary, if the incident P wave is homogeneous then
  (a) the reflected SI wave is homogeneous if and only if $Q_{HS}^{-1} = Q_{HP}^{-1}$,
  (b) the transmitted P wave is homogeneous if and only if

$$Q_{HP}^{-1} = Q_{HP}'^{-1} \qquad \text{and} \qquad \sin^2 \theta_{\phi_1} \leq \frac{k_P'^2}{k_P^2} = \frac{\rho'\left(K_R + \frac{4}{3} M_R\right)}{\rho\left(K_R' + \frac{4}{3} M_R'\right)} = \frac{v_{HP}^2}{v_{HP}'^2},$$

  (c) the transmitted SI wave is homogeneous if and only if

$$Q_{HP}^{-1} = Q_{HS}'^{-1} \qquad \text{and} \qquad \sin^2 \theta_{\phi_1} \leq \frac{k_S'^2}{k_P^2} = \frac{\rho'\left(K_R + \frac{4}{3} M_R\right)}{\rho\, M_R'} = \frac{v_{HP}^2}{v_{HS}'^2}.$$

(5) For the problem of a general (homogeneous or inhomogeneous) SII wave incident on a welded viscoelastic boundary, prove
  (a) if the incident medium is elastic and the transmitted medium is anelastic, then no critical angles exist for either the reflected or transmitted waves,
  (b) if both media are anelastic and the angle of incidence θ_{u_1} $\left(\theta_{u_1} \neq \pi/2\right)$ is a critical angle for the transmitted SII wave, then

$$\tan \gamma_{u_1} = \left(\sin^2 \theta_{u_1} - \frac{k'_{S_R} k'_{S_I}}{k_{S_R} k_{S_I}}\right) \Big/ \left(\sin \theta_{u_1} \cos \theta_{u_1}\right).$$

(6) Describe why there can be a net flow of energy due to interaction of superimposed waves in anelastic media but not in elastic media.

(7) For the problem of a general (homogeneous or inhomogeneous) SII wave incident on a welded viscoelastic boundary, show that interaction of the incident and reflected SII waves results in
  (a) net flow of energy normal to the boundary given by

$$\vec{x}_3 \cdot \left(\left\langle \vec{\mathcal{J}}_{12} \right\rangle + \left\langle \vec{\mathcal{J}}_{21} \right\rangle\right) = |D_1||D_2| \exp\left[-\left(\vec{A}_{u_1} + \vec{A}_{u_2}\right) \cdot \vec{r}\,\right] \omega \sin \Omega \left(M_I \vec{P}_{u_1} - M_R \vec{A}_{u_1}\right) \cdot \hat{x}_3,$$

  and
  (b) a mean rate of energy dissipation given by

$$\langle \mathcal{D}_{12} \rangle + \langle \mathcal{D}_{21} \rangle = |D_1||D_2| \exp\left[-\left(\vec{A}_{u_1} + \vec{A}_{u_2}\right) \cdot \vec{r}\,\right] 2\omega \cos[\Omega] \left(M_R \vec{P}_{u_1} \cdot \vec{A}_{u_1} + M_I \left(\left(\vec{A}_{u_1} \cdot \hat{x}_1\right)^2 - \left(\vec{P}_{u_1} \cdot \hat{x}_3\right)^2\right)\right)$$

in terms of the amplitudes and propagation and attenuation vectors for the waves.

6

Numerical Models for General Waves Reflected and Refracted at Viscoelastic Boundaries

Theoretical results in the previous chapter predict that plane harmonic waves reflected and refracted at plane anelastic boundaries are in general inhomogeneous with the degree of inhomogeneity dependent on the angle of incidence, the degree of inhomogeneity of the incident wave, and properties of the viscoelastic media. As a result physical characteristics of the waves such as phase velocity, energy velocity, phase shifts, attenuation, particle motion, fractional energy loss, direction and amplitude of maximum energy flow, and energy flow due to wave interaction vary with angle of incidence. Consequently, these physical characteristics of inhomogeneous waves propagating in a stack of anelastic layers will not be unique at each point in the stack as they are for homogeneous waves propagating in elastic media. Instead these physical characteristics of the waves will depend on the angle at which the wave entered the stack and hence the travel path of the wave through previous layers. Towards understanding the significance of these dependences of the physical characteristics on angle of incidence and inhomogeneity of the incident wave, numerical models for general SII and P waves incident on single viscoelastic boundaries are presented in this chapter. Study of this chapter, especially the first three sections, provides additional insight into the effects of a viscoelastic boundary on resultant reflected and refracted waves.

A computer code (WAVES) is used to calculate reflection–refraction coefficients and the physical characteristics of reflected and refracted general waves for the problems of general (homogeneous or inhomogeneous) plane P, SI, and SII waves incident on a plane boundary between viscoelastic media (Borcherdt *et al.*, 1986). The code utilizes general solutions in media V and V' as specified by (4.2.1) through (4.2.45), Generalized Snell's Law as specified by (5.2.22), (5.3.16), and (5.4.15), and solutions to the boundary condition equations (5.2.27), (5.3.21), or (5.4.18) and (5.4.19) depending on the type of incident wave being considered. Physical characteristics of the reflected and refracted waves are specified in terms of the given parameters of the incident wave and the parameters of the media using equations (3.6.1) through (3.6.83).

Viscoelasticity of each medium is characterized by material density ρ, wave speeds v_{HS}, v_{HP}, and reciprocal quality factors Q_{HS}^{-1}, Q_{HP}^{-1} for homogeneous S and P waves as discussed following (3.5.7). This characterization permits interpretation of results in terms of any chosen linear viscoelastic model such as those specified in Table (1.3.30), those derivable from an infinite combination of springs and dashpots (Bland, 1960), or those proposed by other investigators, for example Minster and Anderson (1981) or Futterman (1962). It permits straightforward incorporation of any chosen frequency-dependent behavior corresponding to a particular viscoelastic model by specification of wave speeds and reciprocal quality factors from corresponding dispersion and absorption curves for homogeneous waves. Normalization of certain wave parameters by the appropriate power of angular frequency allows the calculations to be applicable to any viscoelastic model and frequency to which the chosen material parameters are appropriate.

Results predicted numerically by the exact theoretical model are compared with those predicted by incorrectly assuming that anelastic waves reflected and refracted at angles less than an "elastic critical angle" are homogeneous. This comparison helps quantify the errors associated with estimates of the physical characteristics of the waves based on low-loss models which do not account correctly for inhomogeneity of anelastic waves. This part of the computer code assumes the physical characteristics of the resultant homogeneous waves are described by low-loss expressions (3.7.2) through (3.7.43). This assumption, referred to here as the homogeneous low-loss, "hll", assumption, incorrectly assumes that conditions of homogeneity and inhomogeneity of the reflected and refracted waves in anelastic media are the same as those for elastic media. It assumes the wave speeds that characterize the material are those derived from the parameters specified for the material using the low-loss expressions for homogeneous waves, namely (3.7.2) and (3.7.3). The values for Q^{-1} used to characterize the material are assumed for the "hll" model to be those specified without adjustment for low-loss materials.

6.1 General SII Wave Incident on a Moderate-Loss Viscoelastic Boundary (Sediments)

The theoretical solution for the problem of a general (homogeneous or inhomogeneous) Type-II S wave incident on an arbitrary plane welded viscoelastic boundary is provided by (5.4.2) through (5.4.19). Media parameters chosen for a corresponding initial numerical model describe either water-saturated sediments (Hamilton *et al.*, 1970) or fractured near-surface materials (Newman and Worthington, 1982). The parameters for the media are given in Table (6.1.1). These parameters characterize media with a moderate amount of intrinsic absorption for which the low-loss assumption is not valid. Choice of this model serves to illustrate several

Table (6.1.1). *Material parameters for the problems of incident Type-II S waves as illustrated in Figures (6.1.2) and (6.1.7).*

Medium	ρ (g/cm^3)	v_{HS} (km/s)	Q_{HS}^{-1}
V: Sediment 1	1.5	0.2	0.5
V': Sediment 2	2	0.4	0.33

physical characteristics of the reflected and refracted waves that are not readily apparent for low-loss media.

6.1.1 Incident Homogeneous SII Wave

To simplify initial considerations, the incident Type-II S wave is assumed to be homogeneous. Notation for the problem of a homogeneous Type-II S wave incident on a plane boundary between viscoelastic media is shown in Figure (6.1.2). Numerical estimates for the physical characteristics of the waves reflected and transmitted at the specified anelastic boundary computed as a function of angle of incidence are shown in Figures (6.1.3) through (6.1.5).

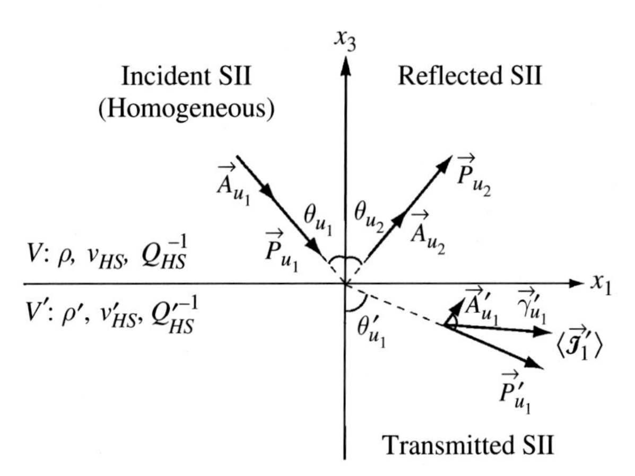

Figure (6.1.2). Notation for the problem of a homogeneous Type-II S wave incident on a plane viscoelastic interface across which the material velocity and intrinsic absorption increase and decrease, respectively. Directions illustrated for phase propagation $\vec{P}$, maximum attenuation $\vec{A}$, and maximum mean energy flux are consistent with numerical results (see text and Figure (6.1.3)).

Figure (6.1.3)a shows that the degree of inhomogeneity of the transmitted wave, γ'_{u_1}, continuously increases with angle of incidence from that of the incident homogeneous wave ($\gamma'_{u_1} = 0$) to an asymptotic value near a physical limit of $\gamma'_{u_1} = -\pi/2$. Figure (6.1.3)c shows that the direction of phase propagation of the transmitted wave asymptotically approaches a value of 78°. It shows that phase propagation is not parallel to the boundary for any angle of incidence, which is consistent with results concerning critical angles as stated in Theorem (5.4.26). In contrast, an elastic model for the boundary or the incorrect "hll" model implies that the degree of inhomogeneity of the transmitted wave is either zero or 90° depending on whether the angle of incidence is less than or greater than the minimum elastic critical angle. For angles of incidence greater than the minimum elastic critical angle, the "hll" model incorrectly assumes that the direction of phase propagation is parallel to the boundary (thin-dashed curves, Figures (6.1.3)a and c). Implications of the incorrect "hll" assumption are apparent upon comparison of the results indicated by the thin-dashed curves with those predicted on the basis of the exact theory as indicated by the bold curves.

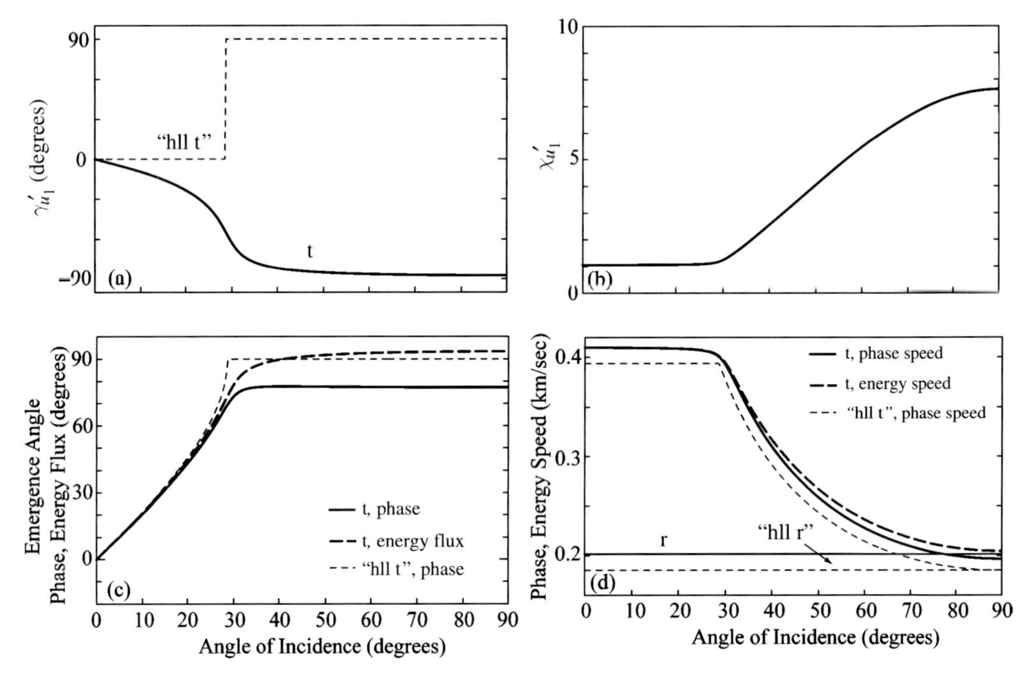

Figure (6.1.3). Inhomogeneity (a), χ'_{Su_1} (b), emergence angles for phase propagation and maximum energy flux (c), and phase and energy speeds (d) for Type-II S waves transmitted (t) and reflected (r) at a sediment–sediment interface (Table (6.1.1)) by an incident homogeneous Type-II S wave.

The exact anelastic model predicts that the direction of maximum mean energy flux for the transmitted wave increases with respect to that of phase propagation as the angle of incidence increases (Figure (6.1.3)c). The direction of maximum energy flux increasingly deviates to a fixed value of 16° from that of phase propagation. For angles of incidence less than about 40° the direction of maximum energy flux is away from the boundary into the transmission medium. For angles of incidence greater than about 40° the direction of the maximum energy flux for the transmitted wave varies from being parallel to the boundary to about 4° toward the boundary

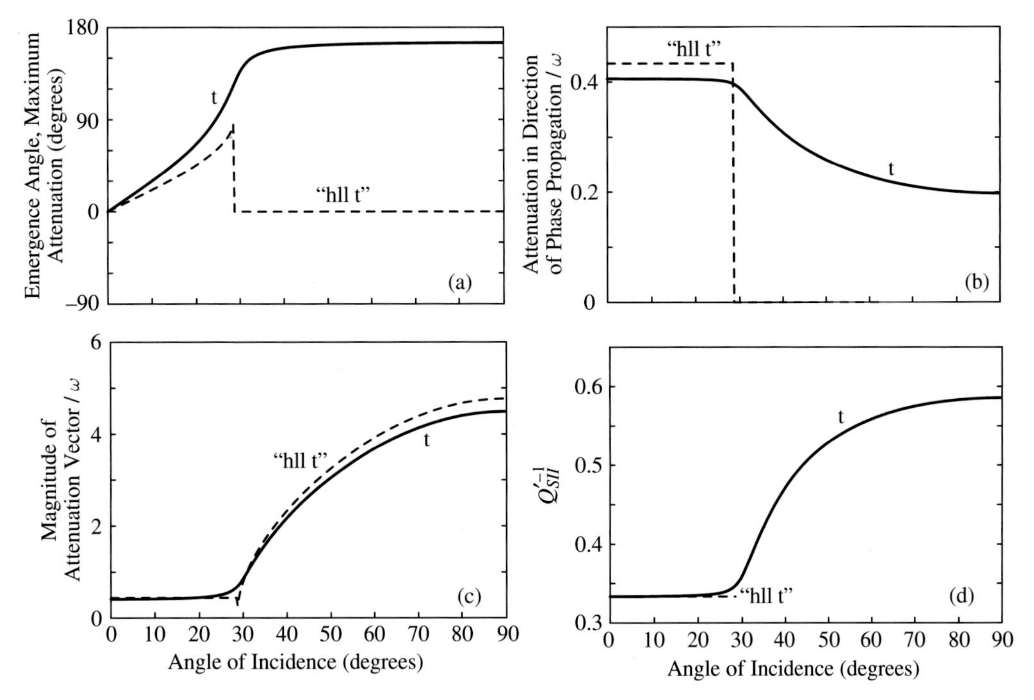

Figure (6.1.4). Emergence angle (a), magnitude for attenuation in direction of propagation (b), maximum attenuation (c), and $Q_{SII}^{\prime -1}$ for the Type-II S wave transmitted at a sediment–sediment interface (see Table (6.1.1)) by an incident homogeneous Type-II S wave.

into the incidence medium (see the bold dashed curve in Figure (6.1.3)c). This negative normal component of energy flux for the transmitted wave is required to maintain continuity of the normal component of energy flow across the boundary as established by Theorem (5.4.27). In contrast, an elastic model and the "hll" assumption predict that the direction of maximum energy flux for the transmitted wave coincides with the direction of phase propagation.

The parameter $\chi'_{Su_1} = \sqrt{1 + Q_{HS}^{\prime -2} \sec^2 \gamma'_{u_1}}$ and the physical characteristics of the transmitted wave undergo rapid transitions for angles of incidence in a range containing the minimum elastic critical angle (see Figures (6.1.3) through (6.1.5)). For convenience, this range in angles of incidence shall be referred to as the "*SII′ Transition Window*". (The concept of this window is introduced to indicate qualitatively a range in angles of incidence over which the rapid transitions occur. A more precise definition could be given in terms of a set of angles of incidence for which χ'_{Su_1} differs from unity by a selected amount, but such a precise definition is not needed here for general identification purposes.)

The magnitudes of the phase and energy velocities of the transmitted wave decrease slightly for angles of incidence less than those in the *SII′ transition window*. They undergo a rapid decrease toward non-zero values near 50 percent of that for a corresponding homogeneous wave (Figure (6.1.3)d) as the angle of

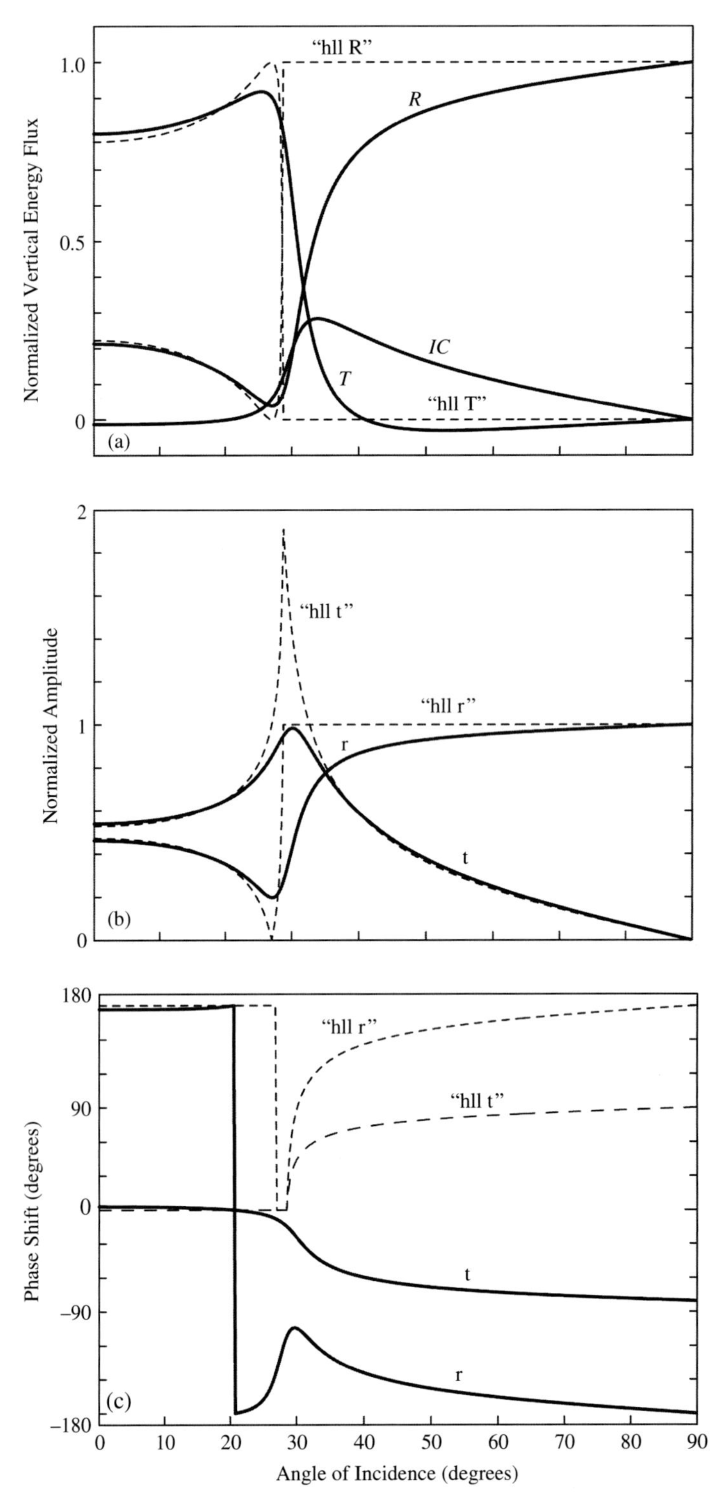

Figure (6.1.5). Energy reflection *R*, transmission *T*, and interaction coefficients *IC* (a), amplitude reflection r and transmission t coefficients (b), and phase reflection r and transmission t coefficients for the Type-II S waves transmitted and reflected at a sediment–sediment interface (Table (6.1.1)) due to an incident homogeneous Type-II S wave.

incidence increases through the *SII′ transition window* toward grazing incidence. The phase speed of the reflected wave is constant and equal to that of the incident wave for each angle of incidence. The "hll" model incorrectly assumes that the phase speed of the transmitted wave is constant for angles of incidence less than the minimum elastic critical angle (~29°) and equal to the apparent phase speed along the interface of the incident wave for larger angles of incidence.

The exact anelastic model shows that the attenuation vector for the transmitted wave is directed into the transmission medium for angles of incidence less than about 25° and away for larger angles of incidence (see Figure (6.1.4)a). The exact model indicates that amplitudes of the transmitted wave decrease with increasing vertical distance from the boundary for angles of incidence less than about 25°. For larger angles of incidence it shows the amplitudes of the transmitted wave increase with vertical distance from the boundary. The fact that γ'_{u_1} does not exceed its physical limit implies the transmitted wave attenuates in the direction of phase propagation so no physical principles are violated. The "hll" model incorrectly assumes the direction of attenuation for the transmitted wave is the same as the direction of phase propagation for angles of incidence less than the elastic critical angle and perpendicular to the boundary directed into the transmission medium for larger angles of incidence (Figure (6.1.4)a).

The phenomenon that amplitudes may increase with increasing vertical distance from a HILV boundary for the harmonic plane-wave problem being considered is implied rigorously by the specification of the problem and the boundary conditions. For the specific problem under consideration, this phenomenon follows from Generalized Snell's Law for anelastic media (Theorem (5.4.15)) for boundaries with contrasts in material properties similar to those chosen here, namely an increase in material velocity and a decrease in the amount of material absorption at the boundary. This situation is typical of that encountered by a seismic wave propagating downward in a layered anelastic model of the Earth.

The maximum attenuation in the direction of phase propagation of the transmitted wave asymptotically decreases to a non-zero value that is about 48 percent of its initial value for a homogeneous wave (Figure (6.1.4)b). The magnitude of the maximum attenuation of the transmitted wave increases asymptotically to a value that is about 11 times larger than its initial value for a homogeneous wave (Figure (6.1.4)c). The "hll" model incorrectly assumes the maximum attenuation in the direction of phase propagation is constant for angles less than the minimum elastic critical angle and zero for larger angles (Figures (6.1.4)b).

Q'^{-1}_{SII} for the transmitted wave shows only a slight increase for angles of incidence less than the minimum elastic critical angle, but then increases rapidly from its value for a homogeneous wave of $Q'^{-1}_{HS} = 0.33$ toward an asymptotic value that is about 176 percent larger near grazing incidence (Figure (6.1.4)d). The "hll" model assumes that Q'^{-1}_{SII} for the transmitted wave does not vary for angles of incidence less than the minimum elastic critical angle and that beyond this angle it vanishes.

Energy, amplitude, and phase reflection and transmission coefficients, as specified for the problem of an incident SII wave by (5.4.18), (5.4.19), (5.4.49), (5.4.50), and (5.4.51), are shown in Figure (6.1.5). The energy coefficients, calculated using the exact anelastic formulation, provide considerable insight into the nature of energy flow across the boundary. The sum of the energy reflection (*R*), transmission (*T*), and interaction (*IC*) coefficients as shown in Figure (6.1.5)a for each angle of incidence is unity as implied by (5.4.48) to insure conservation of energy at the boundary.

The energy coefficients (Figure (6.1.5)a) show that the net normalized vertical energy flux due to interaction of the velocity and stress fields of the incident and reflected waves (*IC*) becomes a significant portion of the energy budget at the boundary for angles of incidence within the *SII′ transition window*. The normalized energy flow due to interaction is a few percent and directed into the transmission medium for angles of incidence near normal. For an angle of incidence near 20°, the net vertical energy flow at the boundary due to interaction reverses direction into the incidence medium. It increases to a maximum of 28 percent of that for the incident wave then asymptotically decreases to zero as grazing incidence is approached. At an angle of incidence near 33° the normal component of energy carried away from the boundary by the transmitted wave equals that flowing away from the transmission medium due to interaction. For angles of incidence greater than about 40° the normal component of the mean energy flux associated with the transmitted wave reverses direction into the incidence medium (Figure (6.1.5)a). The "hll" model incorrectly assumes that the net energy flow due to interaction of the incident and reflected waves is zero across the boundary. It assumes that no energy is transmitted across the boundary for angles of incidence greater than the minimum elastic critical angle (≈29°).

The amplitude reflection and transmission coefficients are shown in Figure (6.1.5)b. The coefficients for the exact anelastic model show gradual transitions for angles of incidence in and beyond the *SII′ transition window*. The normalized displacement amplitude of the transmitted wave increases to a value near that of the incident wave for an angle of incidence within the window, then gradually decreases to zero for grazing incidence (Figure (6.1.5)b). The normalized displacement amplitude of the reflected wave decreases to about 20 percent of that of the incident wave then asymptotically approaches unity near grazing incidence. The "hll" model predicts abrupt changes in the amplitude reflection and transmission coefficients near the elastic critical angle.

Phase shifts for the reflected and transmitted waves with respect to that for the incident wave are specified by (5.4.18) and (5.4.19). They are shown for the anelastic model under consideration in Figure (6.1.5)c. They show that the transmitted and reflected waves are out of phase with the incident wave by amounts of about 4° and 180°, respectively, for angles of incidence less than those in the *SII′ transition window*. For larger angles of incidence the phase shift for the transmitted wave rapidly tends toward −80° and that for the reflected wave abruptly changes to −180° then increases to about −90° before decreasing to about −180° near grazing incidence. Determination of the correct values for the phase shifts requires inference of the correct values for d_β and d'_β as discussed for the inference of the correct direction for the attenuation vector for the transmitted wave. The "hll" model predicts phase shifts consistent with its assumed direction for the attenuation vector that differ significantly from those for the exact anelastic model for angles of incidence greater than the elastic critical angle.

6.1.2 Incident Inhomogeneous SII Wave

Theorem (5.4.21) establishes that a homogeneous SII wave transmitted across a boundary between anelastic media with different amounts of intrinsic absorption will, in general, be inhomogeneous for all angles of incidence except normal. Consequently, to consider wave propagation in layered media, the inhomogeneity of the transmitted wave must be taken into account in order to consider its propagation through a stack of anelastic layers. Notation for the problem of an incident inhomogeneous SII wave is shown in Figure (6.1.7). Corresponding numerical estimates for the physical characteristics of the reflected and refracted waves are shown in Figures (6.1.8) through (6.1.10).

To ensure that the incident inhomogeneous wave will carry energy toward the boundary, the direction of energy flux for the incident wave is constrained to be toward the boundary. Considering that the direction of energy flux for an inhomogeneous wave in anelastic media is at an intermediate angle between that of phase propagation and that of maximum attenuation, this constraint can be imposed by considering only angles of phase incidence for which corresponding angles of energy flux are toward the boundary. The angle between the direction of phase propagation and that of energy flux for the incident SII wave, as implied by (3.4.6), (3.6.17), (3.6.18), and (3.6.19), is given by

$$\angle(\vec{P}_{u_1}, \vec{\mathcal{I}}_{u_1}\rangle) = \arccos\left[\frac{\chi_{Su_1} + \chi^2_{HS}}{\chi_{HS}\left(1 + \chi_{Su_1}\right)}\right]. \qquad (6.1.6)$$

Inhomogeneity of the incident wave is chosen to be −60°. This degree of inhomogeneity corresponds to that of the SII wave transmitted at an angle of incidence of 29° for the problem of an incident homogeneous wave considered in the previous section. For this degree of inhomogeneity of the incident wave the mean energy flux for the incident wave is directed at an angle of about −9° with respect to the direction of phase propagation as implied by (6.1.6). Material

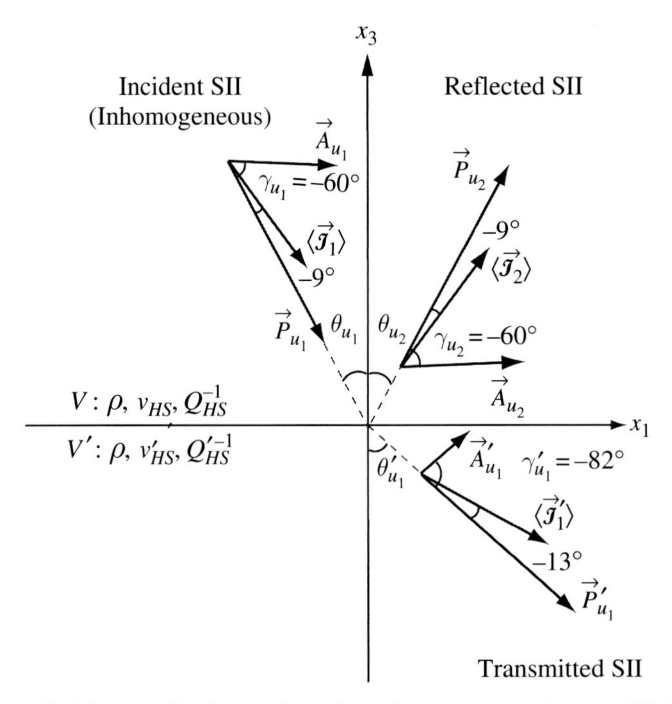

Figure (6.1.7). Notation for the problem of an inhomogeneous ($\gamma_{u_1} = -60°$) Type-II S wave incident on a plane viscoelastic interface with material parameters corresponding to water-saturated sediments (Table (6.1.1)). Directions of phase propagation $\vec{P}$ with respect to maximum attenuation $\vec{A}$ and mean energy flux $\langle \vec{\mathcal{J}} \rangle$ as predicted by the numerical model are indicated.

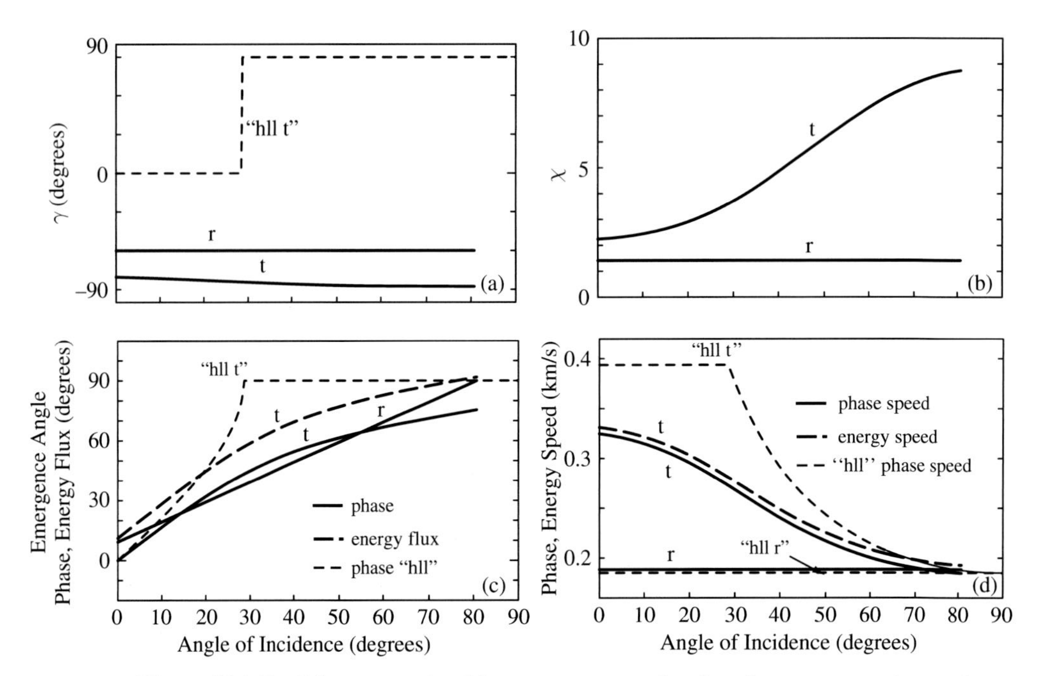

Figure (6.1.8). Inhomogeneity (a), emergence angles for phase propagation and maximum energy flux (b), phase and energy speeds (c), and $1/Q$ for reflected and refracted Type-II S waves at a sediment–sediment interface (Table (6.1.1)) generated by an incident inhomogeneous ($\gamma_{u_1} = -60°$) Type-II S wave.

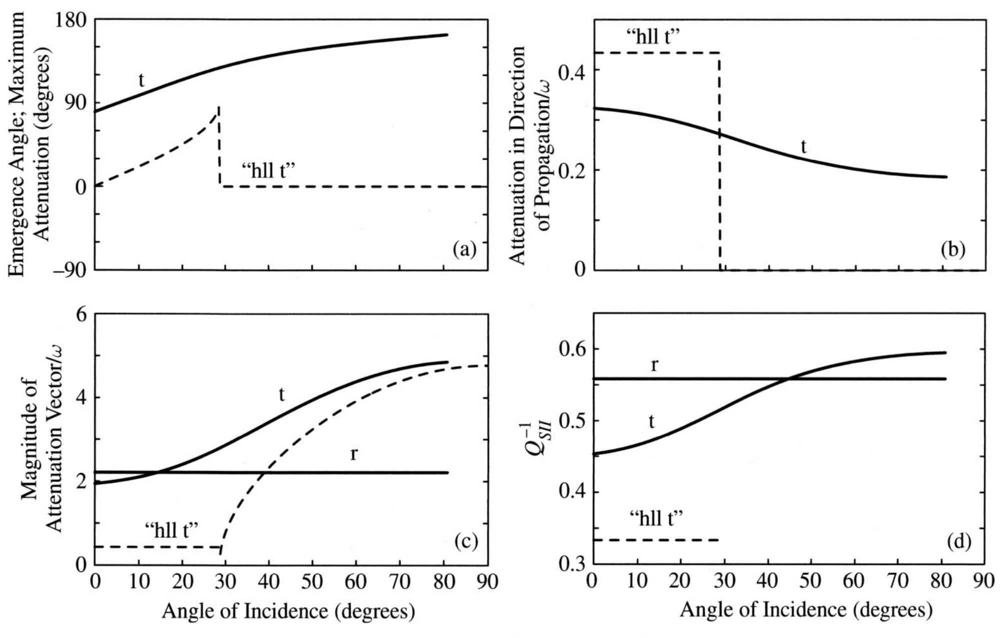

Figure (6.1.9). Emergence angle (a) and magnitude for attenuation in direction of propagation (b), maximum attenuation (c), and $1/Q$ for reflected and refracted Type-II S waves at a sediment–sediment interface (Table (6.1.1)) generated by an incident inhomogeneous ($\gamma_{u_1} = -60°$) Type-II S wave.

parameters were chosen to be the same as those for the problem in the preceding section (Table (6.1.1)) in order to facilitate comparison of results for an incident homogeneous wave with those for an incident inhomogeneous wave.

Quantitative characteristics of the reflected and refracted inhomogeneous waves are shown in Figures (6.1.8) through (6.1.10). Calculations for the "hll" model assume the incident SII wave is homogeneous. The direction of the mean energy flux for the incident inhomogeneous SII wave with $\gamma_{u_1} = -60°$ is parallel to the boundary for an angle of phase incidence $\theta_{u_1} = 81°$ as indicated in Figure (6.1.8)c. Negative components of vertical energy flux for the incident wave for larger angles of phase incidence are not of physical interest. Hence, calculations based on the exact anelastic model are shown only for angles of phase incidence satisfying $0 \le \theta_{u_1} \le 81°$.

The physical characteristics of the transmitted SII wave generated by an incident inhomogeneous SII wave differ significantly from those generated by an incident homogeneous SII wave (compare Figures (6.1.8) through (6.1.10) with Figures (6.1.3) through (6.1.5)). Inhomogeneity of the transmitted wave increases from near $-80°$ towards its physical limit of $-90°$ as the angle of phase incidence increases (Figure (6.1.8)a). The angle between the directions of phase propagation and maximum energy flux for the transmitted wave varies from near $-11°$ to

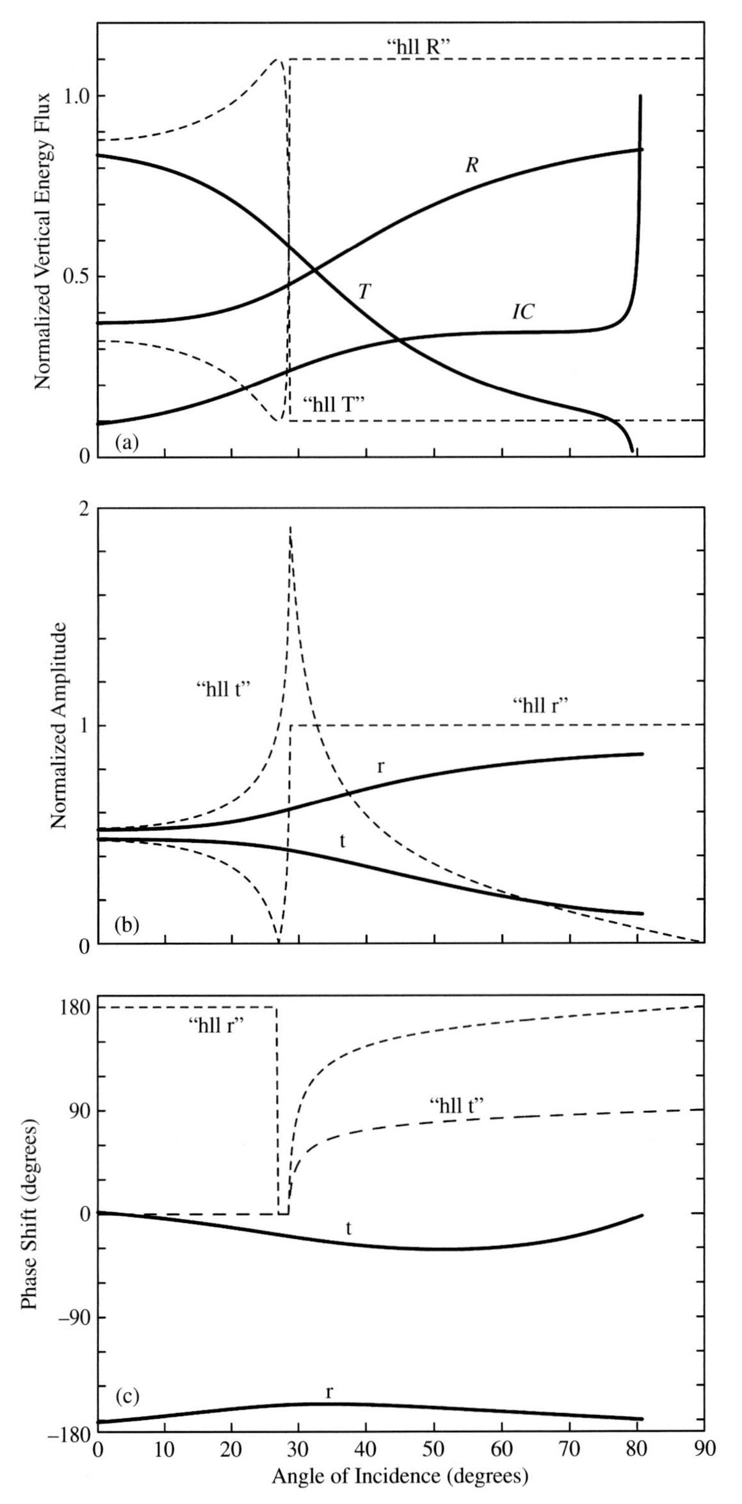

Figure (6.1.10). Energy reflection R, transmission T, and interaction coefficients IC (a), amplitude reflection r and transmission t coefficients (b), and phase reflection r and transmission t coefficients (c) for Type-II S waves reflected and transmitted at a sediment–sediment interface (Table (6.1.1)) due to an incident inhomogeneous Type-II S wave with $\gamma_{u_1} = -60°$.

about −16° (Figure (6.1.8)c). The energy speed for the wave transmitted by an incident inhomogeneous SII wave exceeds the phase speed and both speeds are less than those of a corresponding homogeneous wave by amounts up to 20 and 30 percent (Figure (6.1.8)d). The energy speed is less than the phase speed for the corresponding incident and reflected waves.

The direction and magnitude of the maximum attenuation for the transmitted wave show significant variations with increasing angle of incidence (Figures (6.1.9) a, b, c). Q^{-1} for the transmitted wave (Figure (6.1.9)d) increases gradually from a value that exceeds that of a homogeneous wave by 36 percent for angles of incidence near normal to 79 percent for angles of incidence near grazing.

The energy reflection, transmission, and interaction coefficients for the assumed incident inhomogeneous SII wave (Figure (6.1.10)a) differ significantly in amplitude from those for an incident homogeneous SII wave (Figure (6.1.5)a). The normal component of energy flux due to interaction of the velocity and stress fields of the incident and reflected inhomogeneous SII waves varies gradually from less than 1 percent at normal incidence to 25 percent of the total incident energy for angles of incidence near 76°, then undergoes a rapid increase towards 100 percent for angles of incidence between 76° and 81°. The rapid increase in the interaction coefficient at these angles of incidence corresponds to those angles for which the direction of mean energy flux for the transmitted wave is reversed and into the incidence medium (Figure (6.1.8) c). Amplitude and phase reflection and refraction coefficients also show gradual variations with angle of incidence that differ significantly from those for the corresponding problem of an incident homogeneous wave (Figures (6.1.10)b and (6.1.10)c).

The quantitative results as reviewed for the example of an incident inhomogeneous wave (Figures (6.1.8) through (6.1.10)) illustrate several physical phenomena for the reflection–transmission problem not encountered for problems concerned only with incident homogeneous waves (Figures (6.1.3) through (6.1.5)). The results differ significantly from those predicted by the "hll" model (thin-dashed curves in Figures (6.1.8) through (6.1.10)). The results indicate that inhomogeneity of the incident wave plays a significant role in determining the physical characteristics of the transmitted and reflected waves at an anelastic boundary. They imply that the physical characteristics of a wave in a stack of anelastic layers will depend on the angle of incidence and inhomogeneity of the initial wave and in turn its previous travel path.

6.2 P Wave Incident on a Low-Loss Viscoelastic Boundary (Water, Stainless-Steel)

Quantitative results for the problem of a homogeneous P wave incident on a water, viscoelastic-solid boundary will provide insight into the anelastic characteristics of reflected P and SI waves. They will provide results for comparison with experimental measurements as evidence in confirmation of the theory for viscoelastic wave propagation. The model also will provide quantitative estimates of reflection coefficients for ocean, solid-Earth interfaces of interest in seismology.

6.2.1 Reflected and Refracted Waves

The numerical model considered here assumes a plane ultrasonic wave in water incident on an immersed plane stainless-steel interface of type 304. Notation for the problem is illustrated in Figure (6.2.1).

Material parameters for the water and stainless steel modeled as viscoelastic media are specified in Table (6.2.2). Physical characteristics of the reflected and transmitted waves as computed using the computer code WAVES based on these equations are shown in Figures (6.2.3) through (6.2.5). Examination of these characteristics provides insight into the nature of the incident P wave problem.

The assumption of homogeneity for the incident P wave implies that the reflected P wave in water is homogeneous for all angles of incidence (Theorem (5.3.22)). Generalized Snell's Law (Theorem (5.3.16)) implies that the angle of reflection equals the angle of incidence for both the directions of phase propagation and maximum attenuation.

The contrast in anelastic properties at the interface implies that both the P wave and the Type-I S wave transmitted into the stainless steel are inhomogeneous for all angles of incidence except normal (Theorem (5.3.24); Figure (6.2.3)a). The degree of inhomogeneity of the transmitted P and SI waves increases to a value near 90° for angles of incidence within the P' and SI' *transition windows*. A rapid change in $\chi'_{P\phi_1}$

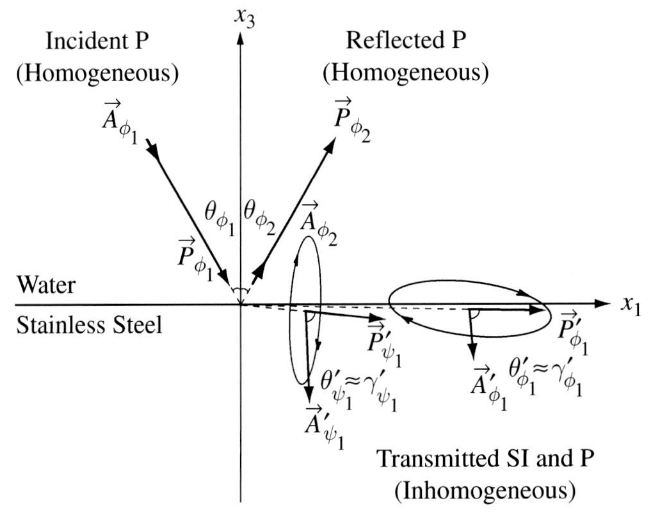

Figure (6.2.1). Notation for a homogeneous P wave incident in water on a plane, anelastic stainless-steel interface with material parameters as indicated in Table (6.2.2).

Table (6.2.2). *Material parameters for a water–stainless-steel boundary for the problem of an incident P wave illustrated in Figure (6.2.1).*

Medium	ρ (g/cm^3)	v_{HS} (km/s)	Q_{HS}^{-1}	v_{HP} (km/s)	Q_{HP}^{-1}
V: Water	1			1.49	0.00012*
V': Stainless steel	7.932	3.142	0.0127*	5.74	0.0073*

*As measured at 10 MHz by Becker and Richardson (1969, 1970).

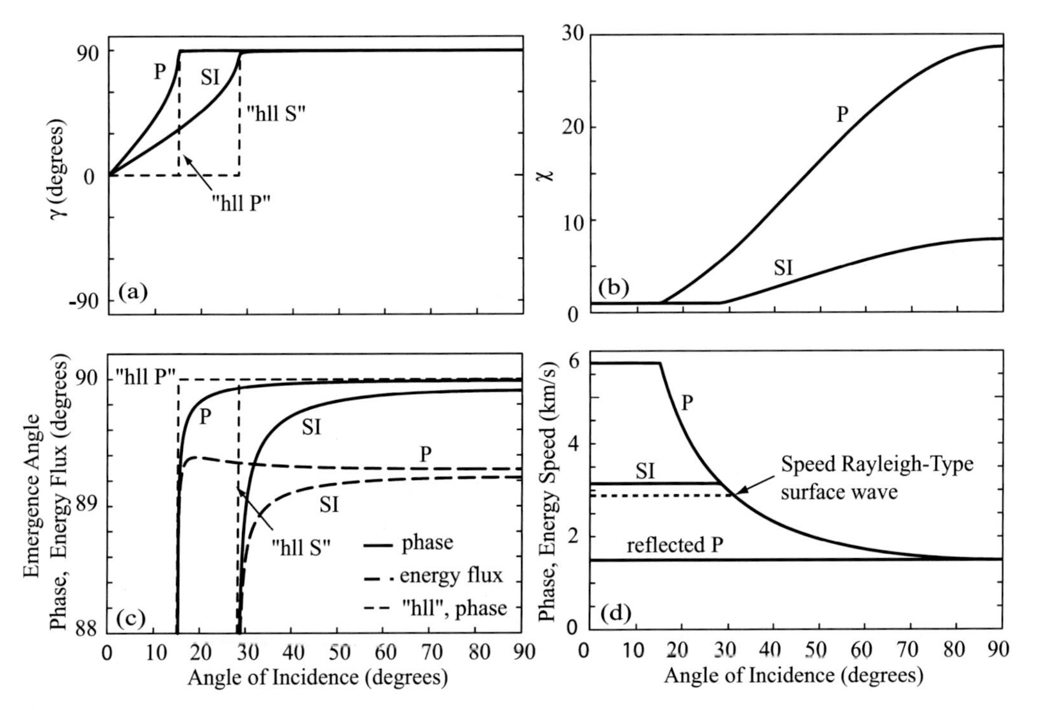

Figure (6.2.3). Inhomogeneity (a), χ (b), emergence angles for phase propagation and maximum energy flux (c), and phase and energy speeds (d) for inhomogeneous transmitted P and Type-I S waves generated by a plane acoustic P wave in water incident on an anelastic plane stainless-steel boundary.

and $\chi'_{S\psi_1}$ for angles of incidence in these windows suggests a similar rapid change in the physical characteristics of the transmitted waves (Figure (6.2.3)b).

The directions of phase and energy propagation for the transmitted P and SI waves are predicted for all angles of incidence to be at some non-zero angle into the stainless steel (Figure (6.2.3)c). In the case of the P wave, energy is carried closer to the boundary than for the S wave; however, neither wave carries energy back into the incident medium as occurred in previous examples (compare Figures (6.1.3)c and (6.2.3)c). It is interesting to note that the P wave carries energy closer to the boundary for angles of incidence within the P' *transition window* than it does for

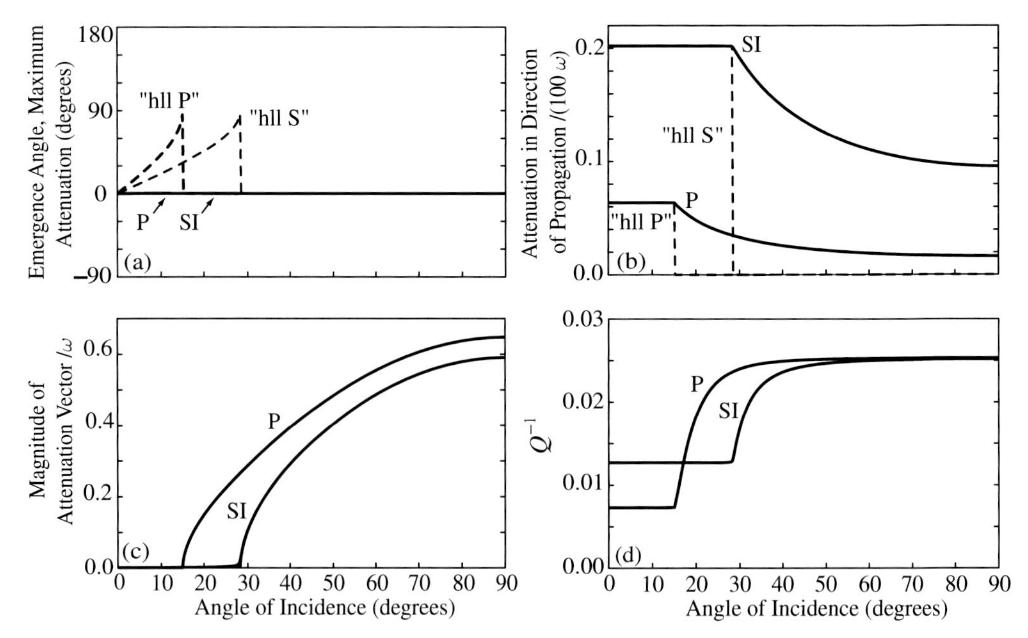

Figure (6.2.4). Emergence angle for direction of maximum attenuation (a), attenuation in direction of propagation (b), maximum attenuation (c), and Q^{-1} for the inhomogeneous transmitted P and Type-I S waves generated by a plane acoustic P wave in water incident on a plane anelastic stainless-steel boundary.

grazing incidence. The mean energy flux for each of the transmitted waves is inclined at a small but finite angle greater than 0.6° away from the boundary for all angles of incidence. The "hll" model incorrectly assumes that phase propagation for each transmitted wave becomes exactly parallel to the boundary for angles greater than the respective minimum elastic critical angles.

The small amount of attenuation specified for the incident P wave in water results in a small component of apparent attenuation along the interface. As a consequence, Generalized Snell's Law (5.3.16) implies that the maximum attenuation vectors for the transmitted waves are directed nearly vertically into the stainless steel (Figure (6.2.4) a). As a result, the amplitudes of each wave decrease away from the boundary for all angles of incidence and the degree of inhomogeneity γ for each transmitted wave increases as the corresponding emergence angle increases (Figures (6.2.4)a and (6.2.3) a)). The component of attenuation in the direction of phase propagation (Figure (6.2.4) b) and the magnitude of the maximum attenuation (Figure (6.2.4)c) of each wave abruptly decrease and increase, respectively, for angles of incidence in the corresponding transition windows for each wave. Q^{-1} for each transmitted wave (Figure (6.2.4) d) abruptly increases in its respective transition widow and asymptotically approaches a value equal to twice that of a corresponding homogeneous SI wave, which is consistent with results in Theorem (3.6.92) and inequalities (3.6.89). The Q^{-1} values for the inhomogeneous transmitted P and SI waves increase by 400 and 200 percent, respectively to the value of $2Q'^{-1}_{HS} = 0.0254$. The "hll" model incorrectly assumes the direction of maximum attenuation for each wave type is parallel to that of phase propagation for angles of incidence less than corresponding elastic critical angles and normal to the boundary for angles beyond critical (Figures (6.2.4)a and (6.2.4)b).

Energy reflection, transmission, and interaction coefficients are shown in Figure (6.2.5). They indicate important characteristics of the reflected and refracted waves that are not predicted by elastic models. Inspection of the normalized mean energy flux

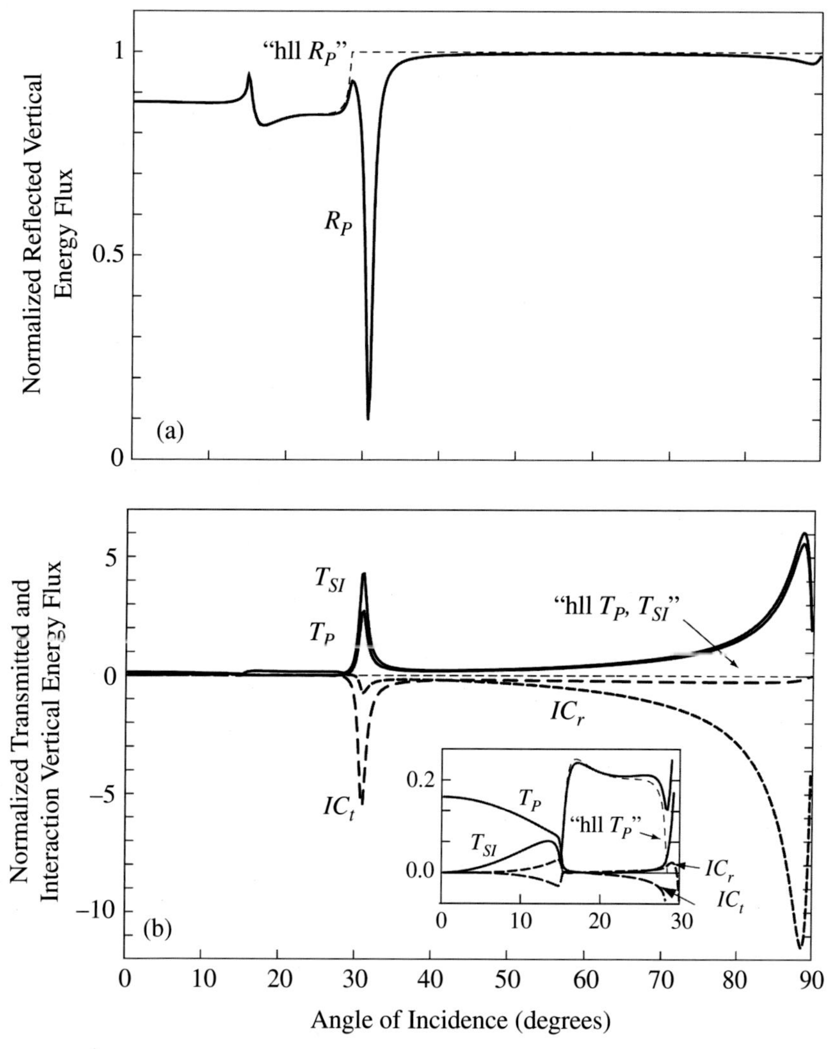

Figure (6.2.5). Energy reflection (a), transmission (b), and interaction (b) coefficients for a plane acoustic P wave incident on a plane water-stainless steel interface. The coefficients show the existence of an *SI′ transition window* in which the reflected and transmitted energy flow differ significantly from corresponding predictions of an elastic model or the "hll" assumption (thin-dashed curves).

normal to the boundary as carried by the reflected acoustic wave (Figure (6.2.5)a) shows a significant reduction in amplitude for a range of angles of incidence near the minimum elastic critical angle for the transmitted SI wave. For this range in angles of incidence the "hll" model assumes total reflection of the incident energy.

Examination of the normalized mean energy flux coefficients shows that in the *SI′ transition window* the normal component of energy flux associated with the reflected P wave decreases substantially and that associated with that the of P and SI waves transmitted into the stainless steel shows a significant increase (see Figures (6.2.5)a and (6.2.5)b). In the *SI′ transition window* the normal component of energy flow due to interaction of the transmitted P and SI waves in the stainless steel increases to an amount larger than that carried by either the transmitted SI or P wave. This normal component of the energy flow due to interaction is opposite to that associated with the transmitted waves and is directed into the water. The normal component of energy flow due to interaction of the incident and reflected waves is also directed into the water for angles of incidence in the *SI′ transition window*. It increases to a maximum of about 10 percent of that due to interaction of the transmitted waves in the transition window. The normal component of energy flow associated with the transmitted wave increases significantly for a range of larger angles of incidence near grazing. This increase is offset by a corresponding increase in the energy flow due to interaction of the incident and reflected waves giving rise to a mean energy flow at the boundary in the opposite direction into the water.

The *SI′ transition window* has also been referred to as the Rayleigh window because the apparent phase velocity for the incident P wave in this window is near that of a Rayleigh wave on stainless steel if modeled as an elastic half space (see Figure (6.2.3)d). Considering that an elastic model of the water, stainless-steel boundary does not predict the significant differences in reflected amplitude for angles of incidence in this window, reference to the window as an *SI′ transition window* seems preferable.

The amplitude of the reflected acoustic wave (Figure (6.2.6)a) and the amplitudes of the transmitted P and SI waves (Figure (6.2.6)b) show significant reductions and increases, respectively, in the *SI′ transition window*. The amplitude of the reflected wave reaches a minimum value that is about 30 percent of that of the incident wave. The transmitted P and SI waves reach maxima that are 2.3 and 2.8 times, respectively, as large as those of the incident wave.

Some characteristics of the transmitted P wave in the *SI′ transition window* that provide additional insight are (1) its degree of inhomogeneity is near its upper limit, calculated to be 89.91° for the chosen material parameters (Figure (6.2.3)a); (2) its phase and energy velocities decrease to nearly equal values at an angle of incidence corresponding to an apparent phase speed near that of a Rayleigh-Type surface wave (see Chapter 8) on a half space of anelastic stainless steel (Figure (6.2.3)c); (3) its directions of phase and energy propagation are near their

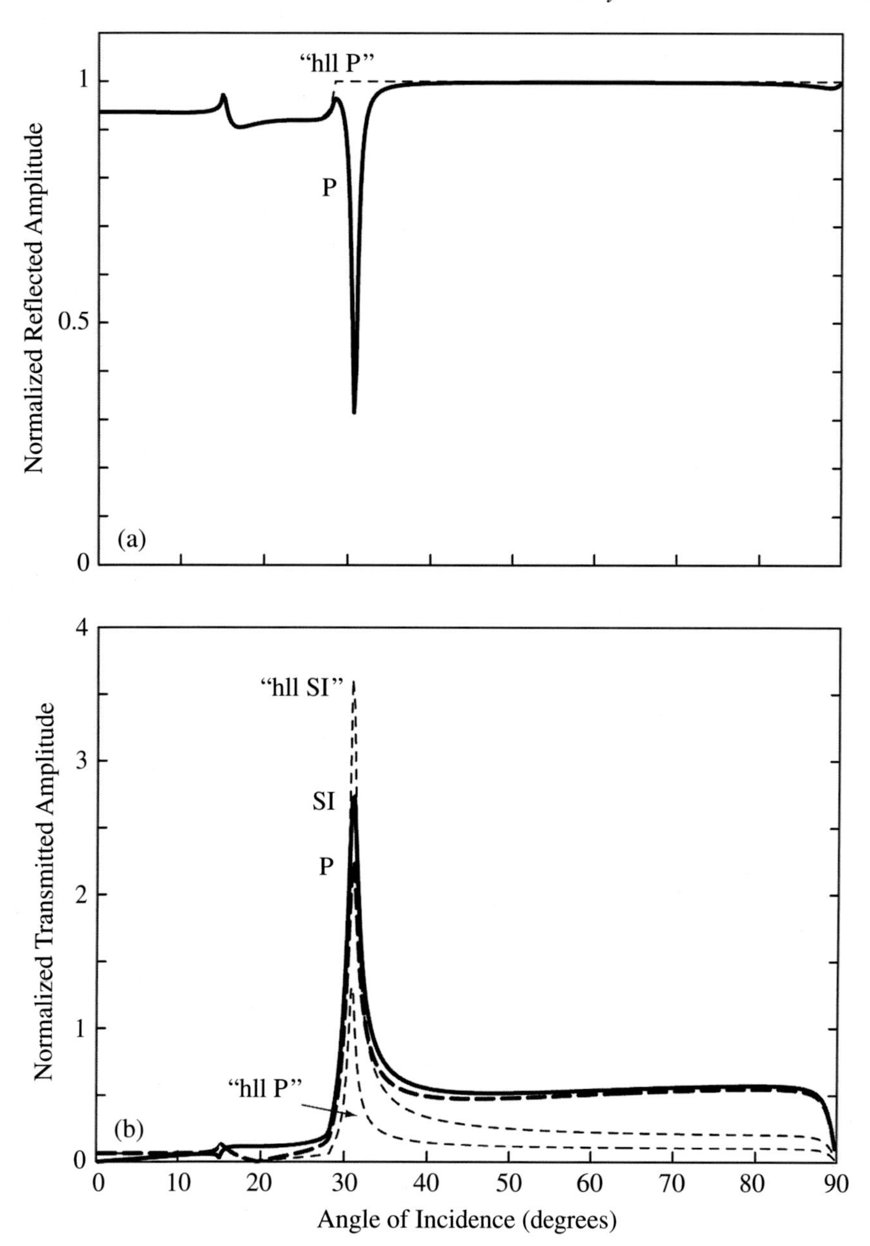

Figure (6.2.6). Amplitude reflection (a) and transmission (b) coefficients characteristics for a plane acoustic P wave incident on a plane water–stainless-steel interface. The coefficients show that the reflected and transmitted amplitudes in the *SI′ transition window* differ significantly from those predicted using an elastic model or the "hll" assumption (thin-dashed curves).

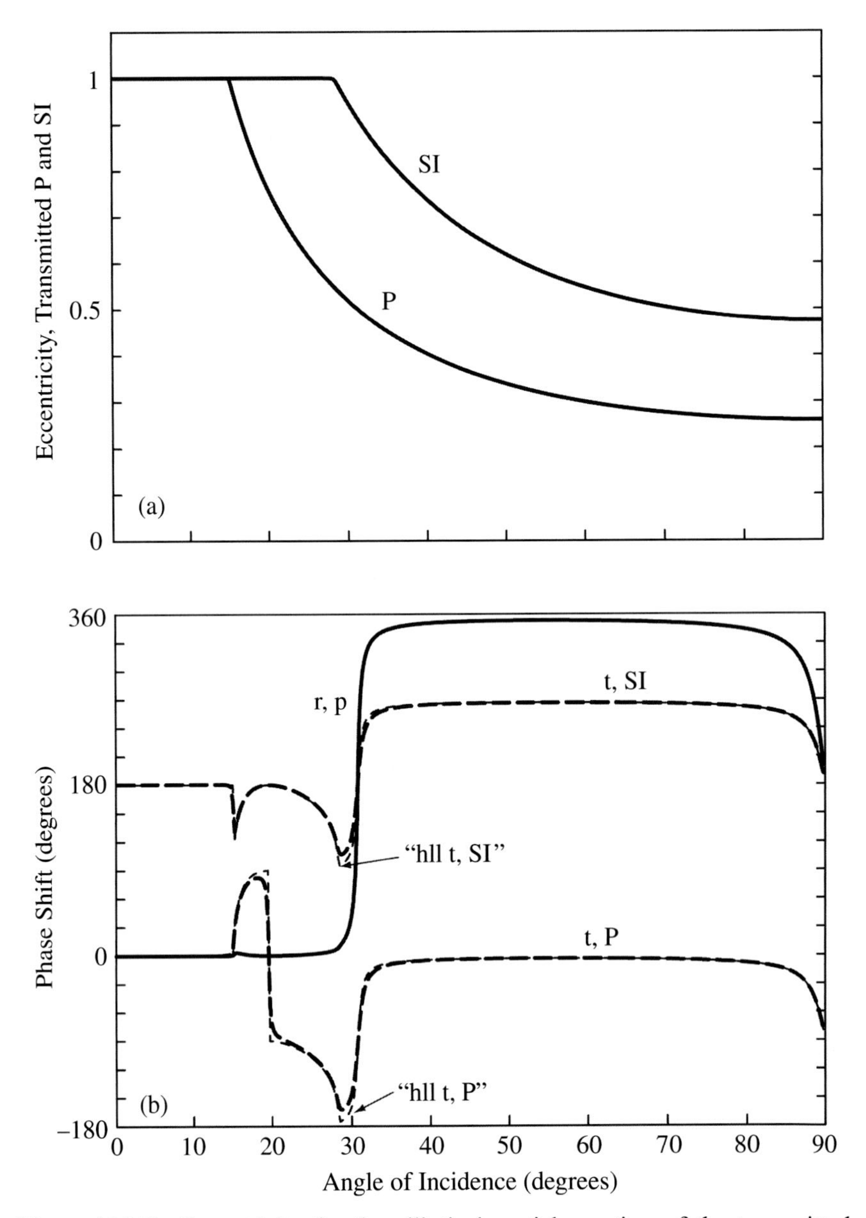

Figure (6.2.7). Eccentricity for the elliptical particle motion of the transmitted P and SI waves (a) and phase shifts for the reflected and transmitted waves (b) for a plane acoustic P wave incident on a plane water, stainless-steel interface.

asymptotic values, which are near but away from the boundary (Figure (6.2.3)b); (4) its $Q_P'^{-1}$ rapidly approaches its limiting value which is within a few percent of $2Q_{HS}'^{-1}$ (Figure(6.2.4)d); and (5) its eccentricity reaches a value near 0.4, indicating the particle motion deviates significantly from linearity (Figure (6.2.7)a).

Characteristics of the transmitted SI wave in the *SI′ transition window* include (1) its directions of phase and maximum energy flux (Figure (6.2.3)c) rapidly approach a value near but away from the boundary; (2) its phase speed decreases with that of the P wave to values less than that of a Rayleigh-Type surface wave (Figure (6.2.3)d); (3) its $Q_{SI}^{\prime -1}$ increases to within about 75 percent of its limiting value near $2Q_{HS}^{\prime -1}$ (Figure (6.2.4)d); (4) its attenuation in the direction of phase propagation decreases (Figure (6.2.4)b) while its maximum attenuation increases (Figure (6.2.4)c); and (5) its particle motion becomes elliptical (Figure (6.2.7)a) with the eccentricity changing from a value near unity to a value near 0.7. Rapid shifts in phase are apparent in the *SI′ transition window* for each of the reflected and refracted waves (Figure (6.2.7)b).

6.2.2 Experimental Evidence in Confirmation of Theory for Viscoelastic Waves

Experimental evidence observed in the laboratory in the 1960s was not consistent with predictions of elastic reflection–refraction theory. Measurements of acoustic amplitudes reflected from water–aluminum interfaces in the laboratory were found to be significantly less than elastic-model predictions for a range of angles of incidence greater than the minimum elastic critical angle for the transmitted S wave. This lack of agreement (e.g. see Brekhovskikh, 1960, p. 34) raised issues regarding the adequacy of plane-wave theory as then only developed for elastic media. Subsequently, considerable experimental evidence concerning the nature of the reflection and refraction of low-loss anelastic waves at water–metal boundaries has been acquired for purposes of discovering material imperfections. These data, some of which have been collected under carefully controlled laboratory conditions provide an established experimental data set for comparison with results predicted on the basis of the theory presented herein.

Experiments to determine the reflective characteristics of a liquid–metal interface provide measurements of the amplitude and phase shift of the reflected acoustic wave generated by a P wave incident on the metal at a known angle of incidence (Becker and Richardson, 1969). Amplitude measurements at 10 MHz and phase measurements at 5 MHz (solid squares) and 16 MHz (open squares) (Becker and Richardson, 1970, pp. 116, 119) for a range of angles of incidence spanning the *SI′ transition window* are shown in Figure (6.2.8). To approximate plane-wave conditions, transducers 2.54 cm in diameter were used with corrections for diffraction effects and lateral beam displacement. (An account of these effects and corresponding references are available in Borcherdt *et al.* (1986). Theoretical predictions for amplitude and phase reflection coefficients corresponding to the measured values (Table (6.2.2)) and to other indicated values of Q_{HS}^{-1} are shown in Figure (6.2.8). The reflection coefficients calculated for various values of Q_{HS}^{-1} correspond approximately to frequencies of 1, 10, 15, 35, and 70 MHz.

The measured amplitude and phase coefficients for the reflected acoustic wave are in excellent agreement with those predicted theoretically as shown in Figures

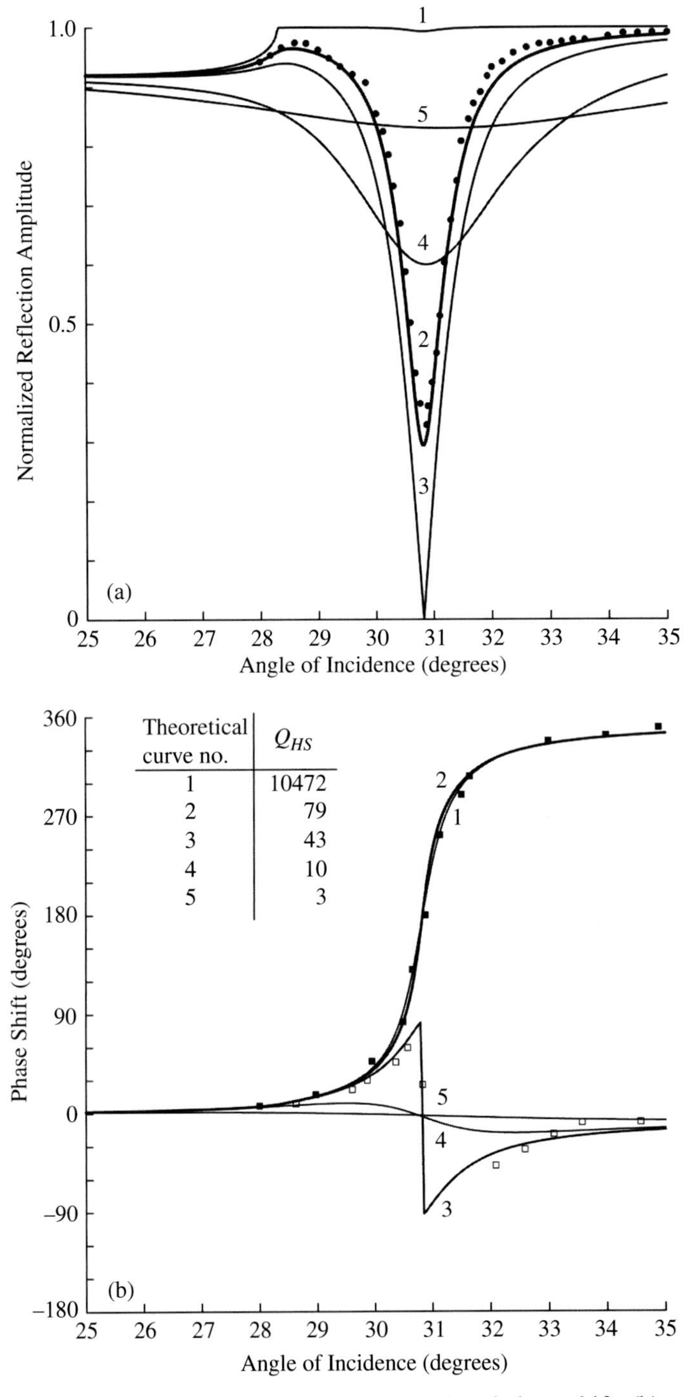

Figure (6.2.8). Amplitude reflection coefficients (a) and phase shifts (b) measured in the laboratory by Becker (1971) and Becker and Richardson (1969, 1970) and predicted theoretically for five water, stainless-steel interfaces with material parameters as indicated. The excellent agreement between measured and theoretically predicted results provides evidence in confirmation of the theory and the empirical measurements.

(6.2.8)a and (6.2.8)b. The measured and calculated amplitude reflection coefficients show a significant dependence on the amount of intrinsic absorption in the stainless steel (Figures (6.2.8)a and (6.2.8)b). The theoretical curves indicate the minimum in the amplitude reflection coefficient decreases to zero then increases as the amount of intrinsic absorption is increased. The theoretical curves predict that a reversal in the phase shift occurs for an amount of intrinsic absorption in the solid that corresponds to the amplitude of the reflected P wave being zero ($Q_{HS}^{-1} = 43$; $f = 15$ MHz). The empirical measurements of phase shifts at 10 and 16 MHz confirm this prediction (see Figures (6.2.8)a and (6.2.8)b).

The excellent agreement between theoretical and laboratory results provides empirical evidence in confirmation of the theory of viscoelastic wave propagation presented herein. It illustrates the significant effect of intrinsic material absorption on the physical characteristics of body waves in layered low-loss anelastic media for certain ranges in angles of incidence. It explains the amplitude reflection observations reported by Brekhovskikh (1960) and the lack of agreement he found with results predicted by an elastic model. It provides evidence for the validity of the physical characteristics of the reflected and refracted waves in low-loss anelastic media as predicted from theoretical considerations for this problem in Figures (6.2.3) through (6.2.8).

6.2.3 Viscoelastic Reflection Coefficients for Ocean, Solid-Earth Boundary

A problem analogous to the water, stainless-steel problem is the problem of reflection of acoustic energy at the ocean floor. This problem is of interest in underwater acoustics and geophysics.

Amplitude and phase reflection coefficients for the P wave generated by a homogeneous P wave incident on various water, solid-Earth interfaces are shown for a range of viscoelastic models in Figures (6.2.10) and (6.2.11). Three distinct material shear-wave speeds as indicated by Ewing and Houtz (1979) and a range in intrinsic absorption values in shear were chosen for the calculations (see Table (6.2.9) and Figures (6.2.10) and (6.2.11)). Intrinsic absorption in bulk for the solid

Table (6.2.9). *Material parameters for a water–solid interface with parameters corresponding to those of ocean, solid-Earth interfaces.*

Medium	ρ (g/cm^3)	v_{HS} (km/s)	v_{HP} (km/s)
V: Ocean	1.03		1.5
V': Sedimentary crust	2.75	2	3.46
V': Basaltic crust	2.85	2.8	4.85
V': Gabbro transition	2.85	3.7	6.41

From Ewing and Houtz (1979).

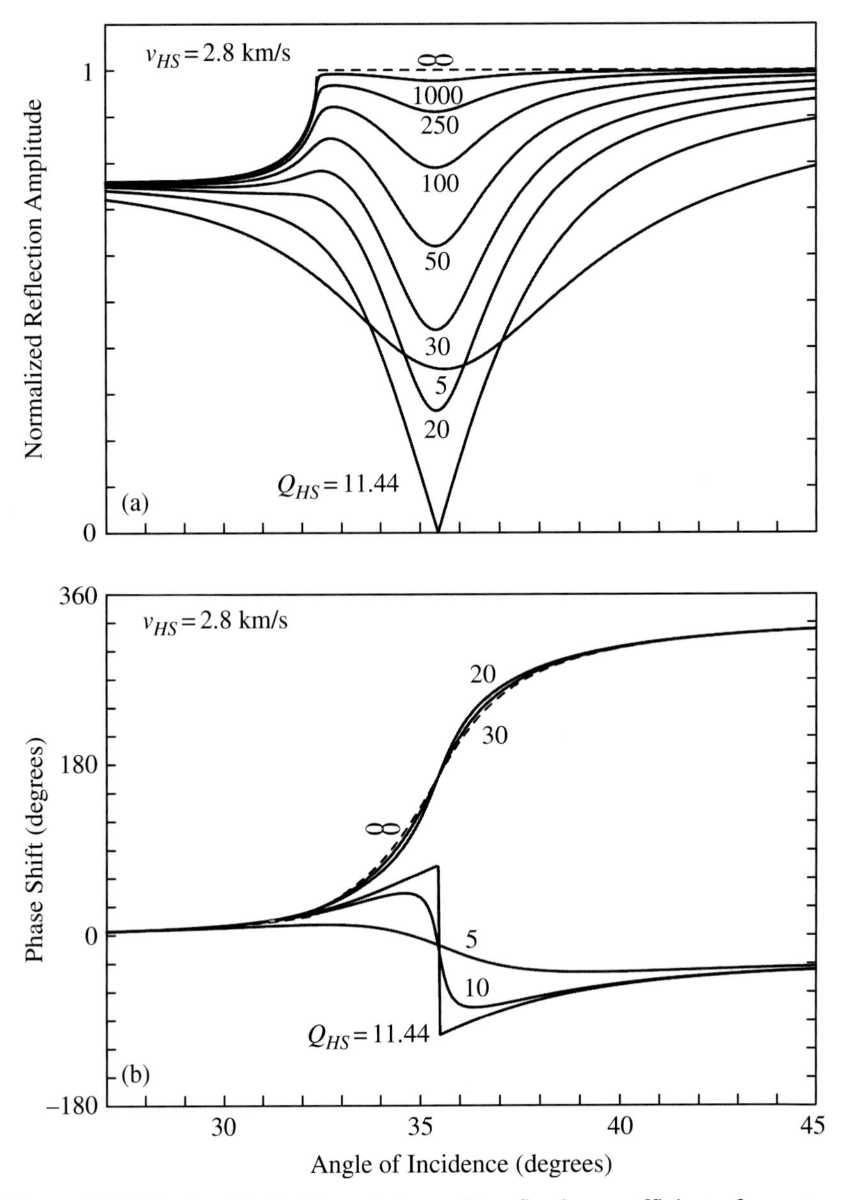

Figure (6.2.10). Amplitude (a) and phase (b) reflection coefficients for water–solid interfaces with shear and P wave speeds corresponding to ocean–basaltic crust (b), and indicated amounts of intrinsic absorption in shear (see Table (6.2.9)).

Earth is assumed negligible with the reciprocal quality factor for a homogeneous P wave given by $Q_{HP}^{-1} = \frac{4}{9}Q_{HS}^{-1}$.

The calculated reflection coefficients (Figures (6.2.10) and (6.2.11)) show a significant dependence on the amount of intrinsic absorption in the solid similar to that observed for the water, stainless-steel boundary. They indicate that the

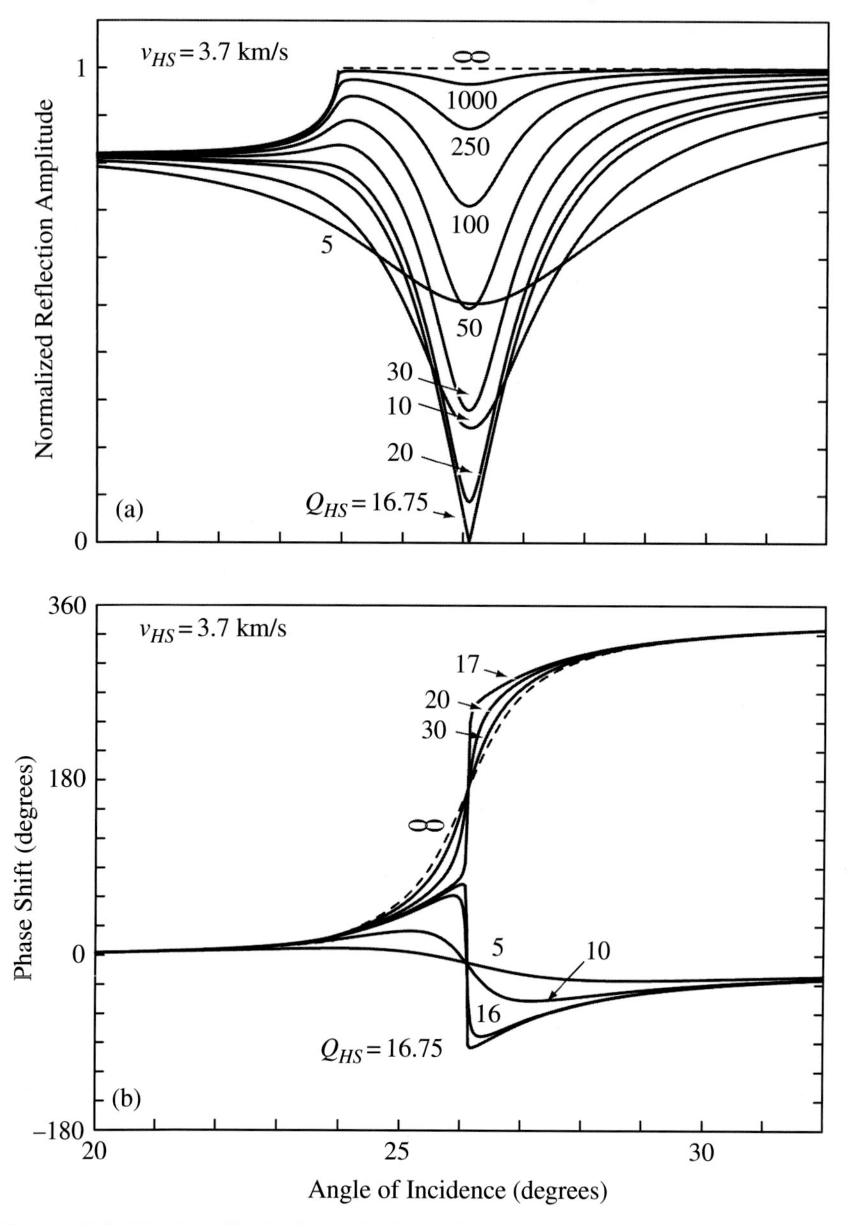

Figure (6.2.11). Amplitude (a) and phase (b) reflection coefficients for water–solid interfaces with shear and P wave speeds corresponding to an ocean–gabbro transition crust computed for indicated amounts of intrinsic absorption in shear (see Table (6.2.9)).

amplitude of the reflected P wave vanishes and a phase-shift reversal occurs for a specific amount of intrinsic absorption at a particular angle of incidence for each model. Calculations show that the local minimum in the amplitude reflection coefficient for each contrast in wave speed decreases to zero then

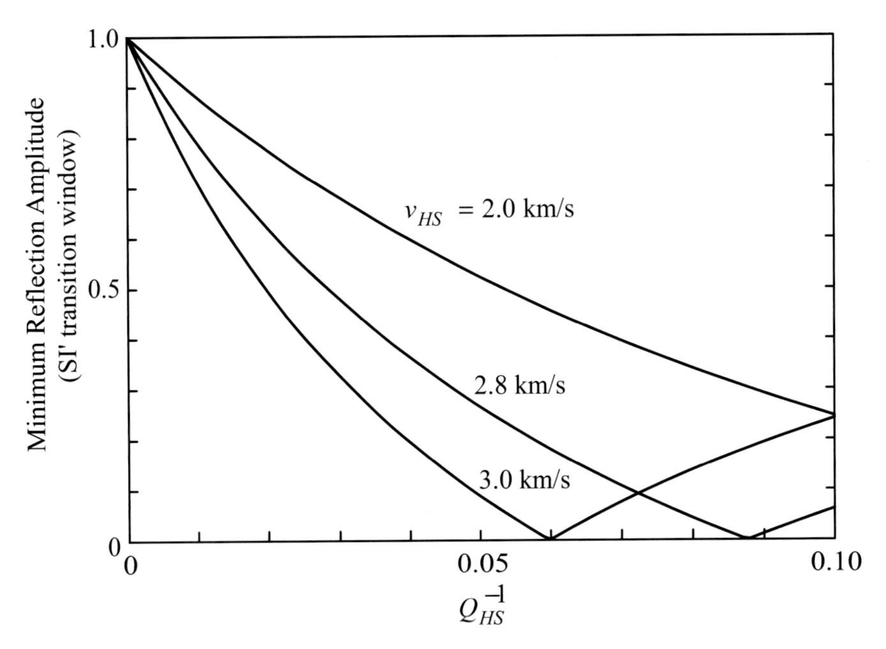

Figure (6.2.12). Amplitude of local minimum in reflected P wave amplitude in the *SI' transition window* as a function of intrinsic absorption in shear for three ocean–solid-Earth interface models as characterized by indicated shear wave speeds (Table (6.2.9)).

increases as the amount of intrinsic absorption increases (Figures (6.2.10)a, (6.2.11)a, and (6.2.12)).

The amplitude reflection coefficient varies with material shear-wave speed of the solid (compare Figures (6.2.10)a and (6.2.11)a). As the shear velocity of the solid approaches the P velocity of the water, the angle of incidence corresponding to the local minimum in the reflected P wave amplitude increases and the amplitude of the local minimum in the reflection coefficient decreases.

The distinguishing characteristics of the amplitude and phase reflection coefficients for the ocean, solid-Earth boundary suggest that their measurement could be useful for inferring the intrinsic absorption characteristics of various crustal materials. The local minima in amplitude reflection data and phase shifts as inferred from wide-angle geophysical reflection data could be useful in developing Q_{HS}^{-1} profiles perpendicular to ocean ridges using techniques such as those described by Stoffa and Buhl (1979). Such profiles if definitive could be useful for inferring geologic age and tectonic spreading rates. The dependence of the reflection coefficients on intrinsic absorption of the ocean bottom suggests that acoustic transmissions of various frequencies, depths, and distances might be expected to propagate more efficiently than others.

6.3 Problems

(1) For the problem of a homogeneous Type-II S wave incident on a plane viscoelastic sediment boundary with parameters of the material specified in Table (6.1.1), sketch graphs showing the dependence on angle of incidence for
   (a) a transmitted SII wave of
      (i) the degree of inhomogeneity from a value of zero to one near its physical limit of $\pi/2$,
      (ii) emergence angles for phase propagation, direction of maximum attenuation, and energy flux,
      (iii) phase and energy speed,
      (iv) reciprocal quality factor,
   (b) normalized energy flux perpendicular to the boundary associated with
      (i) incident SII wave,
      (ii) reflected SII wave,
      (iii) transmitted SII wave, and
      (iv) interaction of incident and reflected SII waves.

(2) Compare the sketches in Problem 1 with corresponding sketches assuming the media are elastic.

(3) Describe how the graphs in Problem 1 change if the incident wave is inhomogeneous with a fixed degree of inhomogeneity of $-60°$

(4) For the problem of a homogeneous P wave incident in water on a viscoelastic stainless-steel boundary,
   (a) sketch graphs of the amplitude reflection coefficient as a function of angle of incidence for various amounts of intrinsic absorption in the stainless steel,
   (b) describe how the reflection coefficient for an elastic model of the stainless steel differs from that for a linear anelastic model with $Q^{-1} \approx 1/80$, and
   (c) describe why the dependence of the reflection coefficient in the *SI′ transition window* on the amount of intrinsic absorption in shear has been useful for nondestructive testing of materials.

7

General SI, P, and SII Waves Incident on a Viscoelastic Free Surface

Solutions for problems involving wave propagation in a semi-infinite half space are of interest for interpreting measurements of radiation fields at locations near or on the free surface. Solutions to these problems as derived for elastic media have formed the basis for the initial interpretation of seismograms and resultant inferences concerning the internal structure of the Earth.

Analytic solutions and corresponding numerical examples for problems involving general SI, P, and SII waves incident on the free surface of a viscoelastic half space are presented in this chapter (Borcherdt, 1971, 1988; Borcherdt and Glassmoyer, 1989; Borcherdt *et al.*, 1989). Closed-form expressions for displacement and volumetric strain are included to facilitate understanding and interpretation of measurements as might be detected on seismometers and volumetric strain meters at or near the free surface of a viscoelastic half space.

The procedures to solve the reflection–refraction problems for a general SI, P, or SII wave incident on a free surface are analogous to those for the corresponding problems for a welded boundary. For brevity, many of the expressions and results in medium V for a welded boundary applicable to the free-surface problems will be referred to here, but not rewritten.

7.1 Boundary-Condition Equations

Solutions of the equations of motion for problems of general P, SI, and SII waves incident and reflected from the surface of a viscoelastic half space are specified by (4.2.1) through (4.2.45) with respect to the coordinate system illustrated in Figure (4.1.3), where medium V' is assumed to be a vacuum. The plane boundary of medium V for a half space is assumed to be stress free.

Steady-state propagation in the $+x_1$ direction of the x_1x_3 plane implies the components of the stress tensor may be expressed in terms of the displacement potentials, using equations (3.1.1), (2.3.5), and (2.3.12). The resulting equations expressing the stresses acting on planes parallel to the free surface are

$$p_{31} = M\left(2\phi_{,13} + \psi_{,11} - \psi_{,33}\right), \tag{7.1.1}$$

$$p_{32} = Mu_{2,3}, \tag{7.1.2}$$

and

$$p_{33} = \left(K + \frac{4}{3}M\right)\left(\phi_{,11} + \phi_{,33}\right) + 2M\left(\psi_{,13} - \phi_{,11}\right), \tag{7.1.3}$$

where the notation $\phi_{,ij}$ denotes the partial derivative of $\phi = \phi_1 + \phi_2$ with respect to the ith and j^{th} component, respectively and ψ denotes the $\hat{x}_2$ component of $\vec{\psi} = \vec{\psi}_1 + \vec{\psi}_2$.

The boundary conditions for an infinite half space are that the stresses on the free surface vanish, that is

$$p_{3j} = 0 \qquad \text{for } j = 1,\ 2,\ 3. \tag{7.1.4}$$

Substitution of the solutions (4.2.1), (4.2.2), and (4.2.26) into (7.1.1) through (7.1.3) restricted by (7.1.4) implies that the complex spatially independent amplitudes of the solutions must satisfy

for $p_{31} = 0$,

$$M\left[2d_\alpha k(B_1 - B_2) + \left(d_\beta{}^2 - k^2\right)(C_{12} + C_{22})\right] = 0, \tag{7.1.5}$$

for $p_{32} = 0$,

$$D_1 = D_2, \tag{7.1.6}$$

and for $p_{33} = 0$,

$$M\left[-\left(d_\beta{}^2 - k^2\right)(B_1 + B_2) + 2d_\beta k(C_{12} - C_{22})\right] = 0. \tag{7.1.7}$$

Solution of these three equations involving six complex amplitudes and the complex wave number k can be achieved upon consideration of a particular problem, such as an incident P, SI, or SII wave on the free surface or consideration of disturbances confined primarily to the free surface.

To consider a specific reflection problem, a single incident wave is assumed. This assumption is equivalent to specifying the complex wave number k and the requirement that the complex amplitudes for the other two incident waves are zero. Hence, the resulting three equations in four unknowns can be solved by expressing the complex amplitudes of the reflected waves in terms of the complex amplitude and complex wave number of the incident wave.

To consider a surface-wave problem the steady-state disturbance is assumed to be concentrated near the surface. This assumption is satisfied by setting $B_1 = C_1 = D_1 = 0$ in (7.1.5) through (7.1.7). With these complex amplitudes set to zero, the three equations in four unknowns can be solved for the complex wave number k and one of the complex amplitudes in terms of the other.

7.2 Incident General SI Wave

The problem of a general Type-I S wave incident on the free surface is specified by assuming the parameters of the incident Type-I S wave are given and that the amplitudes of the solutions for the other incident waves as specified in (4.2.1) and (4.2.26) vanish, namely $B_1 = D_1 = 0$. The incident general Type-I S wave as specified by (4.2.2) is given in medium V by equations (5.2.3) through (5.2.11). For brevity, these equations are not rewritten here. Solutions for the reflected P and SI waves in medium V as specified by equations (4.2.1) and (4.2.2) are given by equations (5.2.14) and (5.2.15). Parameters for the incident and reflected waves are illustrated in Figure (7.2.1).

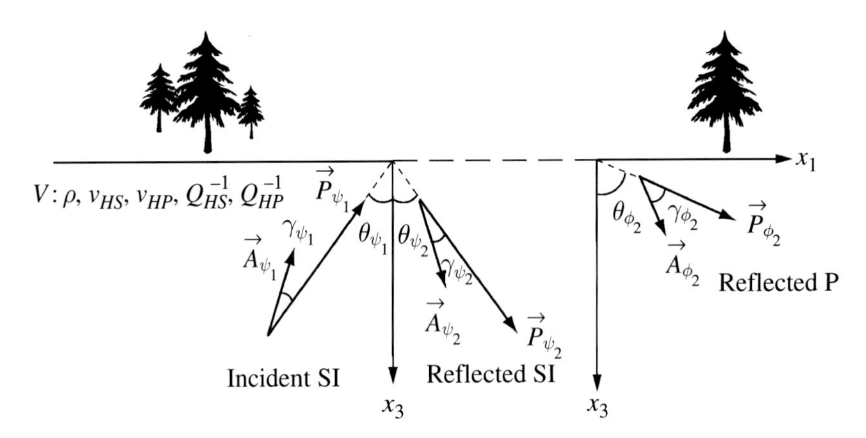

Figure (7.2.1). Diagram illustrating notation for directions and magnitude of propagation and attenuation vectors for the problem of a general SI wave incident on the free surface of a HILV half space.

7.2.1 *Reflected General P and SI Waves*

For the problem of a general Type-I S wave incident on a free surface, the boundary conditions (7.1.5) through (7.1.7) readily admit solutions for the complex amplitudes of the reflected waves in terms of the complex amplitude and complex wave number for the incident wave, namely

$$B_2/C_{12} = 4\, d_\beta\, k\left(d_\beta^2 - k^2\right)/g(k), \tag{7.2.2}$$

$$D_2 = D_1 = 0, \tag{7.2.3}$$

and

$$C_{22}/C_{12} = \left(4\, d_\alpha\, d_\beta\, k^2 - \left(d_\beta^2 - k^2\right)^2\right) \Big/ g(k), \tag{7.2.4}$$

where

$$g(k) = 4\, d_\alpha\, d_\beta k^2 + \left(d_\beta^2 - k^2\right)^2 \tag{7.2.5}$$

for

$$g(k) \neq 0. \tag{7.2.6}$$

The complex wave number k for the incident general SI wave is given by

$$k = \frac{\omega}{v_{HS}} \left(\sqrt{\frac{1 + \chi_{S\psi_1}}{1 + \chi_{HS}}} \sin\theta_{\psi_1} - i\sqrt{\frac{-1 + \chi_{S\psi_1}}{1 + \chi_{HS}}} \sin[\theta_{\psi_1} - \gamma_{\psi_1}] \right), \tag{7.2.7}$$

which simplifies for an elastic half space to

$$k = \frac{\omega}{v_{HS}} \sin\theta_{\psi_1}, \tag{7.2.8}$$

where $\chi_{S\psi_1} = \sqrt{1 + Q_{HS}^{-2} \sec^2 \gamma_{\psi_1}}$ and $\chi_{HS} = \sqrt{1 + Q_{HS}^{-2}}$. The propagation and attenuation vectors as specified by (5.2.5) and (5.2.6) for the incident and reflected waves are completely determined upon specification of the angle of incidence θ_{ψ_1}, the degree of inhomogeneity γ_{ψ_1}, and the circular frequency ω for the incident general SI wave together with the material parameters for V as specified by ρ, k_S or v_{HS} and Q_{HS}^{-1}. Hence, the propagation and attenuation vectors for the reflected general P and SI waves are determined in terms of those for the incident general SI wave, from which it follows that (7.2.2) and (7.2.4) provide the desired solution for the complex amplitude of the reflected general P and SI waves in terms of that given for the incident general SI wave.

For a non-trivial incident wave $C_{12} \neq 0$, so equations (7.1.5) and (7.1.7) imply $g(k) \neq 0$. Roots of equation (7.2.5) will be shown to specify k for a Rayleigh-Type surface wave. Hence, this result shows that a general Type-I S wave incident in a plane perpendicular to the free surface does not generate a Rayleigh-Type surface wave. Also, boundary-condition equation (7.2.3) shows that no Type-II S waves are generated by a Type-I S wave incident in a plane perpendicular to the interface.

The existence of solutions (7.2.2) and (7.2.4) confirms equality of the complex wave number k in the assumed solutions for the incident and reflected waves as specified by (4.2.1) and (4.2.2). Hence, the components of phase propagation and maximum attenuation parallel to the boundary for the incident and reflected waves are equal, that is,

$$k_R = \mathrm{Re}\left[K_{\psi_1 x}\right] = \left|\vec{P}_{\psi_1}\right| \sin\theta_{\psi 1} = \left|\vec{P}_{\psi_2}\right| \sin\theta_{\psi_2} = \left|\vec{P}_{\phi_2}\right| \sin\theta_{\phi_2} \tag{7.2.9}$$

and

$$\begin{aligned} -k_I = -\mathrm{Im}\left[K_{\psi_1 x}\right] &= |\vec{A}_{\psi_1}| \sin[\theta_{\psi_1} - \gamma_{\psi_1}] = |\vec{A}_{\psi_2}| \sin[\theta_{\psi_2} - \gamma_{\psi_2}] \\ &= |\vec{A}_{\phi_2}| \sin[\theta_{\phi_2} - \gamma_{\phi_2}], \end{aligned} \tag{7.2.10}$$

establishing Generalized Snell's Law for the problem of a general Type-I S wave incident on the free surface of a viscoelastic half space, as stated for medium V by Theorem (5.2.22).

It follows, as for the corresponding welded-boundary problem, that $\theta_{\psi_1} = \theta_{\psi_2}$ and $\gamma_{\psi_1} = \gamma_{\psi_2}$. Hence, for the incident and reflected SI waves, the reflection angle equals the angle of incidence and their degrees of inhomogeneity are equal.

The conditions of homogeneity and inhomogeneity for the waves reflected from the free surface for an incident general SI wave are as stated for the reflected waves in medium V by (5.2.28) through (5.2.43). In brief, the reflected SI wave is homogeneous if and only if the incident SI wave is homogeneous. Theorem (5.2.34) indicates that the P wave reflected from the free surface of a HILV half space is homogeneous if and only if $Q_{HS}^{-1} = Q_{HP}^{-1}$ and $\sin^2\theta_{\psi 1} \le v_{HS}^2/v_{HP}^2$. These results indicate that for elastic media the reflected P wave will be homogeneous for angles of incidence less than some elastic critical angle, but for anelastic Earth-type media where $Q_{HS}^{-1} \neq Q_{HP}^{-1}$, the reflected P wave, in general, will be inhomogeneous for all angles of incidence.

The degree of inhomogeneity of the reflected P wave in anelastic media will be shown to increase with angle of incidence and hence the phase velocity of the reflected P wave will decrease, the maximum attenuation will increase, and fractional energy loss and other characteristics will be shown to vary with angle of incidence. As $Q_{HS}^{-1} \neq Q_{HP}^{-1}$ in an anelastic Earth the preceding results show that P waves generated by a SI body wave incident on the free surface, in general, are inhomogeneous with the degree of inhomogeneity and the physical characteristics of the wave dependent on the angle of incidence and the degree of inhomogeneity of the incident wave.

The conditions for propagation of the reflected P wave parallel to the free surface are similar to those for the reflected P wave generated by an SI wave incident on a welded boundary (see the section of Theorem (5.2.45)). The results as stated in Theorems (5.2.45) through (5.2.47) for the P wave reflected in medium V are the same with similar proofs. Hence, for brevity they are not restated in this section. In brief, the results show that for elastic media a critical angle or phase propagation parallel to the boundary exists whenever the apparent phase velocity along the interface of the incident SI wave is less than or equal to that of a homogeneous P wave. In contrast, for anelastic Earth materials, the results indicate that phase propagation for the reflected P wave is away from the boundary for all angles of incidence.

In the case of an elastic medium, a normally incident SV (SI) wave does not generate a dilatational disturbance upon interacting with the free surface. However, in the case of a vertically incident inhomogeneous SI wave in anelastic media, a dilatational disturbance is reflected from the free surface. To show this result, suppose no dilatational disturbance is generated, that is $B_2 = 0$, then equation (7.2.2) implies $k = 0$, $d_\beta = 0$, or $d_\beta^2 - k^2 = 0$. The first alternative is not possible, because if the incident wave is vertically incident then $k = i|\vec{A}_{\psi_1}| \sin \gamma_{\psi_1} \neq 0$ in order for the wave to propagate. The two other alternatives are not possible because the medium is anelastic. Hence, the amplitude of the reflected P wave is not zero and a dilatational disturbance is reflected from the free surface. If the normally incident SI wave is homogeneous, then $k = k_S \sin \theta_{\psi_1} = 0$ and equation (7.2.2) shows the amplitude of the reflected P wave vanishes.

For an elastic half space, angles of incidence exist such that the incident SV (SI) wave is entirely reflected as a dilatational disturbance. For anelastic media it will be evident that such angles in general do not exist. The following theorem indicates that such angles exist for only a restricted class of viscoelastic solids of which elastic is a special case.

Theorem (7.2.11). If the incident SI wave is homogeneous and a non-zero angle of incidence exists for which the amplitude of the reflected SI wave is zero, then the solid is such that $Q_{HS}^{-1} = Q_{HP}^{-1}$.

The contrapositive of this theorem shows that because $Q_{HS}^{-1} \neq Q_{HP}^{-1}$ for anelastic Earth materials, the amplitude of the reflected SI wave is non-zero for every non-vertical angle of incidence for an incident homogeneous SI wave.

To prove the theorem suppose the angle of incidence $\theta_{\psi_1} \neq 0$ and the amplitude of the reflected SI wave $C_{22} = 0$ from which (7.2.4) implies

$$16 \frac{d_\alpha^2}{k^2} \frac{d_\beta^2}{k^2} = \left(\frac{d_\beta^2}{k^2} - 1 \right)^4. \tag{7.2.12}$$

By assumption the incident SI wave is homogeneous, hence $k = k_S \sin \theta_{\psi_1}$ and (4.2.10) implies

$$\frac{d_\beta^2}{k^2} = \tan^{-2} \theta_{\psi_1}. \tag{7.2.13}$$

The preceding two equations imply that d_β^2/k^2 and d_α^2/k^2 are real numbers. Definition (4.2.9) of d_α implies

$$\text{Im}\left[\frac{d_\alpha^2}{k^2}\right] = \frac{1}{\sin^2\theta_{\psi_1}} \text{Im}\left[\frac{k_P^2}{k_S^2}\right] = 0, \tag{7.2.14}$$

from which it follows that k_P^2/k_S^2 is a real number. Hence, Lemma (5.2.35) implies the desired result namely, $Q_{HS}^{-1} = Q_{HP}^{-1}$.

7.2.2 Displacement and Volumetric Strain

Consideration of the effects of the free surface on the displacement fields and volumetric strain associated with an incident general SI wave at or near the free surface is of special interest for interpreting recordings of signals from three-component seismometers and volumetric strain transducers (Borcherdt *et al.*, 1989). Comparison of simultaneous displacement fields and corresponding volumetric strain is of interest, because, theoretically, volumetric strain is associated only with the P wave reflected from the free surface, but not with either the incident or the reflected SI waves. Consequently, simultaneous measurements from corresponding sensors suggest that characteristics of the wave fields might be inferred that cannot be inferred from either measurement alone.

To further investigate the effect of the free surface on the general reflected P wave, equations (7.2.9) and (7.2.10) allow the reflection angle for the P wave to be expressed in terms of the given incidence angle for the incident SI wave, using identities (3.6.21) as

$$\sin\theta_{\phi_2} = \frac{|\vec{v}_{\phi 2}|}{|\vec{v}_{\psi 1}|}\sin\theta_{\psi_1} = \frac{v_{HP}}{v_{HS}}\sqrt{\frac{1+\chi_{HP}}{1+\chi_{HS}}}\sqrt{\frac{1+\chi_{S_{\psi_1}}}{1+\chi_{P_{\phi_2}}}}\sin\theta_{\psi_1} \tag{7.2.15}$$

and the angle that the attenuation vector for the reflected P wave makes with respect to the vertical (3.6.23) in terms of that for the incident SI wave by

$$\begin{aligned}\sin[\theta_{\phi_2} - \gamma_{\phi_2}] &= \frac{|\vec{A}_{\psi_1}|}{|\vec{A}_{\phi_2}|}\sin[\theta_{\psi_1} - \gamma_{\psi_1}] \\ &= \frac{v_{HP}}{v_{HS}}\sqrt{\frac{1+\chi_{HP}}{1+\chi_{HS}}}\sqrt{\frac{-1+\chi_{S_{\psi_1}}}{-1+\chi_{P_{\phi_2}}}}\sin[\theta_{\psi_1} - \gamma_{\psi_1}].\end{aligned} \tag{7.2.16}$$

Substitution of these expressions into the definitions (3.10.12) and (3.10.13) reveals two identities relating parameters of the reflected P wave to those of the incident SI wave, namely

$$F_{P_{\phi_2}}[\sin\theta_{\phi_2}] = \frac{|k_S|}{|k_P|}F_{S_{\psi_1}}[\sin\theta_{\psi_1}] = \frac{v_{HP}}{v_{HS}}\sqrt{\frac{1+\chi_{HP}\chi_{HS}}{1+\chi_{HS}\chi_{HP}}}F_{S_{\psi_1}}[\sin\theta_{\psi_1}] \tag{7.2.17}$$

and

$$\Omega_{P_{\phi_2}}\left[\sin\theta_{\phi_2}\right] = \Omega_{S_{\psi_1}}\left[\sin\theta_{\psi_1}\right]. \tag{7.2.18}$$

These identities when substituted into the expression for the radial or $\hat{x}_1$ (x) component of displacement for the reflected P wave as described by (3.10.10) yield the following expression for the radial component of displacement associated with the general reflected P wave, namely,

$$u_{R\phi_2 x} = |B_2 k_P| \exp\left[-\vec{A}_{\phi_2} \cdot \vec{r}\right] \frac{|k_S|}{|k_P|} F_{S_{\psi_1}}\left[\sin\theta_{\psi_1}\right] \cos\left[\zeta_{P\phi_2}(t) + \psi_P - \Omega_{S_{\psi_1}}\left[\sin\theta_{\psi_1}\right]\right], \tag{7.2.19}$$

where the amplitude $|B_2|$ and the phase term $\zeta_{P_{\phi_2}}(t) \equiv \omega t - \vec{P}_{\phi_2} \cdot \vec{r} + \arg[B_2 k_P] - \pi/2$ from (3.10.2) are given in terms of parameters of the incident wave by (7.2.2). Expression (7.2.19) provides a complete specification of the radial component of displacement for the reflected P wave in terms of the given parameters of the incident general SI wave. It provides an explicit description of the reflected P wave contribution to the radial component of displacement in terms of the given parameters of a general incident SI wave as might be measured on a radial seismometer.

If the incident SI wave is homogeneous, then (3.10.22) and (3.10.23) imply $F_{S_{\psi_1}}\left[\sin\theta_{\psi_1}\right] = \sin\theta_{\psi_1}$ and $\Omega_{S_{\psi_1}}\left[\sin\theta_{\psi_1}\right] = \tan^{-1}\left[Q_{HS}^{-1}/(1+\chi_{HS})\right] = \psi_S$, so the expression (7.2.19) for the radial component of displacement simplifies to

$$u_{R\phi_2 x} = |B_2 k_P| \exp\left[-\vec{A}_{\phi_2} \cdot \vec{r}\right] \frac{|k_S|}{|k_P|} \sin\theta_{\psi_1} \cos\left[\zeta_{P\phi_2}(t) + \psi_P - \psi_S\right]. \tag{7.2.20}$$

If the amount of intrinsic absorption in the medium is assumed to be small, then $\chi_{HS} \approx \chi_{HP} \approx 1$, so (7.2.17) implies the amplitude-modulation factor for the reflected general P wave is approximately that of the incident general SI wave scaled by the ratio of the wave speed of a homogeneous P wave to that of a homogeneous S wave, that is

$$F_{P_{\phi_2}}\left[\sin\theta_{\phi_2}\right] \approx \frac{v_{HP}}{v_{HS}} F_{S_{\psi_1}}\left[\sin\theta_{\psi_1}\right], \tag{7.2.21}$$

and the phase term $\psi_P - \psi_S$ is approximately given by

$$\psi_P - \psi_S \approx \frac{Q_{HP}^{-1}}{2} - \frac{Q_{HS}^{-1}}{2}. \tag{7.2.22}$$

This expression shows that the radial component of displacement for the reflected P wave in a low-loss half space is shifted in phase due to the difference in anelastic response of the medium to homogeneous P and S waves. It shows that the radial component of displacement for the reflected P wave varies sinusoidally with the given incidence angle of the incident SI wave. It provides an explicit description of the reflected P wave contribution to the radial component of displacement as might be measured on a radial seismometer due to a homogeneous SI wave incident on the free surface of a low-loss anelastic half space.

The volumetric strain generated by a general SI wave incident on the free surface results from the generation of the reflected P wave. Equation (3.10.29) implies the volumetric strain as initially derived by Borcherdt (1988) is given by

$$\Delta_R(t) = |B_2 k_P| \exp\left[-\vec{A}_{\phi_2} \bullet \vec{r}\right] |k_P| \cos\left[\zeta_{P_{\phi_2}}(t) - \psi_P - \pi/2\right], \tag{7.2.23}$$

where the amplitude and phase terms are given in terms of those for the incident SI wave by (7.2.2). For media with small amounts of absorption, identities (3.6.16) and (3.10.3) imply

$$|k_P| \approx \omega / v_{HP} \tag{7.2.24}$$

and

$$\psi_P \approx Q_{HP}^{-1}/2, \tag{7.2.25}$$

which in turn imply that the expression (7.2.23) for the volumetric strain of the P wave reflected from the free surface of a low-loss half space may be simplified further.

The times t_1 and t_2 at which the volumetric strain (7.2.23) assumes its zero and maximum values are given by $\zeta_{P_{\phi_2}}(t_2) = \psi_P + \pi$ and $\zeta_{P_{\phi_2}}(t_1) = \psi_P + \pi/2$, respectively. Substitution of these times into the expression for the radial component of displacement (7.2.20) for the P wave reflected from the free surface for an incident homogeneous SI wave yields

$$\frac{u_{R\phi_2 x}(t_2)}{u_{R\phi_2 x}(t_1)} = \tan[2\psi_P - \psi_S], \tag{7.2.26}$$

suggesting that if times t_1 and t_2 can be determined from a volumetric-strain recording and the radial component of displacement is discernible for the reflected P wave then the anelastic characteristics of the material might be inferred. For example, for materials with small amounts of absorption, (7.2.26) implies

$$Q_{HP}^{-1} - Q_{HS}^{-1}/2 \approx \tan^{-1}\left[\frac{u_{R\phi_2 x}(t_2)}{u_{R\phi_2}(t_1)}\right]. \tag{7.2.27}$$

Hence, if an independent estimate of the specific absorption of either P or S can be made then the other can be inferred or if the amount of absorption in P is much less than that in S then (7.2.27) might be used to infer the specific absorption in S.

Similar inferences regarding intrinsic absorption might be made by inferring phase differences between collocated measurements of inferred displacement and volumetric strain using expressions (7.2.19) and (7.2.23).

To consider free-surface amplitude and phase reflection coefficients for a general viscoelastic half space it is useful to extend the definition of free-surface reflection coefficients given for homogeneous waves in elastic media to more general cases that include both homogeneous and inhomogeneous waves. One convention that can be adopted to accommodate both homogeneous and inhomogeneous waves is to define reflection coefficients by considering ratios of the various components of motion and volumetric strain to that of the maximum particle motion displacement amplitude. The maximum displacement amplitude for each general wave is the amplitude of the wave multiplied by the length of the major axis of the particle motion ellipse $|\vec{\xi}_1|$ as given by (3.2.12) and (3.3.24) for P and SI waves. This convention includes as a special case the definition given for homogeneous waves in elastic media, because for homogeneous waves the elliptical particle motion degenerates to linear particle motion either parallel or perpendicular to the direction of phase propagation.

In terms of the notation introduced here the amplitude reflection coefficient for the reflected P wave is defined as the ratio of its maximum particle displacement to that of the incident SI wave, namely

$$r_P \equiv \frac{|B_2 k_P|}{|C_{12} k_S|} \frac{|\vec{\xi}_{1P\phi_2}|}{|\vec{\xi}_{1S\psi_1}|} \exp\left[-\left(\vec{A}_{\phi_2} - \vec{A}_{\psi_1}\right) \cdot \vec{r}\right]. \tag{7.2.28}$$

The phase coefficient for the maximum particle displacement of the reflected P wave is defined as the phase difference between that of the reflected P wave and the incident SI wave at the time of the maximum particle displacement for the incident SI wave. This phase shift, advanced by 180° for convenience, is given from (3.2.11) and (3.3.23), by

$$\phi_P + \pi \equiv \arg\left[\frac{B_2}{C_{12}} \frac{k_P}{k_S}\right] - \left(\vec{P}_{\phi_2} - \vec{P}_{\psi_1}\right) \cdot \vec{r}. \tag{7.2.29}$$

Free-surface reflection coefficients for the radial and vertical components of displacement for the reflected P wave with respect to the maximum particle displacement for the incident SI wave can be defined respectively as

$$r_{Px} \equiv r_P F_{P_{\phi_2}}\left[\sin \theta_{\phi_2}\right] \tag{7.2.30}$$

and

$$r_{Pz} \equiv r_P F_{P_{\phi_2}}\left[\cos \theta_{\phi_2}\right], \tag{7.2.31}$$

which can be rewritten in terms of the modulation factors for the given incident SI wave from (7.2.17) as

$$r_{Px} = r_P \frac{|k_S|}{|k_P|} F_{S_{\psi_1}}[\sin \theta_{\psi_1}] \tag{7.2.32}$$

and

$$r_{Pz} = r_P \sqrt{\frac{\chi_{P_{\phi_2}}}{\chi_{HP}} - \frac{v_{HP}^2}{v_{HS}^2} \frac{1 + \chi_{HP}\chi_{HS}}{1 + \chi_{HS}\chi_{HP}} \left(F_{S_{\psi_1}}[\sin \theta_{\psi_1}]\right)^2}. \tag{7.2.33}$$

Corresponding expressions for the reflected SI wave may be defined and derived in a similar fashion.

Phase coefficients for the horizontal and vertical components of particle motion for the reflected P wave are specified as the phase difference between that of the corresponding component of motion for the reflected P wave and the incident SI wave at the time of the maximum particle motion for the incident SI wave. These phase coefficients, advanced by 180° for convenience, may be defined in terms of the phase difference for the maximum particle motion, namely ϕ_P, and simplified with (7.2.18) for the radial component of motion to yield

$$\phi_{Px} + \pi \equiv \phi_P + \psi_P - \Omega_{P_{\phi_2}}[\sin \theta_{\phi_2}] = \phi_P + \psi_P - \Omega_{S_{\psi_1}}[\sin \theta_{\psi_1}] \tag{7.2.34}$$

and for the vertical component of motion by

$$\phi_{Pz} + \pi \equiv \phi_P + \psi_P - \Omega_{P_{\phi_2}}[\cos \theta_{\phi_2}] = \phi_P + \psi_P - \Omega_{S_{\psi_1}}[\cos \theta_{\psi_1}]. \tag{7.2.35}$$

If the incident SI wave is homogeneous then the difference between the phase coefficients for the radial and maximum motions, advanced by 180°, is given by

$$\phi_{Px} - \phi_P + \pi = \psi_P - \psi_S. \tag{7.2.36}$$

For low-loss media it simplifies to

$$\phi_{Px} - \phi_P + \pi \approx \left(Q_{HP}^{-1} - Q_{HS}^{-1}\right)/2. \tag{7.2.37}$$

These equations show that the difference of the phase coefficients for the radial and maximum particle motions is dependent on the difference in intrinsic material absorption for homogeneous P and S waves.

Following the convention of normalizing by the maximum displacement amplitude the amplitude reflection coefficient for volumetric strain normalized by ω may be defined by

$$r_\Delta \equiv \frac{|B_2 k_P|}{|C_{12} k_S|} \frac{1}{|\vec{\xi}_{1_{S\psi_1}}|} \frac{|k_P|}{\omega} \exp\left[-\left(\vec{A}_{\phi_2} - \vec{A}_{\psi_1}\right)\cdot\vec{r}\right]. \quad (7.2.38)$$

The corresponding phase coefficient for volumetric strain can be defined as the phase of the volumetric strain at the time of the maximum displacement amplitude of the incident wave, advanced by 180° . The resulting phase shift is defined by

$$\phi_\Delta + \pi \equiv \arg\left[\frac{B_2\ k_P}{C_{12}\ k_S}\right] - (\vec{P}_{\phi_2} - \vec{P}_{\psi_1})\cdot\vec{r} - \left(\psi_P + \frac{\pi}{2}\right). \quad (7.2.39)$$

The difference between the phase coefficient for the volumetric strain and that for the radial component of particle motion is

$$\phi_\Delta - \phi_{Px} = -2\psi_P + \psi_S + \frac{\pi}{2}. \quad (7.2.40)$$

The corresponding difference for the vertical component of particle motion is

$$\phi_\Delta - \phi_{Pz} = -2\psi_P + \Omega_{P_{\phi 2}}\left(\cos\theta_{P_{\phi 2}}\right) + \pi/2. \quad (7.2.41)$$

These relations show that volumetric strain is shifted in phase with respect to the radial component of displacement by a fixed amount dependent on the intrinsic absorption characteristics of the medium. The vertical component shows a phase shift with respect to volumetric strain that also is dependent on the inhomogeneity of the reflected wave field.

For elastic media $\psi_P = \psi_S = 0$, hence for angles of incidence less than critical

$$\phi_\Delta - \phi_{P_x} = \phi_\Delta - \phi_{Pz} = \frac{\pi}{2}, \quad (7.2.42)$$

so for elastic media the volumetric strain for the reflected P wave is advanced with respect to that of the particle motion components by 90°.

7.2.3 Numerical Model for Low-Loss Media (Weathered Granite)

Quantitative results for the problem of an SI wave incident on the free surface of an anelastic half space are useful in understanding the effect of anelasticity on free-surface reflections. They will provide insight regarding the interpretation of collocated inferences of displacement and volumetric strain as might be detected simultaneously with separate sensors (Borcherdt and Glassmoyer, 1989).

The simplest problem to consider is that in which the incident SI wave is homogeneous. This assumption implies that the reflected SI wave is homogeneous with characteristics of phase and energy speed equal to those of the incident wave. It implies that for media with $Q_{HS}^{-1} \neq Q_{HP}^{-1}$ the reflected P wave is inhomogeneous

Table (7.2.43). *Material parameters for the problem of a Type-I S wave incident on the free surface of a low-loss anelastic half space.*

Medium	ρ (g/cm^3)	v_{HS} (km/s)	Q_{HS}^{-1}	v_{HP} (km/s)	Q_{HS}^{-1}
Pierre Shale or weathered granite	2.5	0.802	0.1	2.16	0.03

From McDonal *et al.* (1958) and Fumal (1978).

for all angles of incidence except normal with phase and energy speeds and other physical characteristics that depend on angle of incidence (Theorem (5.2.34)).

Material parameters chosen to represent the viscoelastic half space (Table (7.2.43)) correspond to those measured for Pierre Shale (McDonal *et al.*, 1958) and weathered granite (Fumal, 1978). Physical characteristics of the reflected P wave calculated on the basis of (7.2.2) through (7.2.42) are shown in Figures (7.2.45) through (7.2.52).

The degree of inhomogeneity of the reflected P wave (Figure (7.2.45)a) rapidly increases from zero to a value near its physical limit (90°) for angles of incidence near and beyond an angle near 22°. This inhomogeneity of the reflected P waves gives rise to a number of wave–field characteristics not predicted by conventional elasticity treatments of the free-surface reflection problem.

The direction of phase propagation for the reflected P wave is at some finite angle >3.1° away from the free surface for all angles of incidence (Figure (7.2.45)b). The direction of maximum energy flux is closer to the interface than phase propagation Figure (7.2.45)b. It increases with angle of incidence to a value greater than 90°. For these angles of incidence the normal component of energy flow associated with the reflected P wave is toward the free surface. For the low-loss anelastic half space under consideration, the phase and energy speeds associated with the reflected P wave are approximately equal. Abrupt changes in the speeds are apparent only for angles of incidence in the *P transition window* which includes angles near and greater than 22° (Figure (7.2.45)c).

The direction of maximum attenuation for the reflected P wave shows a dependence on angle of incidence that is significantly different from that of the "hll" assumption, which assumes the direction is parallel to phase propagation for angles of incidence less than critical and directed into the medium perpendicular to the interface for larger angles of incidence. Figure (7.2.46)a shows that the direction of maximum attenuation for the reflected P wave is parallel to that of phase propagation only for normal incidence. As the angle of incidence increases, its deviation from phase propagation increases. At one angle of incidence of about one-half the elastic critical angle ($\sim -12°$), it is parallel to the free surface. For larger angles of incidence, the direction of maximum attenuation increases to a value of 176° which is near

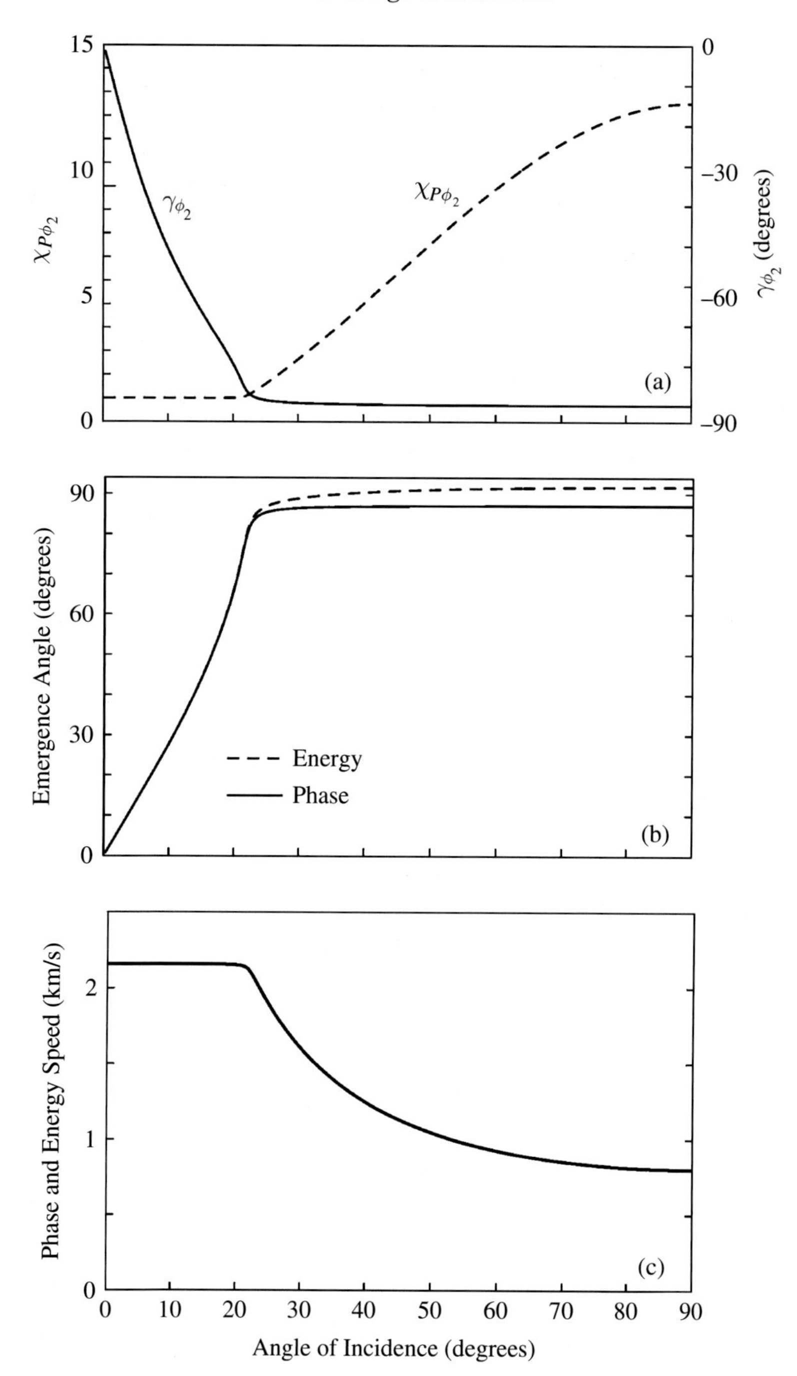

Figure (7.2.45). Inhomogeneity γ_{ϕ_2} and $\chi_{P\phi_2}$ (a), emergence angle for phase propagation and maximum energy flux (b), and phase and energy speeds (c) for the P wave generated by a homogeneous SI wave incident on the free surface of an anelastic half space with parameters corresponding to Pierre Shale (Table (7.2.43)).

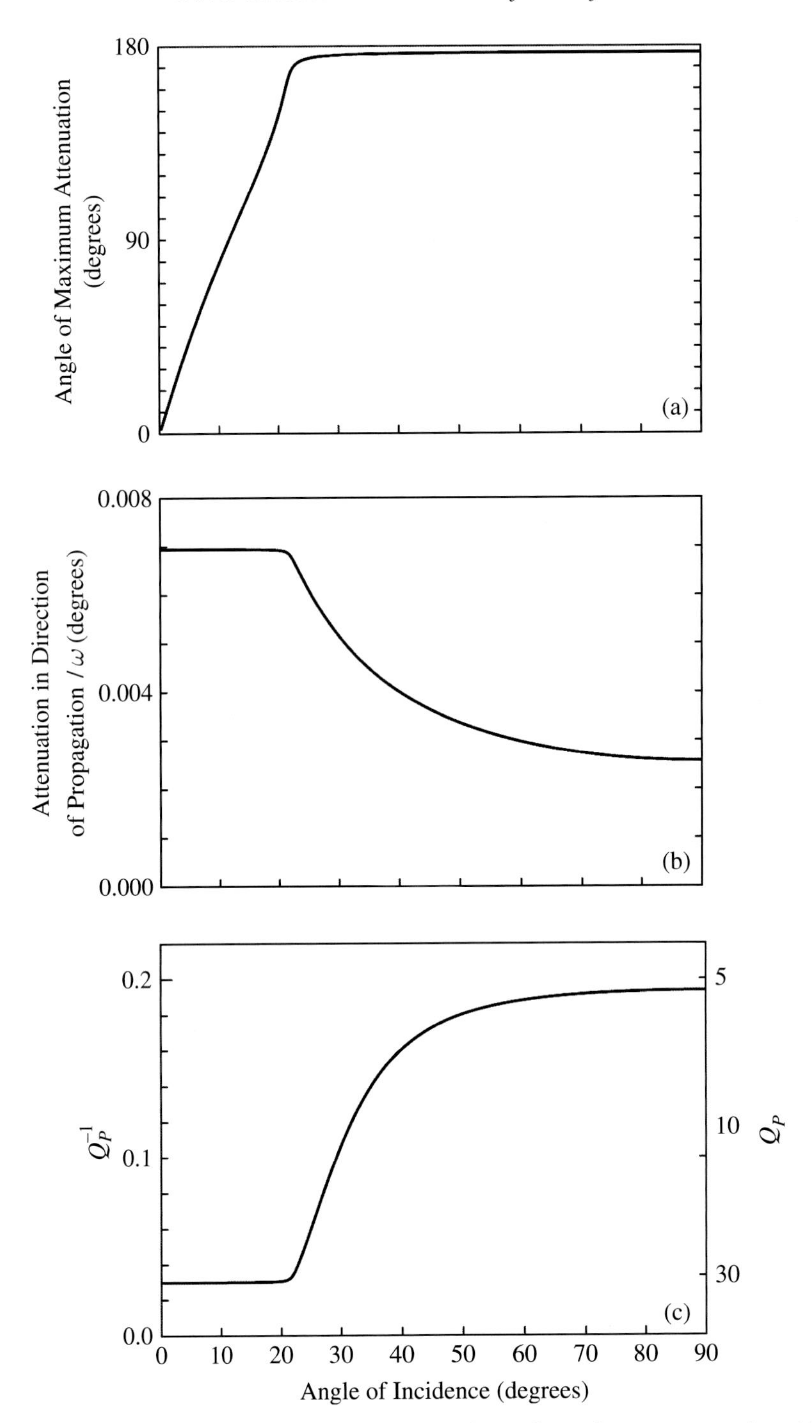

Figure (7.2.46). Emergence angle for direction of maximum attenuation (a), attenuation in direction of propagation (b) and $Q_{\bar{P}}^{-1}$ (c) for the P wave generated by a homogeneous SI wave incident on the free surface of an anelastic half space with parameters corresponding to Pierre Shale (Table (7.2.43)).

vertical. For all angles of incidence greater than $\sim -12°$ the direction of maximum attenuation indicates that amplitudes of the reflected P wave actually increase with depth (Figure (7.2.46)) as measured perpendicular to the free surface. No physical principles are violated. This characteristic of the reflected P wave is required in order that the solutions satisfy the boundary conditions. This result is similar to a previous result encountered for an SII wave transmitted at an anelastic boundary for which the intrinsic material absorption of the transmission medium for the corresponding wave type is greater than that of the incident medium (see Figure (6.1.3)c).

Rapid changes in several of the physical characteristics of the reflected P wave are apparent for angles of incidence in the reflected *P transition window*. For angles of incidence larger than about 22°, the direction of phase propagation decreases more rapidly (Figure (7.2.46)b), $Q_{\bar{P}}^{-1}$increases from its initial value of 1/30 to its limiting value of $2Q_{HS}^{-1} = 0.2$ as implied by Theorem (3.6.92) (Figure (7.2.46)c), the particle motion ellipse increases in tilt with respect to the direction of phase propagation by about a degree at grazing incidence (Figure (7.2.47)a), and the ellipticity of the particle motion tends toward circularity as indicated by the ratio of the minor to major axis which approaches a value near unity (0.929; Figure (7.2.47)b). These physical characteristics of the reflected wave field are in contrast to the familiar characteristics that would be anticipated on the basis of an elastic model.

Consideration of the energy reflection and interaction coefficients provides additional insight into the free-surface reflection problem for a viscoelastic half space. For normal incidence, the energy reflection and interaction coefficients (Figure (7.2.48); see (5.4.59) through (5.4.65) for definitions of the coefficients) indicate that all of the incident SI energy is reflected as SI energy. However, with increasing angles of incidence, the reflection coefficients (Figure (7.2.48)a) indicate that the proportion of energy converted from incident SI energy to reflected dilatational energy increases and reaches a maximum (40 percent for this model) at an angle (~18°) just prior to 22°. For larger and an intermediate range of angles of incidence (~ 40° to about 47° for this model), energy converted to dilatational energy is approximately zero, then for larger angles of incidence the dilatational energy carried by the reflected P wave reverses direction and flows toward the free surface. The proportion of dilatational energy carried by the reflected P wave toward the free surface increases to a maximum of about 28 percent for larger angles of incidence (~78°) then vanishes at grazing incidence. The fact that the reflection coefficient for the P wave reverses sign, indicating that the reflected P wave actually carries energy toward the free surface, is consistent with the earlier observation that the direction of mean energy flux for the reflected P wave is out of the media away from the free surface (Figure (7.2.45)b).

Inspection of the various interaction coefficients (Figure (7.2.48)) shows that for angles of incidence up to about 19°, the amount of energy flow as a result of

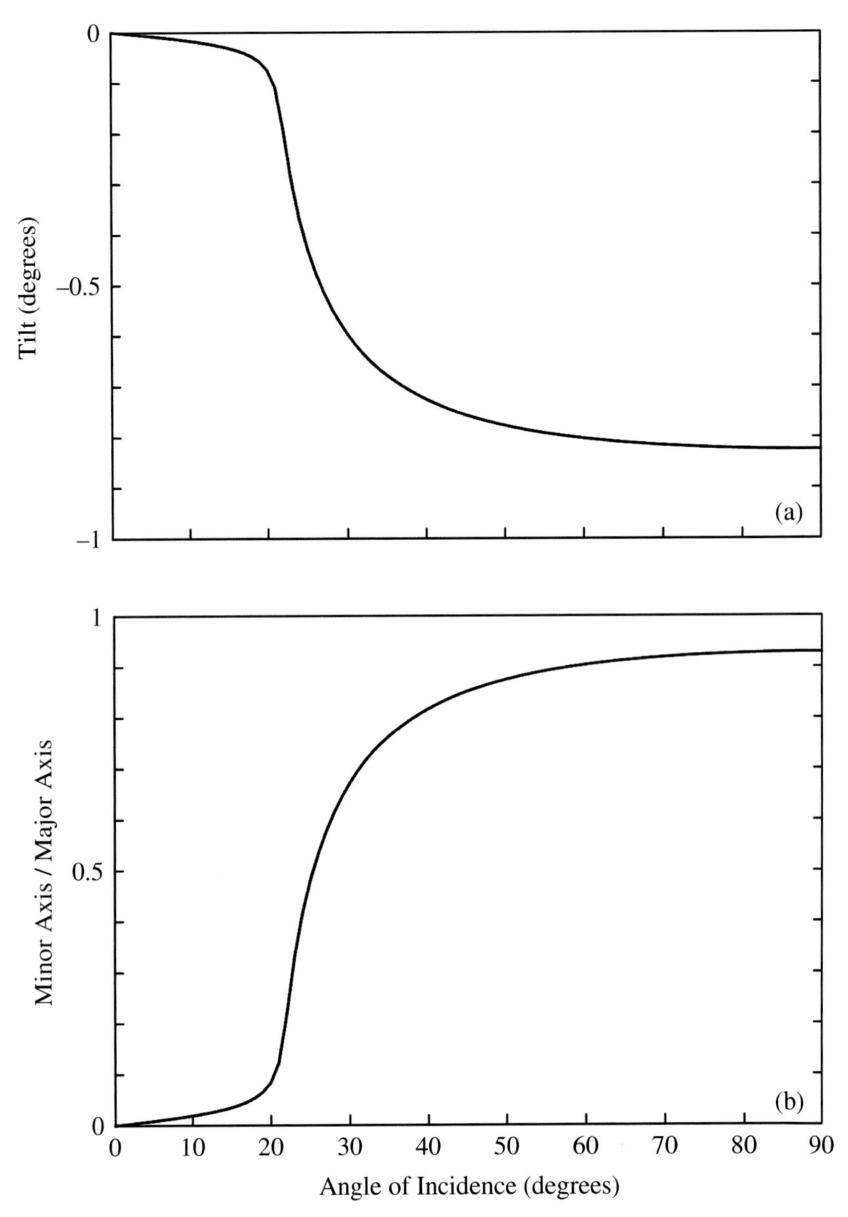

Figure (7.2.47). Tilt (a) and ratio of minor to major axis (b) of the particle motion ellipse for the P wave generated by a homogeneous SI wave incident on the free surface of an anelastic half space with parameters corresponding to Pierre Shale (Table (7.2.43)).

wave–field interaction represents a small proportion of the incident energy. For larger angles of incidence, the coefficient for the total energy flow due to wave–field interaction (IC_{total}; Figure (7.2.48)b) shows that the energy flow due to interaction becomes a significant part of the energy budget at the interface. For the range of

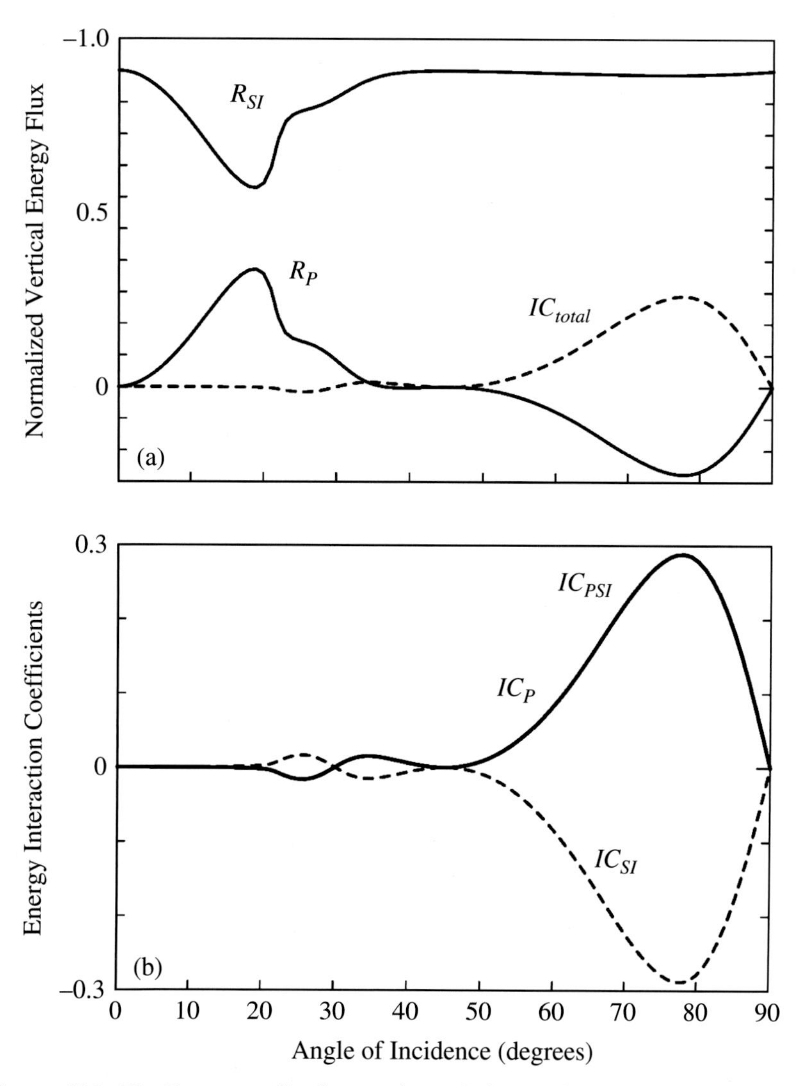

Figure (7.2.48). Energy reflection and total interaction coefficient (a) and individual interaction coefficients for incident and reflected waves (b) generated by a homogeneous SI wave incident on the free surface of an anelastic half space with parameters corresponding to Pierre Shale (Table (7.2.43)).

angles for which the reflected P wave carries energy toward the boundary, the total or net energy flow due to the various wave–field interactions is in the opposite direction into the medium. As a consequence for these larger angles of incidence, the energy flow normal to the boundary due to wave–field interactions counteracts that carried toward the boundary by the reflected P wave. For these larger angles of incidence, most (>99 percent), but not all, of the incident energy that is carried away from the boundary is carried by the reflected SI wave (R_{SI}; Figure (7.2.48)a).

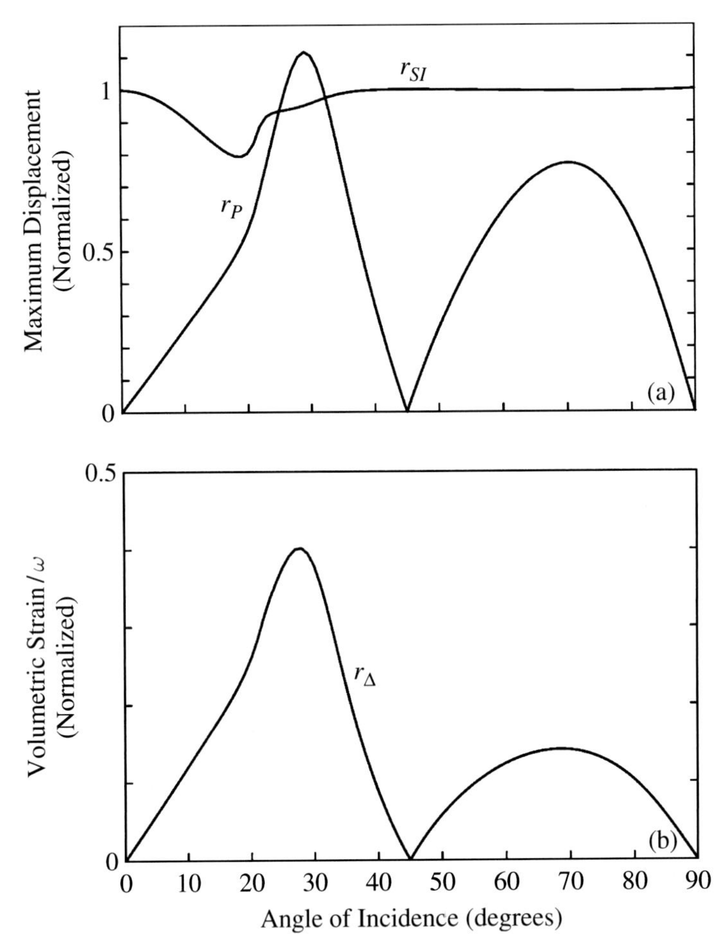

Figure (7.2.49). Reflection coefficients for maximum displacement (a) and volumetric strain (b), for the P and SI waves generated by a homogeneous SI wave incident on the free surface of an anelastic half space with parameters corresponding to Pierre Shale (Table (7.2.43)).

The interactions of the reflected P wave with the incident SI wave (IC_P) and the reflected SI wave (IC_{PSI}) result in a normal component of energy flow at the boundary into the medium (Figure (7.2.48)b). This normal component of energy flow is partially counteracted by the normal component carried by the reflected P wave R_P and that due to interaction of the incident and reflected SI waves (IC_{SI}). The angles of incidence for which the interaction energy coefficients achieve their largest values correspond to a slight decrease (<1 percent for this model) in the energy reflection coefficient for the reflected SI wave.

Amplitude reflection coefficients and phase shifts for the maximum displacement and volumetric strain as defined by (7.2.28), (7.2.29), (7.2.38), and (7.2.39) are shown in Figures (7.2.49) and (7.2.50). They indicate that the maximum amplitude of the

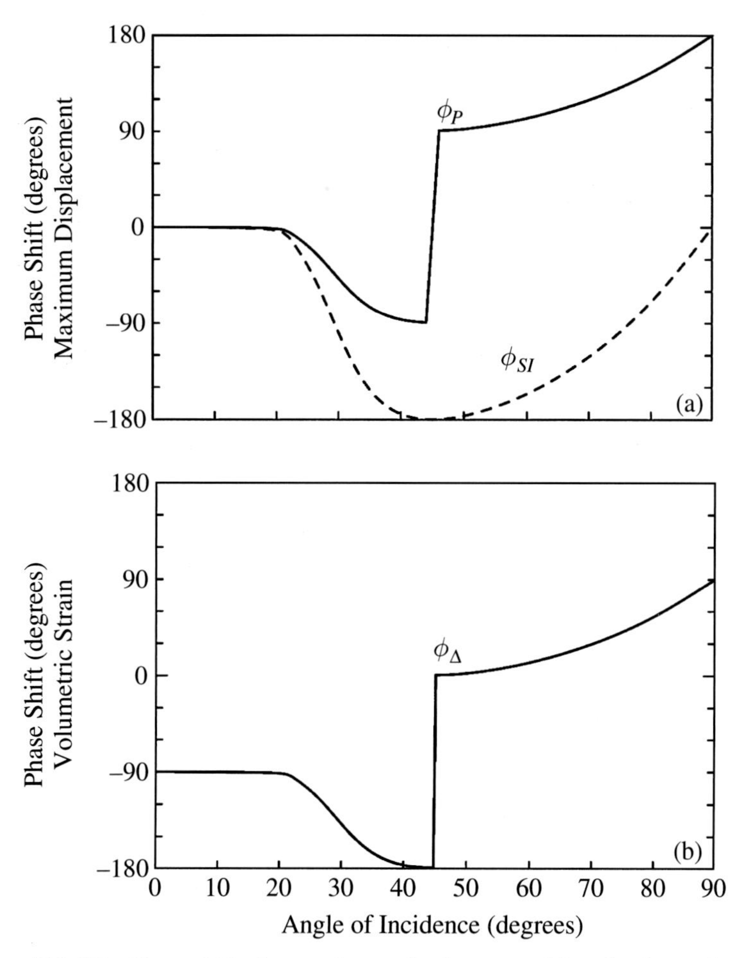

Figure (7.2.50). Phase shifts for maximum displacement (a) and volumetric strain (b) for the P and SI waves generated by a homogeneous SI wave incident on the free surface of an anelastic half space with parameters corresponding to Pierre Shale (Table (7.2.43)).

reflected SI wave decreases to about 80 percent of that of the incident SI wave at an angle of incidence of about 19°, then increases toward the amplitude of the incident wave as grazing incidence is approached (Figure (7.2.49)a). They indicate that the maximum amplitude of the reflected P wave shows maxima at about (29°, 70°) and minima at (0°, 45°, 90°). Corresponding maxima and minima for volumetric strain occur at (28°, 68°) and (0°, 45°, 90°). With the exception of maxima at 68° and 70°, the local maxima and minima in the reflected amplitudes occur at angles of incidence for which the total energy flow due to interaction is near zero. Phase-shift variations occur with increasing angle of incidence for the reflected SI and P waves at about 22°, which correspond to angles for which energy flow due to interaction is apparent.

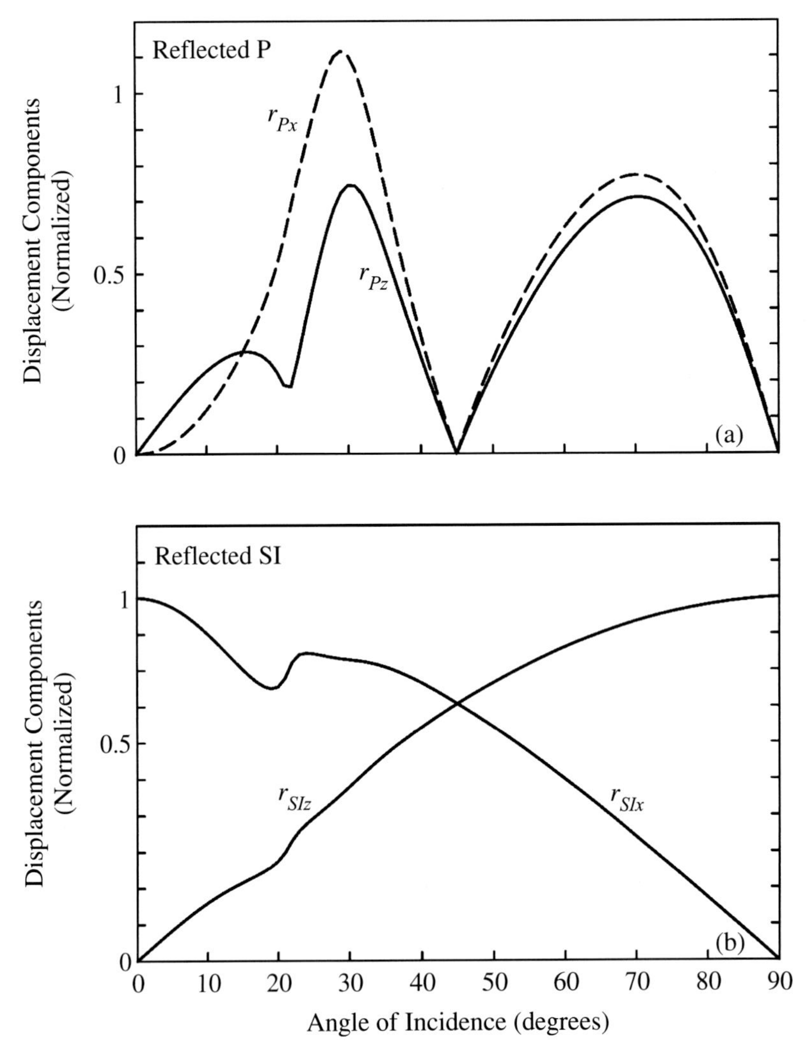

Figure (7.2.51). Reflection coefficients for the radial and vertical components of displacement for the reflected P wave (a) and the reflected SI wave (b) generated by a plane SI wave incident on the free surface of an anelastic half space with parameters corresponding to Pierre Shale (Table (7.2.43)).

The maxima, which occur in the maximum displacement amplitude and volumetric strain for the reflected P wave, occur at angles of incidence for which the degree of inhomogeneity of the reflected P wave is large (>89.5°). As a result the physical characteristics of the reflected P wave differ significantly from those for a corresponding homogeneous P wave. More specifically, at angles of incidence for which the particle displacement and volumetric strain for the reflected inhomogeneous P wave reach local maxima,

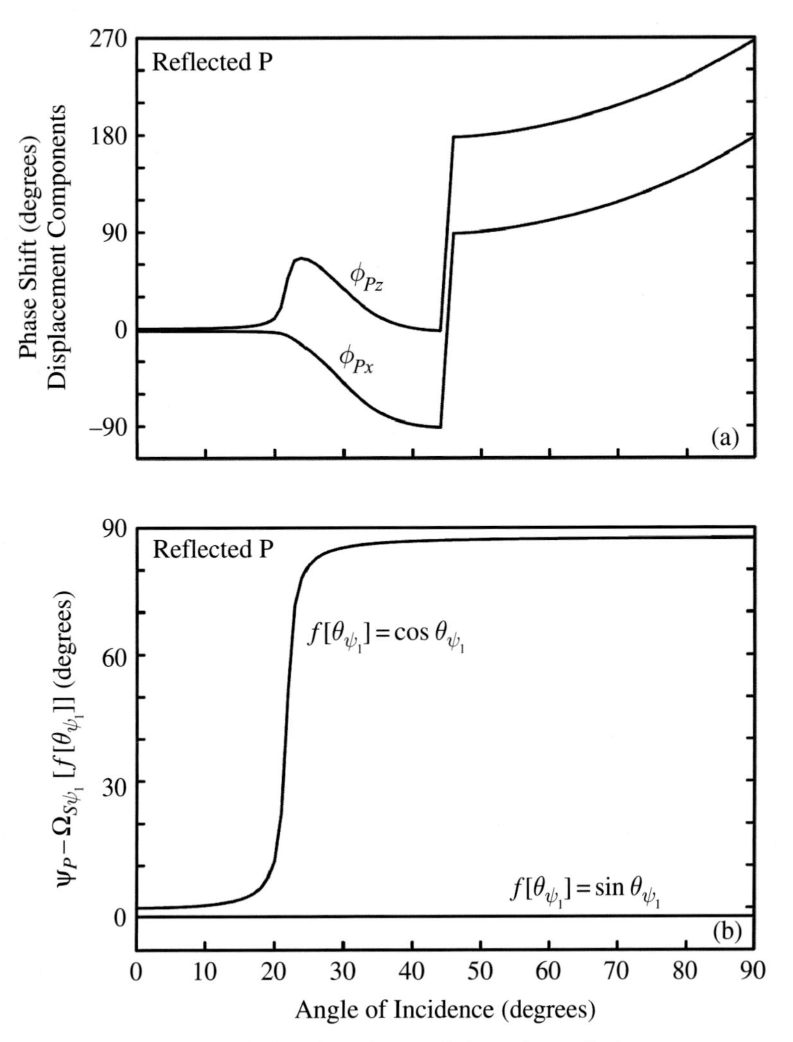

Figure (7.2.52). Phase shifts for the radial and vertical components of displacement for the reflected P wave generated by a plane SI wave incident on the free surface of an anelastic half space with parameters corresponding to Pierre Shale (Table (7.2.43)).

(1) the phase and energy velocities are about 22 and 61 percent less than those for a homogeneous P wave,
(2) Q_P^{-1} is about 233 and 568 percent greater than that for a homogeneous P wave,
(3) the elliptical particle motion of the reflected P wave approaches circularity as indicated by the ratio of minor to major axis with values near 0.6 and 0.9,
(4) the normal component of energy flow carried by the reflected P wave at the first and second local maxima will represent about 11 and 21 percent of the total incident energy, and

(5) the amplitude of the P wave reflected at ~28° increases with depth to a value at a depth of one wavelength that is about 6 percent larger than its amplitude at the free surface ($\exp[0.5854 + 0.997|\vec{A}_P|z]$ for $z = \lambda$).

Amplitude reflection coefficients for the horizontal and vertical components of displacement as implied by (7.2.32) and (7.2.33) are shown in Figure (7.2.51). These coefficients for the reflected P wave (Figure (7.2.51)a) show that the vertical amplitude becomes comparable to the radial amplitude and the ellipticity tends toward circularity as the angle of incidence increases (see Figure (7.2.47)b). The local minimum in the vertical component for the reflected P wave near 22° corresponds to changes in several physical characteristics and a local maximum in the vertical phase shift (see Figures (7.2.51)a and (7.2.52)a). It also corresponds to a local minimum near 19° in the horizontal reflection coefficient for the reflected SI wave.

The amounts that the phase shifts for the horizontal and vertical displacements differ from that for the maximum displacement and volumetric strain as implied by (7.2.36) and (7.2.40) are shown in Figure (7.2.52). For low-loss media corresponding to Pierre Shale, these expressions imply from (7.2.37) that these differences are approximately 2.0° and 90° + 1.1°, respectively. The differences are small and barely discernible at the plotting scales shown in Figure (7.2.52).

The dependence of the vertical phase coefficient on angle of incidence and in turn inhomogeneity of the reflected P wave is evident from Figure (7.2.52)b, where the quantity $\psi_P - \Omega_{P_{\phi_2}}[f[\theta_{\phi_2}]] = \psi_P - \Omega_{S_{\psi_1}}[f[\theta_{\psi_1}]]$ is plotted. The rapid change from 0 to 90°, which takes place in this quantity near 22° with $f[\theta_{\psi_1}] = \cos\theta_{\psi_1}$ (Figure (7.2.52)), manifests itself as a local maximum in the vertical phase coefficient (Figure (7.2.52)a). This local maximum predicted for the reflected inhomogeneous wave field would not be anticipated on the basis of the "hll" model.

7.3 Incident General P Wave

The solution to the problem of a general P wave incident on the free surface of a viscoelastic half space may be derived in a manner analogous to that for a general P wave incident on a welded boundary.

7.3.1 Reflected General P and SI Waves

The problem of a general P wave incident on the free surface of a HILV half space is specified by assuming the parameters of the incident P wave are given and that the amplitudes of the solutions for the incident SI and SII waves vanish, that is

$$\vec{C}_1 = C_{11}\hat{x}_1 + C_{12}\hat{x}_2 + C_{13}\hat{x}_3 = 0. \tag{7.3.1}$$

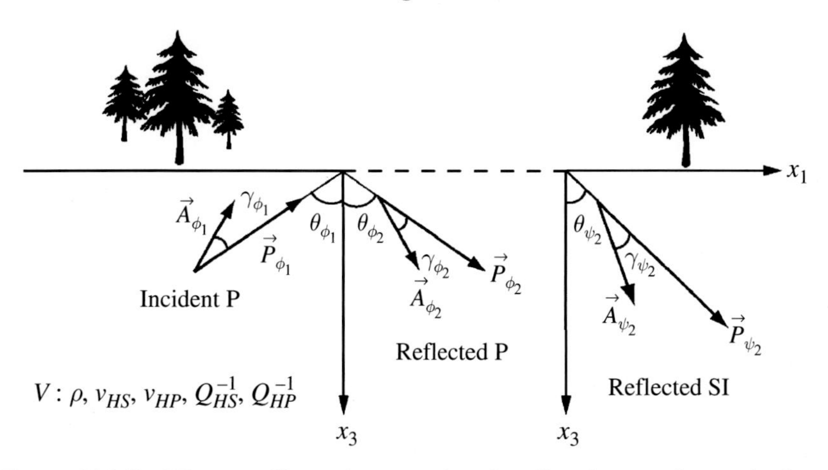

Figure (7.3.2). Diagram illustrating notation for directions and magnitude of propagation and attenuation vectors for the problem of a general P wave incident on the free surface of a HILV half space.

The boundary conditions imply that no SII waves are generated at the boundary, hence without loss of generality $C_{21} = C_{23} = 0$. The incident P wave as specified by (4.2.2) is given in medium V by equations (5.3.3) through (5.3.11). The expressions provide a complete specification of the given general P wave in terms of a given angle of incidence, θ_{ϕ_1}, a given angle between its attenuation and propagation vectors, γ_{ϕ_1} $(0 \leq |\gamma_{\phi_1}| < \pi/2)$, and a given complex amplitude B_1.

The assumed incident general P wave is either an inhomogeneous wave with elliptical particle motion in the $x_1 x_3$ plane or a homogeneous wave with linear particle motion in the direction of propagation, depending on the degree of inhomogeneity, γ_{ψ_1}, chosen for the incident wave. Solutions for the reflected P and SI waves in medium V as specified by equations (4.2.1) and (4.2.2) are given by equations (5.2.14) and (5.2.15). Notation for the problem of a general P wave incident on the free surface of a HILV half space is illustrated in Figure (7.3.2).

For the problem of a general P wave incident on a free surface the boundary conditions (7.1.5) through (7.1.7) permit the complex amplitudes of the displacement potentials for the reflected waves to be solved in terms of the complex amplitude and complex wave number of the incident general P wave, namely

$$B_2/B_1 = \left[4\, d_\beta\, d_\alpha k^2 - \left(d_\beta^2 - k^2\right)^2\right] \Big/ g(k), \tag{7.3.3}$$

$$D_2 = D_1 = 0, \tag{7.3.4}$$

and

$$C_{22}/B_1 = 4k\, d_\alpha \left(d_\beta^2 - k^2\right)/g(k), \tag{7.3.5}$$

where

$$g(k) = 4 d_\alpha d_\beta k^2 + \left(d_\beta^2 - k^2\right)^2 \tag{7.3.6}$$

for

$$g(k) \neq 0. \tag{7.3.7}$$

With the complex wave number k for the incident general P wave given by

$$k = \frac{\omega}{v_{HP}} \left(\sqrt{\frac{1 + \chi_{P_{\phi_1}}}{1 + \chi_{HP}}} \sin \theta_{\phi_1} - i \sqrt{\frac{-1 + \chi_{P_{\phi_1}}}{1 + \chi_{HP}}} \sin \left[\theta_{\phi_1} - \gamma_{\phi_1} \right] \right), \tag{7.3.8}$$

which for an elastic half space reduces to

$$k = \frac{\omega}{v_{HP}} \sin \theta_{\phi_1}, \tag{7.3.9}$$

where $\chi_{P_{\phi_1}} = \sqrt{1 + Q_{HP}^{-2} \sec^2 \gamma_{\phi_1}}$ and $\chi_{HP} = \sqrt{1 + Q_{HP}^{-2}}$, the propagation and attenuation vectors as specified by (5.2.5) and (5.2.6) for the incident and reflected waves are completely determined upon specification of the angle of incidence θ_{ϕ_1}, the degree of inhomogeneity γ_{ϕ_1}, and the circular frequency ω for the incident general P wave together with the material parameters for V as specified by k_P or v_{HP} and Q_{HP}^{-1}. Hence, the propagation and attenuation vectors for the reflected general P and SI waves are determined in terms of those for the incident general P wave, from which it follows that (7.3.3) and (7.3.5) provide the desired solution for the complex amplitude of the reflected general P and SI waves in terms of that given for the incident general P wave.

The existence of solutions (7.3.3) and (7.3.5) confirms equality of the complex wave number k in the assumed solutions for the incident and reflected waves as specified by (4.2.1) and (4.2.2). Hence, the components of phase propagation and maximum attenuation parallel to the boundary for the incident and reflected waves are equal, that is,

$$k_R = \mathrm{Re}\left[K_{\phi_1 x}\right] = |\vec{P}_{\phi_1}| \sin \theta_{\phi_1} = |\vec{P}_{\psi_2}| \sin \theta_{\psi_2} = |\vec{P}_{\phi_2}| \sin \theta_{\phi_2} \tag{7.3.10}$$

and

$$\begin{aligned} -k_I = -\mathrm{Im}\left[K_{\phi_1 x}\right] &= |\vec{A}_{\phi_1}| \sin[\theta_{\phi_1} - \gamma_{\phi_1}] = |\vec{A}_{\psi_2}| \sin[\theta_{\psi_2} - \gamma_{\psi_2}] \\ &= |\vec{A}_{\phi_2}| \sin[\theta_{\phi_2} - \gamma_{\phi_2}], \end{aligned} \tag{7.3.11}$$

establishing Generalized Snell's Law for the problem of a general P wave incident on the free surface of a viscoelastic half space, as stated for medium V by Theorem (5.2.22).

For a non-trivial incident wave $B_1 \neq 0$, so equations (7.1.5) and (7.1.7) imply $g(k) \neq 0$. As for the problem of an incident SI wave, roots of equation (7.3.6) will be shown to specify k for a Rayleigh-Type surface wave. Hence, this result shows that a general P wave incident in a plane perpendicular to the free surface does not generate a Rayleigh-Type surface wave. Also, boundary-condition equation (7.2.3) shows that no Type-II S waves are generated by a P wave incident in a plane perpendicular to the interface.

It follows, as for the corresponding welded-boundary problem, that $\theta_{\phi_1} = \theta_{\phi_2}$ and $\gamma_{\phi_1} = \gamma_{\phi_2}$. Hence, for the incident and reflected P waves, the reflection angle equals the angle of incidence and their degrees of inhomogeneity are equal.

The conditions of homogeneity and inhomogeneity for the waves reflected from the free surface for an incident general P wave are the same and the proofs are similar for the reflected waves in medium V by Theorems (5.3.22) through (5.3.25). In brief, the reflected P wave is homogeneous if and only if the incident P wave is homogeneous. Theorem (5.3.24) indicates that the SI wave reflected from the free surface of a HILV half space is homogeneous if and only if $Q_{HS}^{-1} = Q_{HP}^{-1}$. These results indicate that for elastic media the reflected SI wave will be homogeneous for all angles of incidence, but for anelastic Earth-type media where $Q_{HS}^{-1} \neq Q_{HP}^{-1}$ the reflected SI wave, in general, will be inhomogeneous for all angles of incidence.

The degree of inhomogeneity of the reflected SI wave in anelastic media will be shown to increase with angle of incidence and hence the phase velocity of the reflected SI wave will decrease, the maximum attenuation will increase, and fractional energy loss and other characteristics will be shown to vary with angle of incidence. As $Q_{HS}^{-1} \neq Q_{HP}^{-1}$ in an anelastic Earth the preceding results show that SI waves generated by a P body wave incident on the free surface, in general, are inhomogeneous. As a result the degree of inhomogeneity and physical characteristics of the reflected SI wave are dependent on the angle of incidence and the degree of inhomogeneity of the incident P wave.

In the case of an elastic medium, a normally incident P wave does not generate an S wave upon interacting with the free surface. However, in the case of a vertically incident inhomogeneous P wave in anelastic media, a shear wave is reflected from the free surface. The proof of this result is similar to that for the problem of a vertically incident inhomogeneous SI wave.

A volumetric strain sensor near the surface will detect the incident P wave and the corresponding P wave reflected from the free surface. The fact that $\theta_{\phi_1} = \theta_{\phi_2}$ and $\gamma_{\phi_1} = \gamma_{\phi_2}$ for the incident and reflected P waves implies that the total radial and vertical components of displacement and the total volumetric strain due to the incident and reflected P waves may be written using equations (3.10.10), (3.10.11), and (3.10.29) as

$$\frac{u_{R_{\phi_1}x} + u_{R_{\phi_2}x}}{A\sqrt{a^2+b^2}} = F_{P_{\phi_1}}\left[\sin\theta_{\phi_1}\right] \cos\left[\varsigma(t) + \psi_P - \Omega_{P\phi_1}\left[\sin\theta_{\phi_1}\right] + \varphi\right], \quad (7.3.12)$$

$$\frac{u_{R_{\phi_1}z} + u_{R_{\phi_2}z}}{A\sqrt{a^2+b^2}} = F_{P_{\phi_1}}\left[\cos\theta_{\phi_1}\right] \cos\left[\varsigma(t) + \psi_P - \Omega_{P\phi_1}\left[\cos\theta_{\phi_1}\right] + \varphi\right], \quad (7.3.13)$$

and

$$\frac{\Delta_{R_{\phi_1}} + \Delta_{R_{\phi_2}}}{A\sqrt{a^2 + b^2}} = |k_P| \cos[\varsigma(t) - \psi_P - \pi/2 + \varphi], \tag{7.3.14}$$

where

$$A \equiv |B_1 k_P| \exp[-\vec{A}_{\phi_1} \bullet \vec{r}], \tag{7.3.15}$$

$$\varsigma(t) \equiv \omega t - P_{\phi_1 x} x - \pi/2, \tag{7.3.16}$$

$$\varphi \equiv \tan^{-1}[b/a], \tag{7.3.17}$$

$$a \equiv \cos d_1 + R \cos d_2, \tag{7.3.18}$$

$$b \equiv \sin d_1 + R \sin d_2, \tag{7.3.19}$$

$$R \equiv \frac{|B_2|}{|B_1|} \exp[-(\vec{A}_{\phi_2} - \vec{A}_{\phi_1}) \bullet \vec{r}], \tag{7.3.20}$$

$$d_1 \equiv -P_{\phi_1 z} + \arg[B_1 k_P], \tag{7.3.21}$$

$$d_2 \equiv -P_{\phi_2 z} + \arg[B_2 k_P]. \tag{7.3.22}$$

These expressions show that the total of the components of displacement for the reflected and incident P wave differs from that for the volumetric strain only in an amplitude modulation factor and a phase shift that depend on degree of inhomogeneity and the angle of incidence of the incident P wave. Hence, depending on characteristics of interest simultaneous inferences of displacement (7.3.12), (7.3.13), and volumetric strain (7.3.14) can permit deduction of additional characteristics of the wave fields and local material properties.

7.3.2 Numerical Model for Low-Loss Media (Pierre Shale)

Quantitative results for the problem of a P wave incident on the free surface of an anelastic half space provide insight for interpretation of displacements and volumetric strains as might be inferred from sensors at or near the surface.

The theoretical problem of a general P wave incident on the free surface of an arbitrary viscoelastic half space is illustrated in Figure (7.3.2). Physical characteristics of the P and SI waves reflected from the free surface can be specified by (7.3.3) and (7.3.5) in terms of the given parameters of the incident wave. Reflection coefficients for the problem of an incident P wave can be defined in a similar way to those for the problem of an incident SI wave. For brevity, the definitions are not repeated.

To easily compare numerical results with those previously computed for an incident SI wave, the incident P wave is assumed here to be homogeneous and the material parameters chosen for the half space are those for a low-loss material corresponding to Pierre Shale or weathered granite (see Table (7.2.43)).

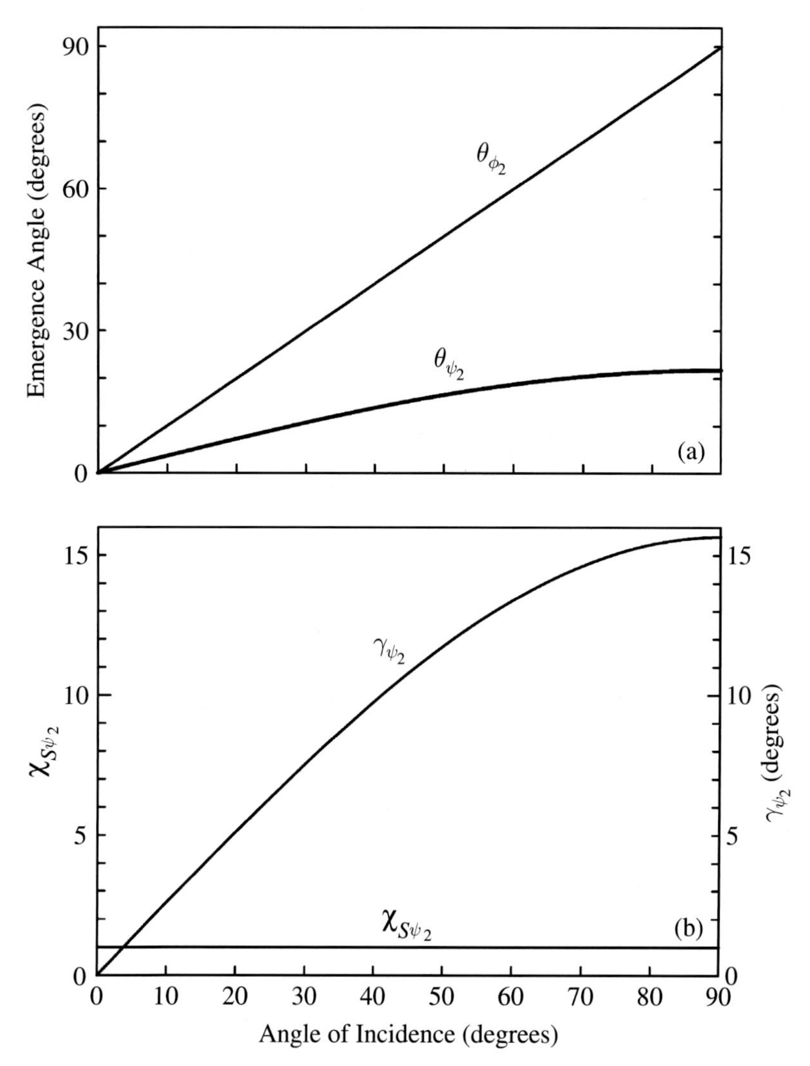

Figure (7.3.23). Emergence angles for phase propagation for reflectied P and SI waves (a), inhomogeneity and parameter χ_S for the reflected SI wave (b) due to a homogeneous P wave incident on the free surface of a low-loss anelastic half space corresponding to Pierre Shale.

The assumption of homogeneity for the incident P wave implies that the reflected P wave is homogeneous. As a result angles of emergence for phase and energy flux are equal to the angle of incidence for the reflected P wave (Figure (7.3.23)a). In addition, phase and energy velocities are equal to that of the incident homogeneous wave and constant as a function of angle of incidence.

The different amounts of intrinsic material absorption for P and S waves in Pierre Shale imply the reflected SI wave is inhomogeneous for all non-normal angles of incidence. The degree of inhomogeneity, the χ parameter, and the angle of

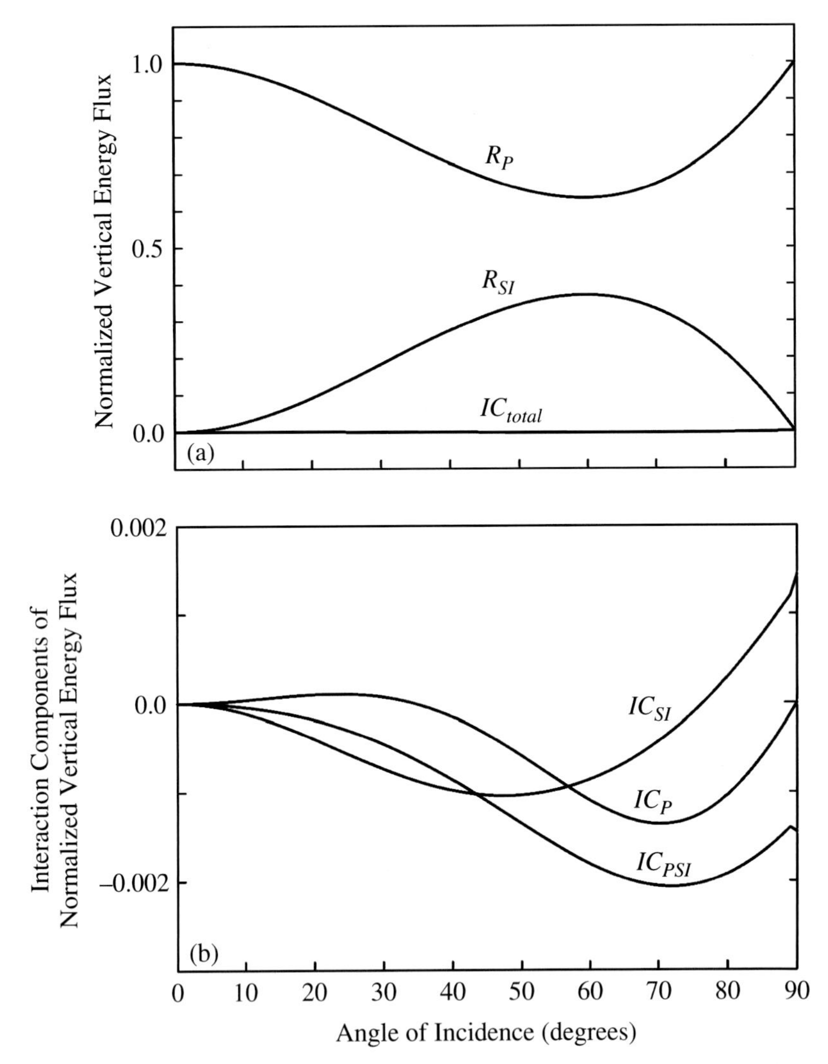

Figure (7.3.24). Energy reflection and total interaction coefficient (a) and individual interaction coefficients (b) for the reflected P and SI waves due to an incident homogeneous P wave on the free-surface of a low-loss anelastic half space with parameters corresponding to Pierre Shale.

emergence for phase and energy, as calculated using the computer code WAVES, are shown for the reflected SI wave in Figure (7.3.23). The degree of inhomogeneity of the reflected SI wave increases asymptotically toward a maximum value less than 16° (Figure (7.3.23)b) and the parameter χ_{ψ_2} does not differ significantly from unity as the angle of incidence approaches grazing incidence (Figure (7.3.23)b). Consequently, for the problem under consideration the physical characteristics of the reflected waves can be described using the low-loss expressions for homogeneous waves as specified by (3.7.6) through (3.7.43).

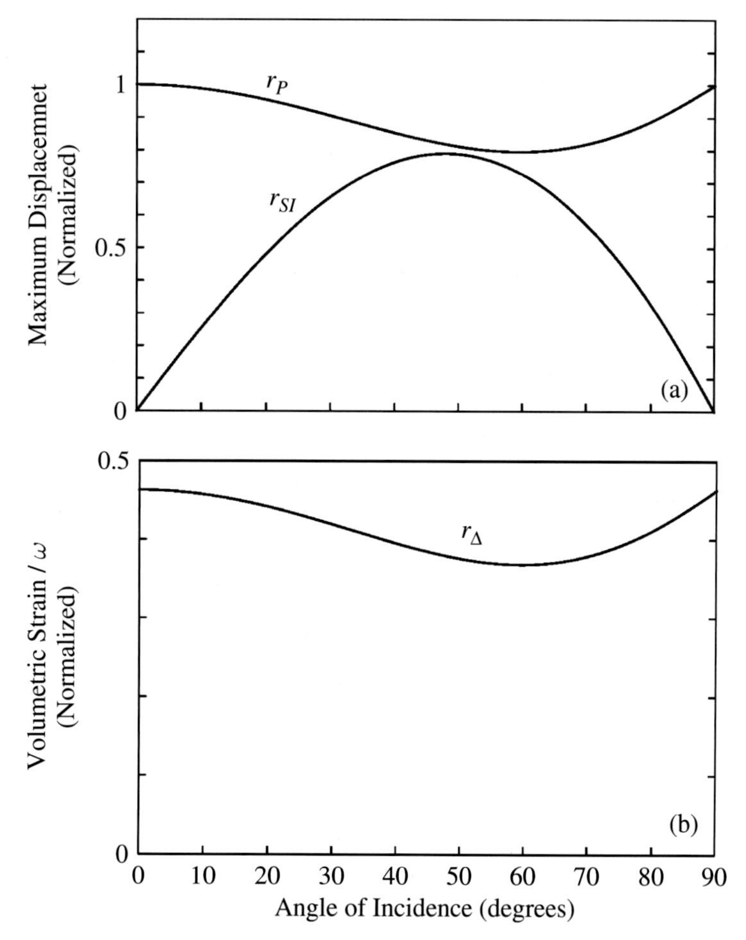

Figure (7.3.25). Reflection coefficients for maximum displacement (a) and volumetric strain (b) for the P and SI waves generated by a homogeneous P wave incident on the free surface of an anelastic half space with parameters corresponding to Pierre Shale (Table (7.2.43)) .

Energy reflection and interaction coefficients are shown in Figure (7.3.24). The coefficients indicate that the largest conversion of incident P to SI energy at the boundary occurs for an angle of incidence of about 60° with no conversion of energy at normal and grazing incidence (see R_{SI}, Figure (7.3.24)a). The coefficients indicate that the energy flow normal to the boundary due to the interaction of the incident P wave with the reflected P and SI waves (IC_P, IC_{SI}) and the interaction of the reflected waves (IC_{PSI}) is small (<0.2%; Figure (7.3.24)b). Hence, the coefficients indicate that for this problem the majority of the incident energy for each angle of incidence is transported away from the boundary by the reflected P and SI waves.

Amplitude and phase reflection coefficients for the maximum displacement and volumetric strain, defined with respect to the maximum particle motion of the incident wave as they were for the incident SI wave problem, are shown in Figures (7.3.25) and

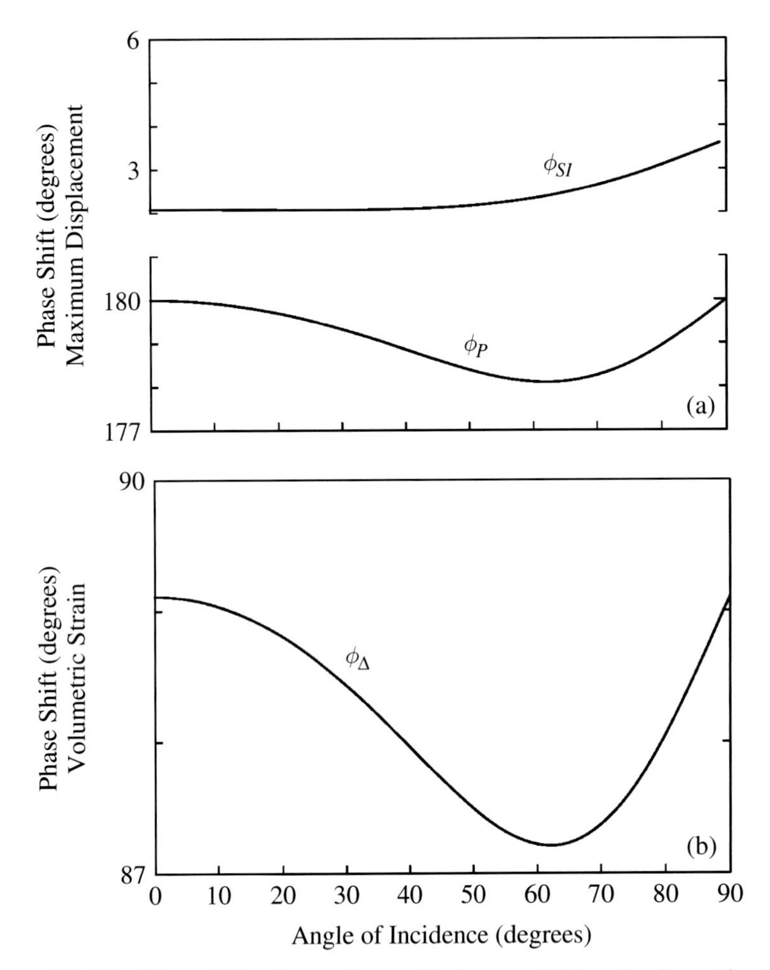

Figure (7.3.26). Phase shifts for maximum displacement (a) and volumetric strain (b) for the P and SI waves generated by a homogeneous P wave incident on the free surface of an anelastic half space with parameters corresponding to Pierre Shale (Table (7.2.43)).

(7.3.26). The reflection coefficients for maximum particle motion (Figure (7.3.25)a) imply that the reflected SI wave vanishes for normal and grazing incidence. The amplitudes of the reflected P and SI waves at an intermediate angle (~50°) are nearly equal and are about 80 percent of those of the incident P wave. The phase shifts for the maximum displacement for the reflected P and SI waves change by less than 2° as the angle of incidence varies from normal to grazing incidence (Figure (7.3.26)a).

The volumetric strain associated with the reflected P wave as normalized by the maximum displacement amplitude of the incident P wave varies by less than 20 percent with angle of incidence (Figure (7.3.25)b). The dependence of the volumetric strain amplitudes on angle of incidence is similar to that of the maximum

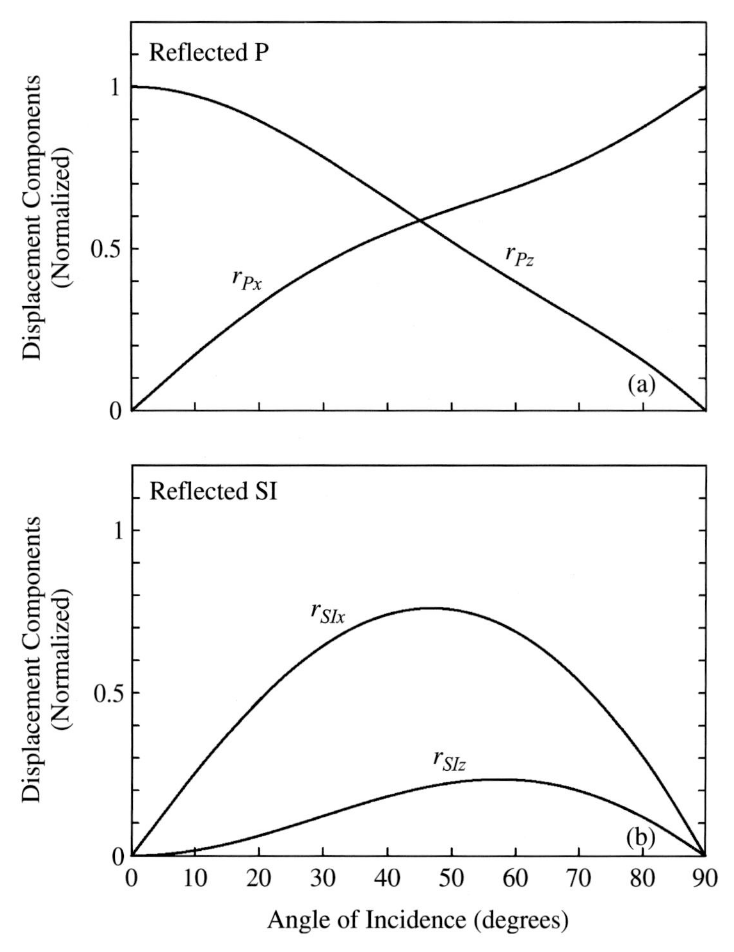

Figure (7.3.27). Reflection coefficients for the radial and vertical components of displacement for the reflected P wave (a) and the reflected SI wave (b) generated by a homogeneous P wave incident on the free surface of an anelastic half space with parameters corresponding to Pierre Shale (Table (7.2.43)).

displacement of the reflected P wave (compare Figures (7.3.25)a and b). This result might be expected because the amount of energy flow due to interaction is small. The phase shift for volumetric strain of the reflected P wave (Figure (7.3.26)b) varies by less than 2° as a function of angle of incidence from the 180° phase shift introduced by reflection from the free surface.

Amplitude and phase reflection coefficients for the horizontal (radial) and vertical components of motion for the reflected P and SI waves are shown in Figure (7.3.27). They indicate the relative contributions of the displacement components of the reflected waves to the total horizontal and vertical components as might be inferred from measurements on radial and vertical seismometers. The coefficients indicate

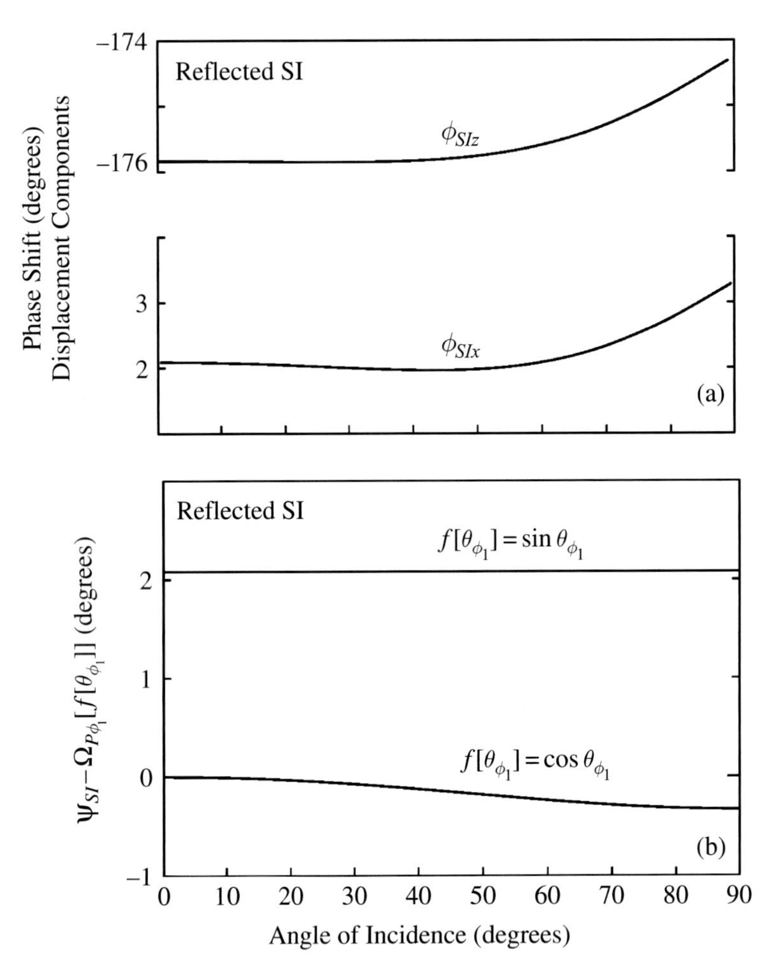

Figure (7.3.28). Phase shifts for the radial and vertical components of displacement for the reflected SI wave generated by a homogeneous P wave incident on the free surface of an anelastic half space with parameters corresponding to Pierre Shale (Table (7.2.43)).

that the horizontal component of motion for the reflected SI wave exceeds the horizontal component for the reflected P wave for angles of incidence less than about 60° (Figures (7.3.27)a and b). Consequently, for such angles of incidence the reflected energy as detected by a horizontal radial seismometer could be expected to be comprised of motion due to both the reflected SI and P waves with the reflected SI being somewhat larger. In contrast, a vertical seismometer would be expected to respond primarily to the reflected P wave with the vertical component of motion for reflected P only approaching the size of that for reflected SI as grazing incidence is approached, where the reflected vertical amplitudes become vanishingly small. As a result of the reflected P wave being homogeneous the horizontal and vertical reflection coefficients for the reflected P wave are simply the maximum reflection coefficient modulated by the sine and cosine functions, respectively. Variations in phase shifts for the horizontal and vertical components of the displacement for the reflected SI wave are less than 2° for the reflected SI wave as a function of angle of incidence (Figure (7.3.28)).

7.4 Incident General SII Wave

The simple problem of a general Type-II S wave incident on the free surface with particle motion parallel to the boundary and perpendicular to the plane of incidence may be readily solved in a manner analogous to that for an SII wave incident on a welded boundary. Solutions for the displacement field of the assumed incident and reflected general SII waves are given by (4.2.26), where the amplitudes of the incident SI and P waves are set to zero. The general incident SII wave is completely specified by (5.4.2) through (5.4.8) for a given angle of incidence, θ_{u_1}, a given angle between its attenuation and propagation vectors, $\gamma_{u_1} (0 \leq |\gamma_{u_1}| < \pi/2)$, a given complex amplitude D_1, and circular frequency ω. The assumed incident general Type-II S wave is either an inhomogeneous wave or a homogeneous wave depending on the degree of inhomogeneity, γ_{u_1}, chosen for the wave. In both cases the particle motion of the incident wave is linear parallel to the interface and perpendicular to the direction of propagation. The solution for the reflected Type-II S wave in medium V is given by (4.2.26).

Parameters for the incident and reflected SII waves are illustrated in Figure (7.4.1).

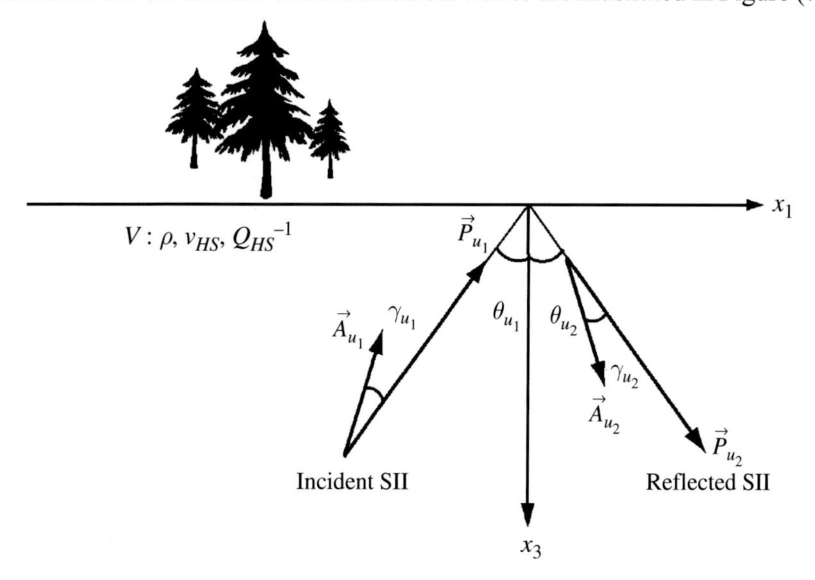

Figure (7.4.1). Diagram illustrating notation for directions and magnitude of propagation and attenuation vectors for the problem of a general Type-II S wave incident on the free surface of a HILV half space.

The boundary condition of vanishing stress at the free surface immediately implies from (7.1.6) that the complex amplitude of the reflected general SII wave equals that of the incident general SII wave, that is,

$$D_1 = D_2. \tag{7.4.2}$$

Hence, the boundary conditions are satisfied with the assumed solutions for the incident and reflected general SII waves showing that no P or Type-I S waves are generated by the incident SII wave. This solution establishes equality of the complex wave number k in the assumed solutions. Hence, with k given by (5.4.9) and (5.4.10) the propagation and attenuation vectors for the reflected general SII wave are completely determined upon specification of the angle of incidence θ_{u_1}, the degree of inhomogeneity γ_{u_1}, and the circular frequency ω for the incident general SI wave together with the material parameters for V as specified by k_S or v_{HS} and Q_{HS}^{-1}. These expressions and equality of the complex wave number k for each of the solutions yields the components of Generalized Snell's Law, namely,

$$k_R = \left|\vec{P}_{u_1}\right| \sin\theta_{u_1} = \left|\vec{P}_{u_2}\right| \sin\theta_{u_2} \tag{7.4.3}$$

or

$$\frac{k_R}{\omega} = \frac{\sin\theta_{u_1}}{\left|\vec{v}_{u_1}\right|} = \frac{\sin\theta_{u_2}}{\left|\vec{v}_{u_2}\right|}. \tag{7.4.4}$$

and

$$-k_I = |\vec{A}_{u_1}| \sin[\theta_{u_1} - \gamma_{u_1}] = |\vec{A}_{u_2}| \sin[\theta_{u_2} - \gamma_{u_2}], \tag{7.4.5}$$

which indicate that the apparent phase velocity and apparent attenuation along the boundary of the general reflected SII wave equal those of the general incident SII wave. It follows that $\theta_{u_1} = \theta_{u_2}$ and $\gamma_{u_1} = \gamma_{u_2}$, hence the reflected SII wave is homogeneous if and only if the incident SII wave is homogeneous.

Results concerning energy flow derived for the problem of an SII wave on a welded boundary ((5.4.27) through (5.4.58)) are valid for the problem of an SII wave incident on a free surface with the understanding that the normal component of energy flux across a free surface is zero. For brevity, the results are not restated here.

7.5 Problems

(1) For the problem of a homogeneous SI wave incident on the free surface of a viscoelastic half space,
   (a) show that the reflected P wave is homogeneous if and only if the amount of intrinsic absorption for a homogeneous shear wave equals that for a homogeneous P wave (i.e. $Q_{HS}^{-1} = Q_{HP}^{-1}$) and the angle of incidence satisfies $\sin^2\theta_{\psi_1} \le v_{HS}^2/v_{HP}^2$,
   (b) describe Generalized Snell's Law for the reflected SI and P waves.

(2) For the problem of a homogeneous SI wave incident on the free surface of a viscoelastic half space with parameters corresponding to Pierre Shale

(Table (7.2.43), sketch graphs showing the dependence on angle of incidence for the reflected P wave of the

(a) degree of inhomogeneity,
(b) emergence angles for directions of phase propagation, maximum attenuation, and energy flux,
(c) reciprocal quality factor,
(d) tilt of the particle motion ellipse,
(e) ratio of minor to major axis of the particle motion ellipse,
(f) normalized vertical energy flux associated with reflected SI, reflected P, and interaction of each of the incident and reflected waves,
(g) normalized maximum displacement amplitude, and
(h) normalized volumetric strain.

(3) For the preceding problem,

(a) find the angles of incidence for which the amplitudes of horizontal and vertical displacement and volumetric strain are the largest,
(b) determine the magnitude of Q_P^{-1}, the ratio of minor to major axis for particle motion ellipse, and normal components of energy flow at the boundary at angles of incidence determined in part (a), and
(c) compare the values determined in part (b) for the reflected inhomogeneous P wave with corresponding quantities for a homogeneous P wave in the same material.

8

Rayleigh-Type Surface Wave on a Viscoelastic Half Space

The problem of a surface wave on a half space is solved by determining if steady-state solutions describing wave fields concentrated near the surface can be chosen such that the boundary conditions are satisfied. This problem was first solved for an elastic half space by Rayleigh (1885). The solution and corresponding numerical results presented here are for a viscoelastic half space (Borcherdt, 1971, 1973b, 1988).

8.1 Analytic Solution

The postulated surface wave on a viscoelastic half space is specified by considering solutions for superimposed inhomogeneous P and SI solutions as specified in (4.2.1) and (4.2.2) with propagation in the $+\hat{x}_1$ direction, with attenuation away from the free surface in the $+\hat{x}_3$ direction, and with the same complex wave number k assumed for each solution. Selection of solutions to represent these assumptions is most easily accomplished by setting

$$B_1 = C_{12} = C_{21} = C_{23} = 0 \tag{8.1.1}$$

in (4.2.1) and (4.2.2) then by defining

$$b_\alpha \equiv \sqrt{k^2 - k_P^2} = i d_\alpha \tag{8.1.2}$$

and

$$b_\beta \equiv \sqrt{k^2 - k_S^2} = i d_\beta, \tag{8.1.3}$$

where "$\sqrt{\ }$" is understood to represent the principal value of the square root, which ensures that the real part of the corresponding complex number is not negative. With these conventions the solutions for the P and SI components of the postulated surface wave are

$$\phi = \phi_2 = B_2 \exp\left[i\left(\omega t - \vec{K}_{\phi_2} \bullet \vec{r}\right)\right] \tag{8.1.4}$$

and

$$\vec{\psi} = \vec{\psi}_2 = C_{22} \exp\left[i\left(\omega t - \vec{K}_{\psi_2} \bullet \vec{r}\right)\right] \hat{x}_2, \tag{8.1.5}$$

where

$$\vec{K}_{\phi_2} = \vec{P}_{\phi_2} - i\vec{A}_{\phi_2} = k\hat{x}_1 - ib_\alpha \hat{x}_3 \tag{8.1.6}$$

$$\vec{K}_{\psi_2} = \vec{P}_{\psi_2} - i\vec{A}_{\psi_2} = k\hat{x}_1 - ib_\beta \hat{x}_3, \tag{8.1.7}$$

and the corresponding propagation and attenuation vectors are given by

$$\vec{P}_{\phi_2} = k_R \hat{x}_1 + b_{\alpha_I} \hat{x}_3, \tag{8.1.8}$$

$$\vec{P}_{\psi_2} = k_R \hat{x}_1 + b_{\beta_I} \hat{x}_3, \tag{8.1.9}$$

and

$$\vec{A}_{\phi_2} = -k_I \hat{x}_1 + b_{\alpha_R} \hat{x}_3, \tag{8.1.10}$$

$$\vec{A}_{\psi_2} = -k_I \hat{x}_1 + b_{\beta_R} \hat{x}_3. \tag{8.1.11}$$

Choice of the principal value for the square root in the definitions of b_α and b_β implies $b_{\alpha_R} \geq 0$ and $b_{\beta_R} \geq 0$. This choice ensures that the direction of attenuation for the chosen solutions is in the $+\hat{x}_3$ direction and hence, the chosen solutions represent a wave concentrated near the surface.

Solutions (8.1.4) and (8.1.5) for viscoelastic media differ from those chosen by Rayleigh for elastic media. The directions of the propagation and attenuation vectors for each of the postulated component solutions for viscoelastic media are not necessarily confined to be parallel and perpendicular respectively to the free surface as they are for elastic media. The directions and magnitude of these vectors for each of the component solutions will be implied by the condition of vanishing stress at the free surface. It will be shown that for anelastic media these vectors are, in general, not parallel and perpendicular to the boundary.

For the postulated surface wave disturbance, the boundary condition of vanishing stress at the free surface may be expressed in terms of the assumed displacement potential solutions with propagation in the $+x_1$ direction as

$$P_{31} = M\left(2\phi_{,13} + \psi_{,11} - \psi_{,33}\right) = 0 \tag{8.1.12}$$

and

$$P_{33} = \left(K + \frac{4}{3}M\right)\left(\phi_{,11} + \phi_{,33}\right) + 2M\left(\psi_{,33} - \phi_{,11}\right) = 0. \tag{8.1.13}$$

Substitution of (8.1.4) and (8.1.5) into these expressions implies the parameters of the assumed solutions must satisfy

$$2 i b_\alpha k B_2 = \left(2k^2 - k_S^2\right) C_{22} \tag{8.1.14}$$

and

$$\left(2k^2 - k_S^2\right) B_2 = -2 i b_\beta k C_{22}. \tag{8.1.15}$$

These two equations involve three complex unknowns, k, B_2, and C_{22}, and hence provide a unique solution for two of the parameters in terms of a third. Assuming $C_{22} \neq 0$ and solving the second equation for B_2/C_{22} implies

$$\frac{B_2}{C_{22}} = \frac{-2 i b_\beta k}{2k^2 - k_S^2}. \tag{8.1.16}$$

Substitution of this result into the first equation yields

$$4 b_\alpha b_\beta k^2 = \left(2k^2 - k_S^2\right)^2. \tag{8.1.17}$$

Solution of (8.1.17) for k will establish that a solution for the postulated surface wave exists. Substitution of the solution for k into (8.1.16) will provide the solution for the amplitude of the P component of the solution in terms of that for the SI component and hence the solution for the physical characteristics of the postulated surface wave. For convenience, the solution for the quantity k for a viscoelastic half space is termed the solution for the complex apparent wave number for a Rayleigh-Type surface wave.

For $k \neq 0$ roots of (8.1.17) are the same as roots of the rationalized equation

$$4\sqrt{1 - \frac{k_P^2}{k^2}}\sqrt{1 - \frac{k_S^2}{k^2}} = \left(2 - \frac{k_S^2}{k^2}\right)^2. \tag{8.1.18}$$

Squaring each side of the equation introduces extraneous roots, but permits the equation to be written as a cubic polynomial, namely

$$\left(\frac{k_S^2}{k^2}\right)^3 - 8\left(\frac{k_S^2}{k^2}\right)^2 + \left(24 - 16\frac{k_P^2}{k_S^2}\right)\left(\frac{k_S^2}{k^2}\right) - 16\left(1 - \frac{k_P^2}{k_S^2}\right) = 0. \tag{8.1.19}$$

This equation may be written in terms of complex velocities by defining $c \equiv \omega/k$ and using definitions (3.1.5) and (3.1.6) as

$$\left(\frac{c^2}{\beta^2}\right)^3 - 8\left(\frac{c^2}{\beta^2}\right)^2 + \left(24 - 16\frac{\beta^2}{\alpha^2}\right)\left(\frac{c^2}{\beta^2}\right) - 16\left(1 - \frac{\beta^2}{\alpha^2}\right) = 0. \tag{8.1.20}$$

This equation for viscoelastic media termed here the complex Rayleigh equation differs from the equation for elastic media derived by Rayleigh (1885, p. 7) in that the velocities are complex. Hence, the equation is a cubic polynomial with complex coefficients. The set of roots of this equation includes those for which the coefficients are real as a special case. It will be shown that complex roots of the equation corresponding to anelastic media imply significantly different physical characteristics for the corresponding surface wave than those for a Rayleigh wave on an elastic half space with roots that are real numbers.

The Fundamental Theorem of Algebra implies the cubic polynomial (8.1.20) has three roots in the complex field, not all of which are necessarily unique. The solution of the equation is provided in Appendix 4. Roots of this cubic polynomial with coefficients in the complex field are given by

$$\begin{aligned} y_j = &\left(\frac{-q}{2} - \sqrt{\left(\frac{q}{2}\right)^2 + \left(\frac{p}{3}\right)^3}\right)^{1/3} u^{j-1} \\ &- \frac{p}{3}\left(\frac{-q}{2} - \sqrt{\left(\frac{q}{2}\right)^2 + \left(\frac{p}{3}\right)^3}\right)^{-1/3} u^{-(j-1)} + \frac{8}{3} \qquad \text{for } j = 1, 2, 3, \end{aligned} \tag{8.1.21}$$

where

$$p \equiv \frac{8}{3} - 16\frac{\beta^2}{\alpha^2}, \tag{8.1.22}$$

$$q \equiv \frac{272}{27} - \frac{80}{3}\frac{\beta^2}{\alpha^2}, \tag{8.1.23}$$

and

$$u \equiv \exp[2\pi i/3]. \tag{8.1.24}$$

The roots specified by (8.1.21), which in turn satisfy the original equation (8.1.18) together with the solutions (8.1.4) and (8.1.5), provide the analytic solution for the postulated surface wave on a HILV half space. Roots that satisfy (8.1.19), but do not satisfy (8.1.18), are extraneous. The wave represented by the analytic solution is referred to as a Rayleigh-Type surface wave on a HILV half space to distinguish it and its distinctly different physical characteristics from that derived by Rayleigh for an elastic half space. Results for an elastic half space will follow as a special case.

A lemma useful in determining which root specified by (8.1.21) is the appropriate root of (8.1.18) for a Rayleigh-Type surface wave is

Lemma (8.1.25). If $y_j = c^2/\beta^2$ for $j = 1, 2, 3$ *as given by (8.1.21) is a root of (8.1.20) and* $|y_j| < 1$, *then* $y_j = c^2/\beta^2$ *is a root of complex Rayleigh equation (8.1.18) and in turn a solution for a Rayleigh-Type surface wave on a HILV half space.*

The proof of this lemma is provided in Appendix 5.

A result which shows that the roots of (8.1.20) as derived by Rayleigh are a special case of a larger class of viscoelastic solids, namely those for which $Q_{HS}^{-1} = Q_{HP}^{-1}$, is

Theorem (8.1.26). If $Q_{HS}^{-1} = Q_{HP}^{-1}$, then the root c^2/β^2 of the Rayleigh-Type surface wave equation (8.1.18) is a real number that satisfies inequalities $0 < c^2/\beta^2 < 1$.

To prove this result, assume $Q_{HS}^{-1} = Q_{HP}^{-1}$, then (3.5.8) and (3.5.9) imply β^2/α^2 may be written as

$$\frac{\beta^2}{\alpha^2} = \left(\frac{v_{HS}^2}{v_{HP}^2}\right) \frac{1 + \sqrt{1 + Q_{HS}^{-2}}}{1 + \sqrt{1 + Q_{HP}^{-2}}} \frac{1 - iQ_{HP}^{-1}}{1 - iQ_{HS}^{-1}} = \frac{v_{HS}^2}{v_{HP}^2}, \tag{8.1.27}$$

which shows that β^2/α^2 is a real number and hence the coefficients of equation (8.1.20) are real. If $f(c^2/\beta^2)$ is used to represent the left-hand side of (8.1.20), it follows that $f(0) < 0$ and $f(1) > 0$, hence equation (8.1.20) has a real root c^2/β^2 satisfying $0 < c^2/\beta^2 < 1$. Hence, Lemma (8.1.25) implies the desired conclusion that c^2/β^2 is a root of equation (8.1.18).

8.2 Physical Characteristics

Physical characteristics of a Rayleigh-Type surface wave on a viscoelastic half space are derived in this section for comparison with those inferred as a special case for an elastic half space.

8.2.1 Velocity and Absorption Coefficient

The magnitude of the velocity (wave speed) and the absorption coefficient for a Rayleigh-Type surface wave along the surface are given by

$$v_B = \omega/k_R \tag{8.2.1}$$

and

$$a_B = -k_I. \tag{8.2.2}$$

The wave speed and absorption coefficient may be written in terms of the appropriate root c/β for a Rayleigh-Type surface wave as

$$v_B = v_{HS}\left(\mathrm{Re}\left[\frac{\beta}{c}\right] + \frac{\sqrt{\chi_{HS} - 1}}{\sqrt{\chi_{HS} + 1}}\mathrm{Im}\left[\frac{\beta}{c}\right]\right)^{-1} \tag{8.2.3}$$

and

$$a_B = \left|\vec{A}_{HS}\right| \left(\mathrm{Re}\left[\frac{\beta}{c}\right] - \frac{\sqrt{\chi_{HS} - 1}}{\sqrt{\chi_{HS} + 1}} \mathrm{Im}\left[\frac{\beta}{c}\right] \right), \tag{8.2.4}$$

where $\chi_{HS} = \sqrt{1 + Q_{HS}^{-2}}$ and v_{HS}, $a_{HS} \equiv \left|\vec{A}_{HS}\right|$, and Q_{HS}^{-1} represent the speed (3.5.5), absorption coefficient (3.6.6), and reciprocal quality factor for a homogeneous S wave (3.5.2). Expressions (8.2.3) and (8.2.4) follow immediately from the relation for any two complex numbers z and z_1 that

$$\frac{z_R}{z_{1_R}} = \mathrm{Re}\left[\frac{z}{z_1}\right] - \frac{z_{1_I}}{z_{1_R}} \mathrm{Im}\left[\frac{z}{z_1}\right] \tag{8.2.5}$$

and the identity

$$\frac{k_{S_I}}{k_{S_R}} = -\sqrt{\frac{\chi_{HS} - 1}{\chi_{HS} + 1}}. \tag{8.2.6}$$

8.2.2 Propagation and Attenuation Vectors for Component Solutions

The solutions (8.1.4) and (8.1.5) suggest that the propagation and attenuation vectors for the component solutions for a Rayleigh-Type surface wave on an anelastic half space form acute angles e_ϕ and e_ψ with respect to the free surface (see Figure (8.2.11)) given by

$$\tan e_\phi = -b_{\alpha_I}/k_R = -b_{\alpha_I} \frac{v_B}{\omega} \tag{8.2.7}$$

and

$$\tan e_\psi = -b_{\beta_I}/k_R = -b_{\beta_I} \frac{v_B}{\omega}. \tag{8.2.8}$$

Similarly, the attenuation vectors for the component solutions form acute angles d_ϕ and d_ψ with respect to the normal to the surface as given by

$$\tan d_\phi = \frac{-k_I}{b_{\alpha_R}} = \frac{a_B}{b_{\alpha_R}}, \tag{8.2.9}$$

$$\tan d_\psi = \frac{-k_I}{b_{\beta_R}} = \frac{a_B}{b_{\beta_R}}. \tag{8.2.10}$$

For an elastic half space $b_{\alpha_I} = b_{\beta_I} = k_I = 0$, hence the propagation vectors (8.1.8) and (8.1.9) are parallel and perpendicular, respectively, to the undisturbed free surface. For an anelastic half space, the propagation and attenuation vectors for the component solutions cannot be perpendicular, because such a wave cannot

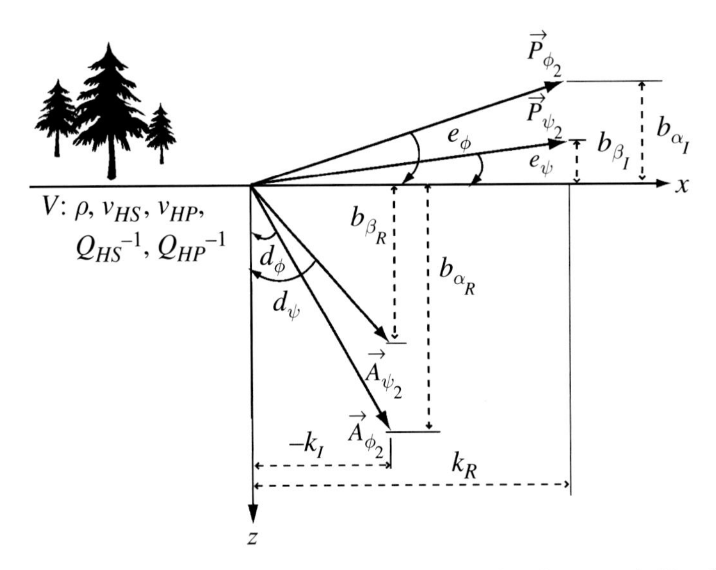

Figure (8.2.11). Notation and relative orientations for the P and SI solutions comprising a Rayleigh-Type surface wave on a viscoelastic half space. Inclinations of the propagation and attenuation vectors with respect to the horizontal and vertical, respectively, result in volumetric strain and displacement components showing an exponentially damped sinusoidal dependence on depth which does not exist for elastic media.

propagate in an anelastic medium as stated in Theorem (3.1.18). In addition, it is possible to show that, for an anelastic medium, both of the propagation vectors cannot be parallel to the free surface. A proof of this result is provided by Borcherdt (1973b).

The directions of the propagation and attenuation vectors as indicated by (8.2.7) through (8.2.10) are indicative of significant distinctions in the displacement field for a surface disturbance on an anelastic half space as opposed to that derived by Rayleigh (1885) for an elastic half space (Borcherdt, 1973b). Inclinations of the various vectors as given by (8.2.7) through (8.2.10) for anelastic media will be shown to imply variations in amplitude with depth, tilt of the particle motion ellipse, and other displacement field characteristics not predicted by elastic models.

8.2.3 Displacement and Particle Motion

The physical displacement field is given from (3.1.1) in terms of the displacement-potential solutions by

$$\vec{u}_R = \mathrm{Re}[\nabla\phi + \nabla \times \vec{\psi}]. \tag{8.2.12}$$

Substitution of solutions (8.1.4) and (8.1.5) into (8.2.12) together with identities in Appendix 2 allow the components of the physical displacement field for a Rayleigh-Type surface wave to be written as

$$u_{Rx_1} = |B_2 k| \exp[k_I x_1] \begin{pmatrix} \exp[-b_{\alpha_R} x_3] \sin\left[\omega t - k_R x_1 - b_{\alpha_I} x_3 + \arg[k B_2]\right] \\ -\dfrac{\left|k^2 + b_\beta^2\right|}{2|k^2|} \\ \exp[-b_{\beta R} x_3] \sin\left[\omega t - k_R x_1 - b_{\beta_I} x_3 + \arg\left[(k^2 + b_\beta^2)/k^2\right] + \arg[B_2 k]\right] \end{pmatrix}, \quad (8.2.13)$$

and

$$u_{Rx_3} = |B_2 k| \exp[k_I x_1] \begin{pmatrix} -\left|\dfrac{b_\alpha}{k}\right| \exp[-b_{\alpha_R} x_3] \sin\left[\omega t - k_R x_1 - b_{\alpha_I} x_3 + \arg[b_\alpha B_2] + \dfrac{\pi}{2}\right] \\ +\dfrac{\left|k^2 + b_\beta^2\right|}{2\left|b_\beta k\right|} \exp[-b_{\beta_R} x_3] \\ \sin\left[\omega t - k_R x_1 - b_{\beta_I} x_3 + \arg\left[\left(k^2 + b_\beta^2\right)\Big/ b_\beta\right] + \arg[B_2] + \dfrac{\pi}{2}\right] \end{pmatrix}, \quad (8.2.14)$$

where k is given by (8.1.17). These equations simplify to

$$u_{Rx_j} = D \exp[k_I x_1]\left(F_j \sin\left[\omega t + f_j + g_{1j}\right] + G_j \sin\left[\omega t + f_j + g_{3j}\right]\right) \quad (8.2.15)$$

for $j = 1, 3$ upon introducing notation

$$\begin{aligned}
&D \equiv |k B_2|, \\
&F_1 \equiv \exp[-b_{\alpha_R} x_3], \qquad F_3 \equiv -\frac{|b_\alpha|}{|k|} \exp[-b_{\alpha_R} x_3] = -\frac{|b_\alpha|}{|k|} F_1, \\
&G_1 \equiv -\frac{\left|k^2 + b_\beta^2\right|}{2|k^2|} \exp[-b_{\beta_R} x_3], \qquad G_3 \equiv \frac{\left|k^2 + b_\beta^2\right|}{2\left|k b_\beta\right|} \exp[-b_{\beta_R} x_3] = -\frac{|k|}{\left|b_\beta\right|} G_1, \\
&f_1 \equiv -k_R x_1 + \arg[k B_2], \qquad f_3 \equiv f_1 + \frac{\pi}{2}, \qquad (8.2.16) \\
&g_{11} \equiv -b_{\alpha_I} x_3, \qquad g_{13} \equiv -b_{\alpha_I} x_3 + \arg\left[\frac{b_\alpha}{k}\right], \\
&g_{31} \equiv -b_{\beta_I} x_3 + \arg\left[\frac{k^2 + b_\beta^2}{k^2}\right], \qquad g_{33} \equiv -b_{\beta_I} x_3 + \arg\left[\frac{k^2 + b_\beta^2}{k b_\beta}\right].
\end{aligned}$$

Equation (8.2.15) may be further simplified to

$$u_{Rx_j} = D \exp[k_I x_1]\left(H_j \sin\left[\omega t + f_j + \varsigma_j\right]\right) \qquad \text{for } j = 1, 3 \quad (8.2.17)$$

with the additional notation defined as follows

$$H_j \equiv \sqrt{F_j^2 + G_j^2 + 2F_jG_j\cos\left[g_{1j} - g_{3j}\right]} \tag{8.2.18}$$

and

$$\varsigma_j \equiv \tan^{-1}\left[\frac{F_j\sin\left[g_{1j}\right] + G_j\sin\left[g_{3j}\right]}{F_j\cos\left[g_{1j}\right] + G_j\cos\left[g_{3j}\right]}\right] \qquad \text{for } j = 1, 3. \tag{8.2.19}$$

The form of equation (8.2.17) shows that the normalized component amplitude distributions at a fixed time "t" when the corresponding amplitude at the surface is a maximum are given by

$$\frac{H_j(x_3)}{H_j(0)}\cos\left[\varsigma_j(x_3) - \varsigma_j(0)\right] \qquad \text{for } j = 1, 3. \tag{8.2.20}$$

For an elastic solid this equation reduces to

$$\frac{H_j(x_3)}{H_j(0)}\cos[\varsigma_j(x_3) - \varsigma_j(0)] = \frac{F_j(x_3) + G_j(x_3)}{F_j(0) + G_j(0)} \qquad \text{for } j = 1, 3. \tag{8.2.21}$$

Equation (8.2.20) shows that, in general, for anelastic solids the normalized amplitude distributions show a superimposed sinusoidal dependence on depth. Equation (8.2.21) shows that for elastic solids the amplitudes do not show this sinusoidal dependence.

The maximum amplitudes during a cycle of oscillation normalized by the corresponding maximum during a cycle of oscillation at the surface are given by

$$\frac{H_j(x_3)}{H_j(0)} \qquad \text{for } j = 1, 3. \tag{8.2.22}$$

For elastic solids, this expression is the same as the amplitude distribution for a fixed time t (8.2.21).

The components of the physical displacement field, (8.2.17), may be written as a pair of simple parametric equations in terms of the absorption coefficient for a Rayleigh-Type surface wave as given by (8.2.2) as

$$u_{Rx_1} = D\exp[-a_Bx_1]H_1(x_3)\sin\vartheta_B(t) \tag{8.2.23}$$

and

$$u_{Rx_3} = D\exp[-a_Bx_1]H_3(x_3)\cos[\vartheta_B(t) + S(x_3)] \tag{8.2.24}$$

upon introducing additional definitions

$$\vartheta_B(t) \equiv \omega t + f_1(x_1) + \varsigma_1(x_3) \qquad \text{and} \qquad S(x_3) \equiv \varsigma_3(x_3) - \varsigma_1(x_3). \tag{8.2.25}$$

These parametric equations describe the motion of a particle in space with respect to the chosen coordinate system as a function of time during the passage of a Rayleigh-Type surface wave. They indicate that the particle motion for a Rayleigh-Type surface wave traces an ellipse with time whose properties are stated explicitly in the following theorem.

Theorem (8.2.26). The motion of a particle as a function of time as described by parametric equations (8.2.23) and (8.2.24) for a Rayleigh-Type surface wave on a HILV half space describes an ellipse with characteristics that,

(1) the direction in which the particle describes the ellipse with time is

(a) retrograde (counter clockwise) if $\cos[S(x_3)] > 0$,

(b) prograde (clockwise) if $\cos[S(x_3)] < 0$, *and*

(c) linear if $\cos[S(x_3)] = 0$,

(2) the tilt η_B *of the major axis of the ellipse with respect to the vertical axis* (x_3) *is given by*

$$\tan[2\eta_B] = \begin{cases} 2\sin S(x_3) \Big/ \left(\dfrac{H_3}{H_1} - \dfrac{H_1}{H_3} \right) & \text{if } H_1 \neq H_3 \\ \pi/4 & \text{if } H_1 = H_3 \end{cases}, \tag{8.2.27}$$

(3) the ellipticity of the ellipse defined as the ratio of the lengths of the principal axis is given by

$$\sqrt{A'/C'}, \tag{8.2.28}$$

where

$$A' \equiv A\cos^2\eta_B + B\sin\eta_B\cos\eta_B + C\sin^2\eta_B, \tag{8.2.29}$$

$$C' \equiv A\sin^2\eta_B - B\sin\eta_B\cos\eta_B + C\cos^2\eta_B, \tag{8.2.30}$$

and

$$A \equiv \frac{1}{(D\exp[-a_B x_1]H_1(x_3)\cos[S(x_3)])^2}, \tag{8.2.31}$$

$$B \equiv \frac{2}{(D\exp[-a_B x_1])^2 H_1(x_3)H_3(x_3)} \frac{\sin[S(x_3)]}{(\cos[S(x_3)])^2}, \tag{8.2.32}$$

$$C \equiv \frac{1}{(D\exp[-a_B x_1]H_3(x_3)\cos[S(x_3)])^2} \tag{8.2.33}$$

The characteristics of the particle motion described by Theorem (8.2.26) are derived in Appendix 6.

An additional parameter useful for describing characteristics of the particle motion for a Rayleigh-Type surface wave as a function of depth is a parameter termed here *Axis Ratio*. It is defined here for later reference as the reciprocal of the ellipticity with the algebraic sign of $\cos[S(x_3)]$ superimposed, specifically the *Axis Ratio* is defined by

$$Axis\ Ratio \equiv \frac{\sqrt{C'}}{\sqrt{A'}} \operatorname{sign}[\cos[S(x_3)]]. \tag{8.2.34}$$

Theorem (8.2.26) indicates that if the *Axis Ratio* is positive, negative or zero, then the corresponding particle motion orbit is retrograde, prograde, or linear. Parameters describing the particle motion ellipse as given by (8.2.18), (8.2.27), and (8.2.28) are illustrated in Figure (8.2.35).

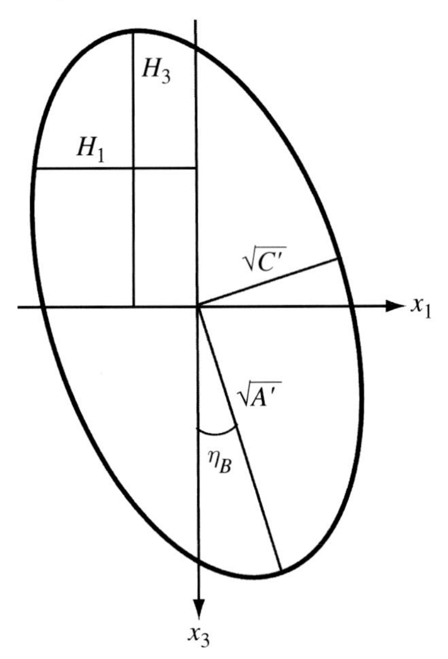

Figure (8.2.35). Diagram illustrating parameters of particle motion ellipse for a Rayleigh-Type surface wave (see text).

The definitions of the parameters in the parametric equations indicate that parameters $H_1 = H_1(x_3)$, $H_3 = H_3(x_3)$, and $S = S(x_3)$ are dependent on depth (x_3). This depth dependence together with (8.2.27) shows that for general viscoelastic media the tilt of the particle motion ellipse varies with depth.

For an elastic half space,

$$g_{ij} = 0, \qquad \varsigma_j = \left\{ \begin{matrix} 0 & \text{for } F_j + G_j > 0 \\ \pi & \text{for } F_j + G_j < 0 \end{matrix} \right\} \text{ for } i = 1,3 \text{ and } j = 1,3, \tag{8.2.36}$$
$$S = \varsigma_3 - \varsigma_1, \qquad \eta_B = 0, \qquad A' = A, \qquad \text{and } C' = C.$$

Hence, an immediate corollary of Theorem (8.2.26) for elastic media is

Corollary (8.2.37). The motion of a particle as a function of time as described by parametric equations (8.2.23) and (8.2.24) for a Rayleigh surface wave on an elastic half space is an ellipse with characteristics that

(1) *the direction in which the particle describes the ellipse with time is retrograde (counter clockwise) at the surface to a depth at which $F_1 + G_1 = 0$ and prograde for all depths below this depth,*

(2) *the tilt η_β of the major axis of the ellipse with respect to the vertical axis is zero for all depths, and*

(3) *the ellipticity of the ellipse defined as the ratio of the lengths of the principal axis is given by*

$$H_3/H_1 = |(F_3 + G_3)/(F_1 + G_1)|. \tag{8.2.38}$$

An important distinction in the characteristics of the particle motions for a Rayleigh-Type surface wave on an anelastic half space as compared to a Rayleigh wave on an elastic half space is that for anelastic media the particle motion ellipse shows tilt with respect to the vertical that varies with depth. For anelastic media tilt of the particle motion ellipse can also be shown to vanish, but only at the free surface for the special case of viscoelastic solids in which the amount of absorption for homogeneous P waves equals that for homogeneous S waves, which is of course a class of viscoelastic solids that includes elastic solids as a special case. Establishment of this result follows from the following result.

Theorem (8.2.39). If $Q_{HS}^{-1} = Q_{HP}^{-1}$ for a HILV half space, then the tilt of the particle motion ellipse of a Rayleigh-Type surface wave is zero at the free surface i.e. $\eta_\beta = 0$ at the free surface.

To prove this result assume $Q_{HS}^{-1} = Q_{HP}^{-1}$, then Theorem (8.1.26) implies

$$g_{ij} = 0 \quad \text{and} \quad F_j + G_j > 0 \qquad \text{for } i = 1, 3 \quad \text{and} \quad j = 1, 3. \tag{8.2.40}$$

Hence, definitions (8.2.17) and (8.2.25) imply the desired result that the tilt vanishes at the free surface, that is $\eta_\beta = 0$ at the free surface.

8.2.4 Volumetric Strain

The volumetric strain associated with the passage of a Rayleigh-Type surface wave must be due to that associated with the P component of the solution, because the requirement that $\nabla \cdot \vec{\psi} = 0$ (3.1.2) immediately implies that the contribution due to the SI component of the solution vanishes. Hence, the expression for the volumetric strain of a Rayleigh-Type surface wave is given by (3.10.29), where k is interpreted as the root of equation (8.1.17). The desired expression for the volumetric strain written in the form of (8.2.23) as initially derived by Borcherdt (1988) is

$$\Delta_R(t) = |D| \exp[-a_B x_1] H_\Delta(x_3) \sin[\vartheta_\Delta(t)] \tag{8.2.41}$$

where

$$H_\Delta(x_3) \equiv \left|\frac{c}{\beta}\right| \left|\frac{\beta}{\alpha}\right| |k_P| \exp[-b_{\alpha_R} x_3], \tag{8.2.42}$$

$$\vartheta_\Delta(t) \equiv \omega t + f_1(x_1) + \varsigma_\Delta(x_3), \tag{8.2.43}$$

$$\varsigma_\Delta(x_3) \equiv \psi_S - 2\psi_P + \arg\left[\frac{c}{\beta}\right] - b_{\alpha_I} x_3 - \pi/2, \tag{8.2.44}$$

$$D \equiv |kB_2|, \qquad f_1 \equiv -k_R x_1 + \arg[kB_2], \tag{8.2.45}$$

and c/β is the appropriate root of (8.1.18) for a Rayleigh-Type surface wave.

Equation (8.2.41) shows that the volumetric strain for a Rayleigh-Type surface wave on a viscoelastic half space is attenuated along the surface with absorption coefficient a_B and confined to the surface by H_Δ. Comparison of (8.2.41) with (8.2.23) shows that the volumetric strain may be represented as the radial component of displacement with a depth-dependent scale factor $H_1(x_3)/H_\Delta(x_3)$ and phase shift $\vartheta_B(t) - \vartheta_\Delta(t)$. Similarly, a depth-dependent scale factor and phase shift are apparent for the vertical component of displacement. Equations (8.2.41) and (8.2.23) provide a complete description of the volumetric strain and displacement components for a Rayleigh-Type surface wave on a HILV half space as might be inferred from corresponding collocated sensors.

The volumetric strain at depth, normalized by the maximum at the surface at time t_m, is given from (8.2.41) by

$$\frac{\Delta_R(t_m)}{\Delta_{R_0}(t_m)} = \exp[-b_{\alpha_R} x_3] \cos\left[b_{\alpha_I} x_3\right], \tag{8.2.46}$$

showing that the normalized volumetric strain has an exponentially damped sinusoidal dependence on depth for anelastic media. This sinusoidal dependence for anelastic media is associated with the inclination of the propagation and attenuation vectors for the component P solution as can be easily seen by rewriting (8.2.46) using definitions (8.2.7) and (8.2.9); that is,

$$\frac{\Delta_R(t_m)}{\Delta_{R_0}(t_m)} = \exp[-a_B x_3/\tan d_\phi] \cos\left[-\frac{\omega}{v_B}(\tan e_\phi) x_3\right]. \tag{8.2.47}$$

For elastic media $b_{\alpha_I} = 0$, so (8.2.46) shows that the volumetric strain has no superimposed sinusoidal dependence on depth. In addition for elastic media, (8.2.25) and (8.2.43) show that

$$\vartheta_\Delta(t) = \vartheta_B(t) - \pi/2 \tag{8.2.48}$$

and

$$S(x_3) = \begin{Bmatrix} 0 & \text{if } F_j + G_j > 0 \\ \pi & \text{if } F_j + G_j < 0 \end{Bmatrix} \quad \text{for } j = 1, 3, \tag{8.2.49}$$

so the phase of the volumetric strain lags the phase of the radial component of displacement by $\pi/2$ down to a depth at which $F_1 + G_1 = 0$. The phase of the volumetric strain lags the phase of the vertical component of displacement by π down to a depth at which $F_3 + G_3 = 0$.

8.2.5 Media with Equal Complex Lamé Parameters ($\Lambda = M$)

The assumption of equal complex Lamé parameters provides considerable simplification in some of the formulae describing a Rayleigh-Type surface wave. The simplification provides additional insight into characteristics of the surface wave on anelastic media versus those of the wave on elastic media.

The assumption of equal complex Lamé parameters implies a number of identities between the various material parameters and the parameters for P and S waves as stated in (3.8.1) through (3.8.14). The identity $\beta^2/\alpha^2 = 1/3$, (3.8.5), when substituted into (8.1.20) shows that coefficients of the equation for a Rayleigh-Type surface wave are real. In fact, the resulting equation is the same as that for elastic media with $\Lambda_R = M_R$. Hence the roots of the equation are the same. Substitution of $\beta^2/\alpha^2 = 1/3$ into the general expressions for the roots of (8.1.21) in the complex field shows that the roots are given by

$$y_j = \begin{Bmatrix} 4 & \text{for } j = 1 \\ 2 + 2/\sqrt{3} & \text{for } j = 2 \\ 2 - 2/\sqrt{3} & \text{for } j = 3 \end{Bmatrix}. \tag{8.2.50}$$

These roots are the same as those derived for elastic media with equal real Lamé parameters (see for example Bullen, 1965, p. 90). Substitution of each value of y_j into (8.1.17) or application of Lemma (8.1.25) shows that $y_3 = 2 - 2/\sqrt{3} = 0.845\,299$ is the desired root of the equation for a Rayleigh-Type surface wave. Substitution of this root into (8.1.17) also shows that for solids with equal Lamé parameters

$$k^2 = \left(2 - \frac{2}{\sqrt{3}}\right)^{-1} k_S^2 = 1.1830\, k_S^2. \tag{8.2.51}$$

Hence, the magnitudes of the velocity and absorption coefficient for a Rayleigh-Type surface wave are given in terms of those for corresponding homogeneous S and P waves to four significant decimals by

$$v_B = \frac{\omega}{k_R} = 0.9194\, v_{HS} = 0.5308\, v_{HP} \tag{8.2.52}$$

and

$$a_B = -k_I = 1.0877\, a_{HS} = 1.8839\, a_{HP}, \tag{8.2.53}$$

where $a_{HS} \equiv |\vec{A}_{HS}|$ and $a_{HP} \equiv |\vec{A}_{HP}|$. These expressions show that the speed and attenuation of a Rayleigh-Type surface wave along the surface of the given half space are less and greater, respectively, than those for corresponding homogeneous S and P waves.

Recalling $\sqrt{}$ is interpreted herein to indicate the principal value of the square root (8.2.51) implies

$$b_\alpha = 0.9218\, k_S \tag{8.2.54}$$

and

$$b_\beta = 0.4278\, k_S. \tag{8.2.55}$$

Hence, the expressions for the propagation and attenuation vectors for the component P and SI solutions of a Rayleigh-Type surface wave may be expressed in terms of those for a corresponding homogeneous S wave to four significant decimals as

$$\vec{P}_{\phi_2} = 1.0877\, k_{S_R} \hat{x}_1 + 0.9218\, k_{S_I} \hat{x}_3, \tag{8.2.56}$$

$$\vec{P}_{\psi_2} = 1.0877\, k_{S_R} \hat{x}_1 + 0.4278\, k_{S_I} \hat{x}_3 \tag{8.2.57}$$

and

$$\vec{A}_{\phi_2} = -1.0877\, k_{S_I} \hat{x}_1 + 0.9218\, k_{S_R} \hat{x}_3, \tag{8.2.58}$$

$$\vec{A}_{\psi_2} = -1.0877\, k_{S_I} \hat{x}_1 + 0.4278\, k_{S_R} \hat{x}_3, \tag{8.2.59}$$

where $k_{S_I} \leq 0$ and $k_{S_R} \geq 0$ implies the propagation vectors are inclined toward the surface and the attenuation vectors are inclined into the solid as intended.

The acute angles that the propagation and attenuation vectors for the component solutions make with respect to the free surface and the vertical, respectively are given from (8.2.7) through (8.2.10) by

$$\tan e_\phi = -b_{\alpha_I}/k_R = -0.8475\, k_{S_I}/k_{S_R}, \tag{8.2.60}$$

$$\tan e_\psi = -b_{\beta_I}/k_R = -0.3933\, k_{S_I}/k_{S_R}, \tag{8.2.61}$$

$$\tan d_\phi = -k_I/b_{\alpha_R} = -1.1800\, k_{S_I}/k_{S_R}, \tag{8.2.62}$$

and

$$\tan d_\psi = -k_I/b_{\beta_R} = -2.5425\, k_{S_I}/k_{S_R}. \tag{8.2.63}$$

These relations show that the angles the propagation and attenuation vectors for the component P and SI solutions of a Rayleigh-Type surface wave make with respect to the surface and the vertical, respectively, are related by

$$\tan e_\phi = 2.1547 \tan e_\psi = 0.7182 \tan d_\phi = 0.3333 \tan d_\psi = -0.8475\, k_{S_I}/k_{S_R}. \tag{8.2.64}$$

Identities (3.6.13), (3.6.14), (3.6.22), and (3.6.25) imply

$$\frac{-k_{S_I}}{k_{S_R}} = \frac{a_{HS} v_{HS}}{\omega} \leq 1. \tag{8.2.65}$$

Hence, for anelastic media with $k_{S_I} \neq 0$ the acute angles specified in (8.2.64) must satisfy

$$0 < e_\psi < 21.47^\circ, \qquad 0 < e_\psi < e_\phi < 40.28^\circ, \tag{8.2.66}$$

and

$$0 < d_\phi < 49.72^\circ, \qquad 0 < d_\phi < d_\psi < 68.53^\circ. \tag{8.2.67}$$

For elastic media $k_{S_I} = 0$, hence the acute angles are given by

$$e_\psi = e_\phi = d_\phi = d_\psi = 0. \tag{8.2.68}$$

With the angles between the corresponding propagation and attenuation vectors for each of the component solutions given in terms of the acute angles by

$$\gamma_\phi = \pi/2 - d_\phi + e_\phi \tag{8.2.69}$$

and

$$\gamma_\psi = \pi/2 - d_\psi + e_\psi, \tag{8.2.70}$$

the inequalities (8.2.66) and (8.2.67) for the acute angles immediately imply for anelastic media

$$\gamma_\psi < \gamma_\phi < \pi/2, \tag{8.2.71}$$

$$40.28^\circ < \gamma_\phi < 90^\circ, \tag{8.2.72}$$

$$21.47^\circ < \gamma_\psi < 90^\circ, \tag{8.2.73}$$

and for elastic media

$$\gamma_\psi = \gamma_\phi = \pi/2. \tag{8.2.74}$$

These results for anelastic media with equal Lamé parameters establish the relative orientations of the propagation and attenuation vectors for the component solutions of a Rayleigh-Type surface wave as illustrated in Figure (8.2.11). For anelastic media they reconfirm that the propagation and attenuation vectors for the component P and SI solutions cannot be parallel and perpendicular to the free surface as for elastic media. They show that the angles the propagation vectors make with respect to the free surface for the component P and SI solutions can be no larger than 40.28° and 21.47°, respectively, for any anelastic solid with equal Lamé parameters. They show the angles the attenuation vectors are inclined with respect to the vertical can be no more than 68.53° and 49.72°, respectively. They show that for such solids the degree of inhomogeneity γ_ϕ of the component P solution exceeds the degree of inhomogeneity γ_ψ of the component SI solution. For anelastic media the degree of inhomogeneity for the component P solution can be no smaller than 40.48° and that for the component SI solution no smaller than 21.47°.

The complex amplitudes of the component P and SI solutions for a Rayleigh-Type surface wave on a HILV half space are related upon substitution of (8.2.51), (8.2.54), and (8.2.55) into (8.1.14) or (8.1.15) by

$$B_2 = -i\,0.6812\,C_{22}. \tag{8.2.75}$$

Explicit expressions for the particle displacements and volumetric strain for a Rayleigh-Type surface wave on a HILV half space with equal Lamé parameters may be expressed in terms of those for a homogeneous S or a homogeneous P wave. Substituting (8.2.51) into (8.2.13) and (8.2.14) then simplification with (3.6.13), (3.6.16), and (3.6.22) yields the following expressions for components of the particle motion in terms of those for a homogeneous S wave:

$$u_{Rx_1} = 1.0877\,|B_2|\frac{\omega}{v_{HS}}\sqrt{\frac{2\chi_{HS}}{1+\chi_{HS}}}\,\exp\left[-1.0877\,a_{HS}x_1\right]$$
$$\left(\begin{array}{l} \exp\left[-0.9218\dfrac{\omega}{v_{HS}}x_3\right] \\ \sin\left[\omega t - 1.0877\dfrac{\omega}{v_{HS}}x_1 + 0.9218\,a_{HS}x_3 - \psi_S + \arg[B_2]\right] \\ -0.5773\,\exp\left[-0.4278\dfrac{\omega}{v_{HS}}x_3\right] \\ \sin\left[\omega t - 1.0877\dfrac{\omega}{v_{HS}}x_1 + 0.4278\,a_{HS}x_3 - \psi_S + \arg[B_2]\right] \end{array}\right) \tag{8.2.76}$$

$$u_{Rx_3} = 1.0877\,|B_2|\frac{\omega}{v_{HS}}\sqrt{\frac{2\chi_{HS}}{1+\chi_{HS}}}\;\exp[-1.0877\,a_{HS}x_1]$$
$$\begin{pmatrix} -0.8475\;\exp\left[0.9218\frac{\omega}{v_{HS}}x_3\right] \\ \sin\left[\omega t - 1.0877\frac{\omega}{v_{HS}}x_1 + 0.9218\,a_{HS}x_3 - \psi_S + \arg[B_2] + \frac{\pi}{2}\right] \\ +1.4679\;\exp\left[-0.4278\frac{\omega}{v_{HS}}x_3\right] \\ \sin\left[\omega t - 1.0877\frac{\omega}{v_{HS}}x_1 + 0.4278\,a_{HS}x_3 - \psi_S + \arg[B_2] + \frac{\pi}{2}\right] \end{pmatrix}, \tag{8.2.77}$$

where $a_{HS} = -k_{S_I} = \frac{\omega}{v_{HS}}\frac{Q_{HS}^{-1}}{1+\chi_{HS}}$.

Similarly, the expression for volumetric strain (8.2.41) of a Rayleigh-Type surface wave on a half space with equal Lamé parameters may be written explicitly in terms of the material parameters for a corresponding homogeneous shear wave as

$$\Delta_R(t) = 0.3333|B_2|\frac{2\omega^2}{v_{HS}^2}\frac{\chi_{HS}}{1+\chi_{HS}}\;\exp\left[-1.0877\,a_{HS}x_1\;\exp - 0.9218\frac{\omega}{v_{HS}}x_3\right]$$
$$\sin\left[\omega t - 1.0877\frac{\omega}{v_{HS}}x_1 + 0.9218\,a_{HS}x_3 - 2\psi_S + \arg[B_2] - \pi/2\right]. \tag{8.2.78}$$

At the free surface the expressions for the components of the particle motions and the volumetric strain simplify to

$$u_{Rx_1} = 0.4597\,|B_2|\frac{\omega}{v_{HS}}\sqrt{\frac{2\chi_{HS}}{1+\chi_{HS}}}\;\exp[-1.0877\;a_{HS}x_1]$$
$$\sin\left[\omega t - 1.0877\frac{\omega}{v_{HS}}x_1 - \psi_S + \arg[B_2]\right], \tag{8.2.79}$$

$$u_{Rx_3} = 0.6748\,|B_2|\frac{\omega}{v_{HS}}\sqrt{\frac{2\chi_{HS}}{1+\chi_{HS}}}\;\exp[-1.0877\;a_{HS}x_1]$$
$$\sin\left[\omega t - 1.0877\frac{\omega}{v_{HS}}x_1 - \psi_S + \arg[B_2] + \frac{\pi}{2}\right], \tag{8.2.80}$$

and

$$\Delta_R(t) = 0.3333|B_2|\frac{2\omega^2}{v_{HS}^2}\frac{\chi_{HS}}{1+\chi_{HS}} \exp[-1.0877 a_{HS} x_1]$$
$$\sin\left[\omega t - 1.0877\frac{\omega}{v_{HS}}x_1 - 2\psi_S + \arg[B_2] - \pi/2\right]. \quad (8.2.81)$$

These expressions immediately imply that at the surface of a HILV half space with equal Lamé parameters passage of a Rayleigh-Type surface wave implies

(1) particles at the free surface describe a retrograde elliptical orbit with time,
(2) the ratio of the major to minor axis of the ellipse is 1.4679,
(3) the vertical axis of the particle motion ellipse is perpendicular to the undisturbed free surface, that is, there is no tilt of the particle motion ellipse at the free surface, and
(4) the amplitude and phase of the volumetric strain are readily related to those of the radial and vertical components of displacement as indicated.

The preceding results are valid for both elastic and anelastic solids with equal Lamé parameters. For low-loss anelastic solids $\chi_{HS} \approx 1$, $\psi_S \approx Q_{HS}^{-1}/2$, and $a_{HS} \approx \omega Q_{HS}^{-1}/(2v_{HS})$. Substitution of these expressions into (8.2.76) through (8.2.81) allows the expressions for the amplitude and phase of the particle motion components and the volumetric strain to be simplified further.

For elastic media $a_{HS} = Q_{HS}^{-1} = 0$, so the components of the particle motion and the volumetric strain simplify to

$$u_{Rx_1} = |B_2| 1.0877\frac{\omega}{v_{HS}}\left(\exp\left[-0.9218\frac{\omega}{v_{HS}}x_3\right] - 0.5773\ \exp\left[-0.4278\frac{\omega}{v_{HS}}x_3\right]\right)$$
$$\sin\left[\omega t - 1.0877\frac{\omega}{v_{HS}}x_1 + \arg[B_2]\right], \quad (8.2.82)$$

$$u_{Rx_3} = |B_2| 1.0877\frac{\omega}{v_{HS}}\left(-0.8475\ \exp\left[0.9218\frac{\omega}{v_{HS}}x_3\right] + 1.4679\ \exp\left[-0.4278\frac{\omega}{v_{HS}}x_3\right]\right)$$
$$\sin\left[\omega t - 1.0877\frac{\omega}{v_{HS}}x_1 + \arg[B_2] + \frac{\pi}{2}\right], \quad (8.2.83)$$

and

$$\Delta_R(t) = 0.3333|B_2|\frac{\omega^2}{v_{HS}^2}\ \exp\left[-0.9218\frac{\omega}{v_{HS}}x_3\right]\ \sin\left[\omega t - 1.0877\frac{\omega}{v_{HS}}x_1 + \arg[B_2] - \pi/2\right]. \quad (8.2.84)$$

The expressions for the particle motion for elastic media agree with those presented by Ewing, Jardetsky, and Press (1957, p. 33).

8.3 Numerical Characteristics of Rayleigh-Type Surface Waves

Quantitative descriptions of the physical characteristics of a Rayleigh-Type surface wave are readily deduced from the analytic expressions derived in the previous section as a function of the appropriate root of the complex Rayleigh equation (8.1.20) and the material parameters of the corresponding HILV half space. The appropriate root of (8.1.20) is the one of the three roots described by (8.1.21) which Lemma (8.1.25) implies must satisfy $0 < |c^2/\beta^2| < 1$.

In choosing parameters to characterize the response of the material it is desirable to choose general parameters so that the results are applicable to any viscoelastic model once the particular frequency dependence of their moduli is specified. Towards this end consideration of (8.1.20) shows that the material response is characterized in the equation by β^2/α^2, which may be written from (3.5.8) through (3.5.11) as

$$\frac{\beta^2}{\alpha^2} = \left(\frac{v_{HS}^2}{v_{HP}^2}\right) \frac{1 + \sqrt{1 + Q_{HS}^{-2}}}{1 + \sqrt{1 + Q_{HP}^{-2}}} \frac{1 - iQ_{HP}^{-1}}{1 - iQ_{HS}^{-1}} \tag{8.3.1}$$

Hence, the material response could be characterized in terms of v_{HS}^2/v_{HP}^2, Q_{HP}^{-1}, and Q_{HS}^{-1}. Alternatively, defining a parameter corresponding to Poisson's ratio for elastic media, namely,

$$\sigma \equiv \frac{3K_R - 2M_R}{2(3K_R + M_R)} \tag{8.3.2}$$

and loss in shear and bulk as $Q_{HS}^{-1} \equiv M_I/M_R$ and $Q_K^{-1} \equiv K_I/K_R$ allows β^2/α^2 to be written as

$$\frac{\beta^2}{\alpha^2} = \frac{1 + iQ_{HS}^{-1}}{\frac{2}{3}\left(\frac{1+\sigma}{1-2\sigma}\right)(1 + iQ_K^{-1}) + \frac{4}{3}(1 + iQ_{HS}^{-1})}. \tag{8.3.3}$$

This equation suggests computation of the roots and hence the physical characteristics of a Rayleigh-Type surface wave as a function of the material parameters σ, Q_{HS}^{-1}, and Q_K^{-1}. An additional useful relation involving the elastic Poisson's ratio in terms of the elastic P and S velocities as specified by (3.5.6) and (3.5.7) is

$$\sigma = \left((v_{HPe}/v_{HSe})^2 - 2\right) \Big/ \left(2(v_{HPe}/v_{HSe})^2 - 1\right). \tag{8.3.4}$$

The Q^{-1} in bulk is related to that for homogeneous S and P waves in terms of σ by

$$Q_K^{-1} = \frac{1-\sigma}{1+\sigma}\left(3Q_{HP}^{-1} - 2\left(\frac{1-2\sigma}{1-\sigma}\right)Q_{HS}^{-1}\right). \tag{8.3.5}$$

For purposes of theoretical considerations there is no a priori reason to assume behavior in bulk is related to behavior in shear. With parameters K_R, K_I, M_R, and M_I defined on the interval $[0, \infty)$, the elastic Poisson's ratio is defined on the interval $[-1, 0.5)$. Negative values of Poisson's ratio are not excluded by material-stability considerations as pointed out by Love (1944, p. 104), however, Poisson's ratio for most materials is non-negative.

Characteristics of the displacement field, particle motion, and volumetric strain associated with a Rayleigh-Type surface wave as described by (8.2.13) through (8.2.49) can be computed efficiently in terms of fractions of a wavelength along the free surface defined by

$$\lambda \equiv \frac{2\pi}{k_R}. \tag{8.3.6.}$$

To derive these expressions a basic term needed for computation of the quantities defined in (8.2.16) is k/k_R. This term is specified in terms of a_B/a_{HS} and v_B/v_{HS} by

$$\frac{k}{k_R} = k\frac{\lambda}{2\pi} = 1 - i\frac{a_B}{a_{HS}}\frac{v_B}{v_{HS}}\sqrt{\frac{\chi_{HS}-1}{\chi_{HS}+1}}, \tag{8.3.7}$$

which in turn can be expressed in terms of the appropriate root of the complex Rayleigh equation by (8.2.3) and (8.2.4). The desired characteristics of a Rayleigh-Type surface wave (8.2.16) expressed in terms of (8.3.7) and fractions of a wavelength λ are

$$F_1 \equiv \exp\left[-\frac{b_{\alpha_R}}{k_R}2\pi\frac{x_3}{\lambda}\right], \qquad F_3 \equiv -\frac{|b_\alpha/k_R|}{|k/k_R|}\exp\left[-\frac{b_{\alpha_R}}{k_R}2\pi\frac{x_3}{\lambda}\right] = -\frac{|b_\alpha/k_R|}{|k/k_R|}F_1, \tag{8.3.8}$$

$$\begin{aligned} G_1 &\equiv -\frac{\left|(k^2+b_\beta^2)/k_R^2\right|}{2\left|k^2/k_R^2\right|}\exp\left[-\frac{b_{\beta_R}}{k_R}2\pi\frac{x_3}{\lambda}\right], \\ G_3 &\equiv \frac{\left|(k^2+b_\beta^2)/k_R^2\right|}{2\left|\frac{k}{k_R}\frac{b_\beta}{k_R}\right|}\exp\left[-\frac{b_{\beta_R}}{k_R}2\pi\frac{x_3}{\lambda}\right] = -\frac{|k/k_R|}{|b_\beta/k_R|}G_1, \end{aligned} \tag{8.3.9}$$

$$f_1 \equiv -k_R x_1 + \arg[kB_2], \qquad f_3 \equiv f_1 + \frac{\pi}{2}, \tag{8.3.10}$$

$$g_{11} \equiv -\frac{b_{\alpha_I}}{k_R}2\pi\frac{x_3}{\lambda}, \qquad g_{13} \equiv -\frac{b_{\alpha_I}}{k_R}2\pi\frac{x_3}{\lambda} + \arg\left[\frac{b_\alpha/k_R}{k/k_R}\right], \tag{8.3.11}$$

$$g_{31} \equiv -\frac{b_{\beta_I}}{k_R} 2\pi \frac{x_3}{\lambda} + \arg\left[\frac{(k^2 + b_\beta^2)/k_R^2}{k^2/k_R^2}\right], \qquad g_{33} \equiv -\frac{b_{\beta_I}}{k_R} 2\pi \frac{x_3}{\lambda} + \arg\left[\frac{(k^2 + b_\beta^2)/k_R^2}{\frac{k}{k_R}\frac{b_\beta}{k_R}}\right], \tag{8.3.12}$$

where

$$\frac{b_\beta}{k_R} = \frac{k}{k_R}\sqrt{1 - \frac{c^2}{\beta^2}} \tag{8.3.13}$$

and

$$\frac{b_\alpha}{k_R} = \frac{k}{k_R}\sqrt{1 - \frac{c^2}{\alpha^2}}. \tag{8.3.14}$$

Setting $x_3 = 0$ in these expressions allows the properties of the waves to be calculated at the surface.

8.3.1 Characteristics at the Free Surface

Ratios of the wave speeds and absorption coefficient for a Rayleigh-Type surface wave on a HILV half space to the corresponding quantity for a homogeneous S wave in the same material are readily calculated using (8.2.3), (8.2.4), and the root c^2/β^2 of the complex Rayleigh equation (8.1.20), which satisfies $0 < |c^2/\beta^2| < 1$. Families of curves for wave speed and absorption coefficient computed over the complete range of theoretically possible values for Poisson's ratio are shown in Figure (8.3.16). The curves are computed for materials with no loss in bulk, $Q_K^{-1} = 0$ and for materials with significant loss in bulk $Q_K^{-1} = 1$.

For elastic and anelastic solids such that $Q_K^{-1} = Q_{HS}^{-1}$, (8.3.5) implies $Q_K^{-1} = Q_{HS}^{-1} = Q_{HP}^{-1}$, from which Theorem (8.1.26) implies the appropriate root of the complex Rayleigh equation is real and hence depends only on σ, but not on the amount of absorption in the material. Consequently, (8.2.3) and (8.2.4) imply the wave-speed ratio v_B/v_{HS} and the absorption coefficient ratio a_B/a_{HS} do not depend on the amount of absorption. As a result the curves in Figure (8.3.16) corresponding to $Q_K^{-1} = Q_{HS}^{-1} = Q_{HP}^{-1}$ are valid for both elastic solids and anelastic solids. Curves for solids with small amounts of intrinsic absorption, say $Q_{HS}^{-1} < 0.1$ for $Q_K^{-1} = 0$ cannot be distinguished at the scale plotted from those for solids with $Q_K^{-1} = Q_{HS}^{-1}$.

Tilt, η_β, of the major axis of the particle motion ellipse for a Rayleigh-Type surface wave is specified by (8.2.27). Corresponding families of curves computed as a function of σ for various amounts of absorption in shear and bulk are shown in Figure (8.3.17). The plots illustrate Corollary (8.2.39), which indicates that there is no tilt of the particle motion ellipse at the free surface if $Q_K^{-1} = Q_{HS}^{-1} = Q_{HP}^{-1}$. They

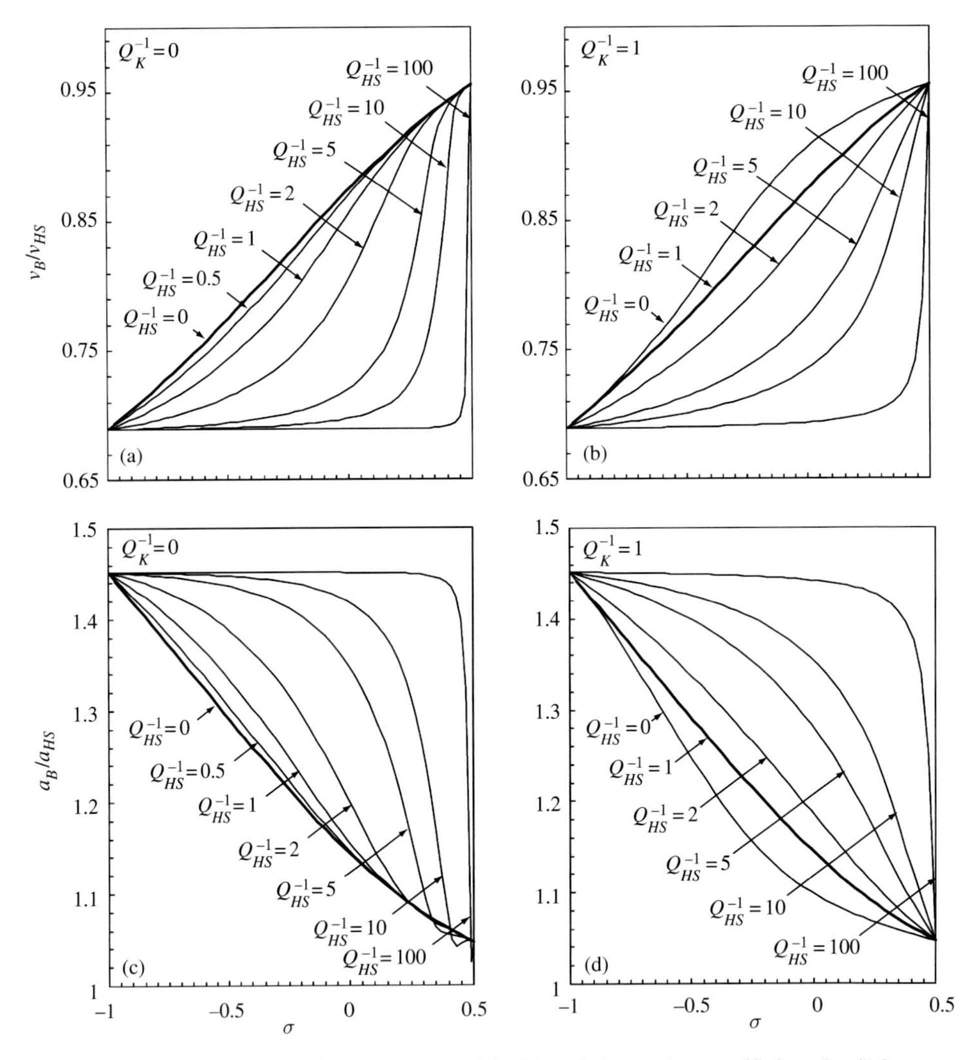

Figure (8.3.16). Ratio of the wave speed (a, b) and absorption coefficient (c, d) for a Rayleigh-Type surface wave to that for a homogeneous S wave as a function of σ and indicated values of Q_{HS}^{-1} and Q_K^{-1}.

also show that there is no tilt of the particle motion ellipse for values of $\sigma = -1$ and $\sigma = 0.5$ regardless of the amounts of absorption in bulk and shear. They indicate that the amount of tilt increases in the counter-clockwise direction as the amount of loss in shear increases relative to that in bulk. They show that the tilt increases in the clockwise direction as the amount of loss in bulk increases with respect to that in shear. They indicate for materials with small amounts of absorption, i.e. $Q^{-1} < 0.1$, that the magnitude of the tilt for materials with non-negative elastic Poisson ratio is small and generally less than 2°.

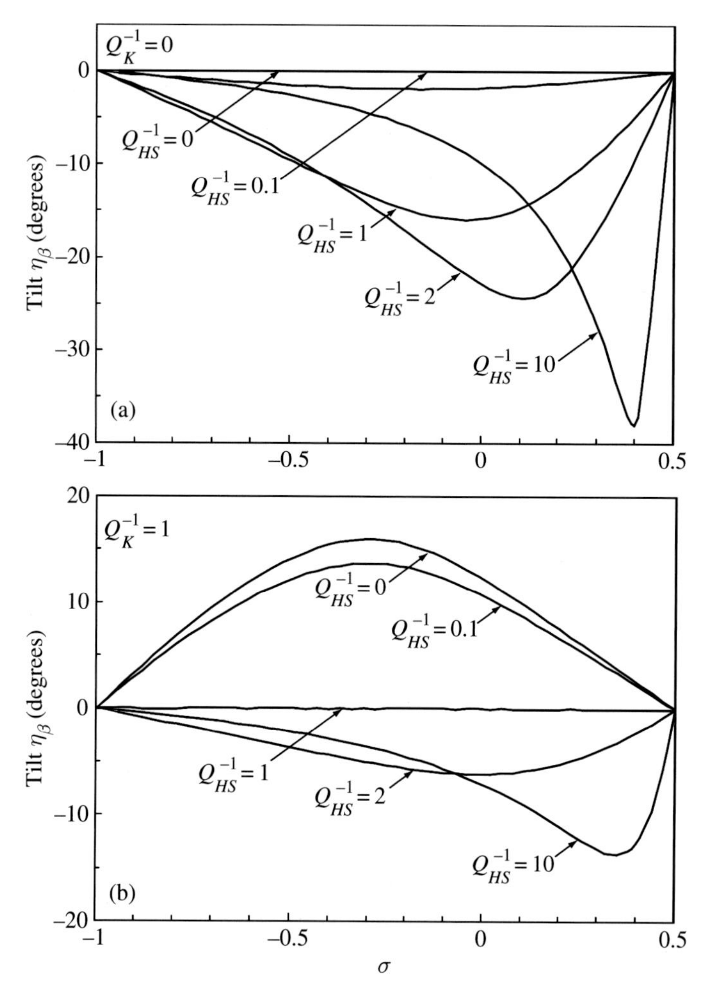

Figure (8.3.17). Tilt of the major axis of the particle motion ellipse at the free surface for a Rayleigh-Type surface wave as a function of σ for materials with the indicated amounts of absorption in shear and bulk.

The ratio of the maximum horizontal (radial) displacement, $H_1(0)$, to the maximum vertical displacement $H_3(0)$ as might be inferred from radial and vertical measurements at the surface is specified by (8.2.18). Families of curves computed as a function of the material parameters are shown in Figure (8.3.19). They indicate that the ratio of maximum horizontal to vertical amplitude generally decreases with increasing σ with the exception of cases in which the amount of absorption in shear is much larger than that in bulk. They indicate that for solids such that $Q_K^{-1} = Q_{HS}^{-1} = Q_{HP}^{-1}$ the ratio is independent of the amount of absorption and depends only on σ. This result is also implied by definitions (8.2.18), Theorem (8.1.26) and (8.3.5) from which it follows that

$$\frac{H_1(0)}{H_3(0)} = \frac{|F_1 + G_1|}{|F_3 + G_3|} = \frac{1}{2}\left|\frac{c^2}{\beta^2}\right| \Bigg/ \left| \frac{2 - \frac{c^2}{\beta^2}}{2\sqrt{1 - \frac{c^2}{\beta^2}}} - \sqrt{1 - \frac{c^2}{\alpha^2}} \right| \tag{8.3.18}$$

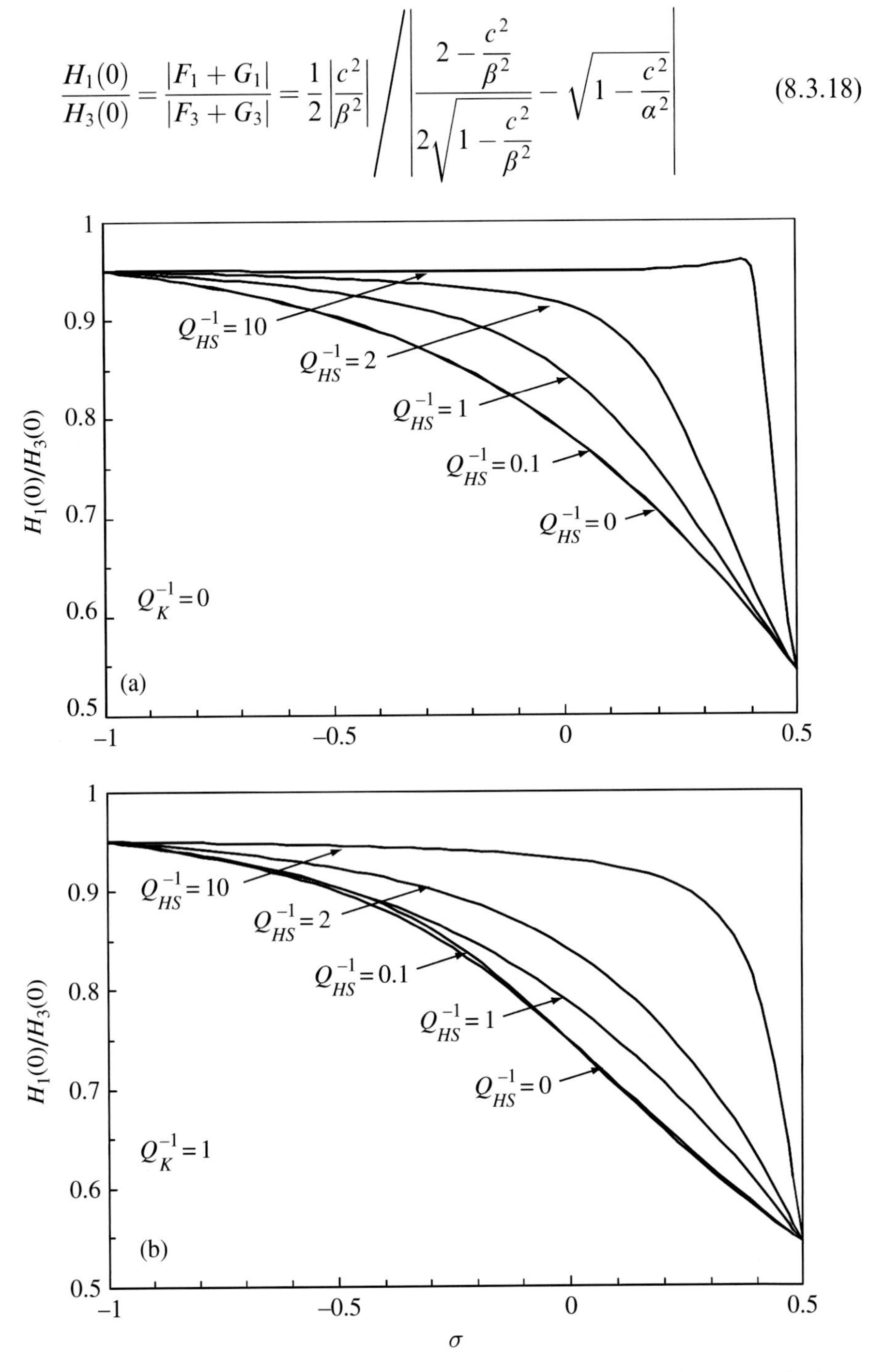

Figure (8.3.19). Ratio of the maximum horizontal to the maximum vertical displacement of the particle motion ellipse at the free surface for a Rayleigh-Type surface wave as a function of σ for materials with the indicated amounts of absorption in shear and bulk.

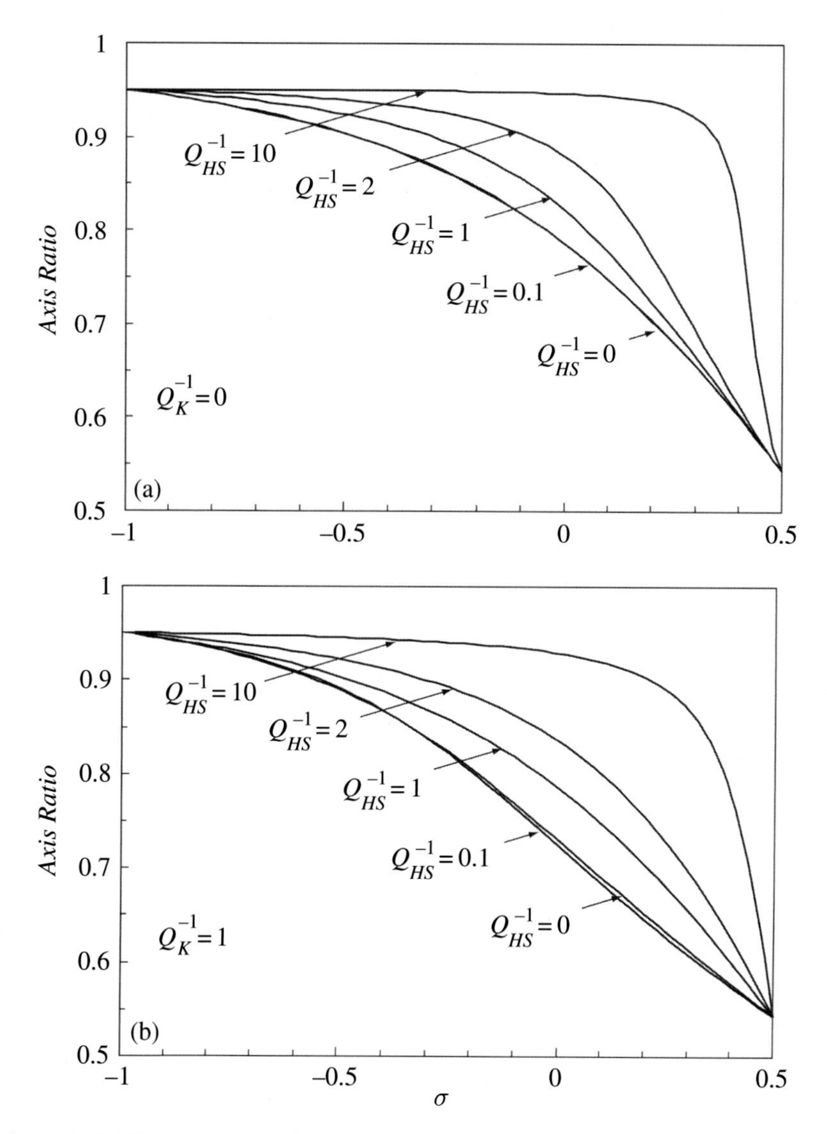

Figure (8.3.20). *Axis Ratio* for a Rayleigh-Type surface wave as a function of σ for materials with the indicated amounts of absorption in shear and bulk. Positive values of the ratio indicate that the elliptical particle motion orbit is retrograde.

and that the ratio depends only on σ for this special class of viscoelastic solids. The curves also show that for solids with no loss in bulk and low loss in shear deviations of the ratios from that for elastic solids are indistinguishable at the scale plotted.

Families of curves for the *Axis Ratio* at the free surface are shown as a function of the material parameters in Figure (8.3.20). The curves indicate that for the material

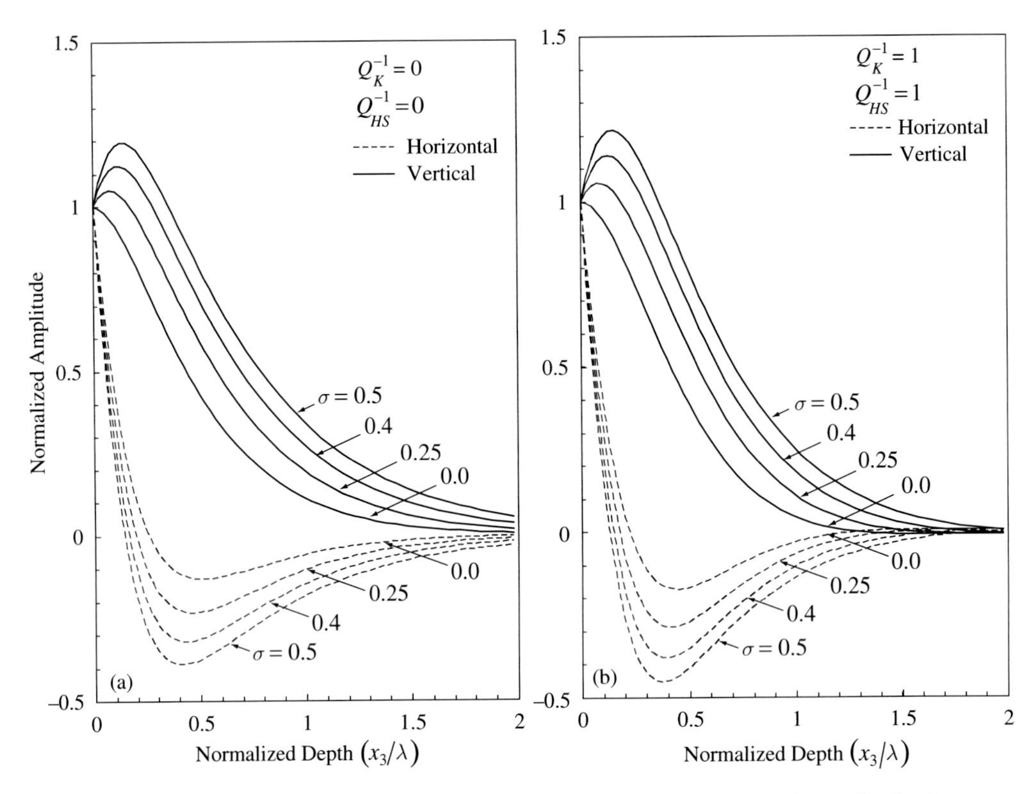

Figure (8.3.21). Normalized maximum horizontal and vertical amplitude for a Rayleigh-Type surface wave as a function of normalized depth for a range of non-negative values of σ and for two fixed equal amounts of absorption in shear and bulk.

parameters considered, the particle motion for a Rayleigh-Type surface wave at the free surface is retrograde. As expected for media with $Q_K^{-1} = Q_{HS}^{-1} = Q_{HP}^{-1}$ from Corollary (8.2.39), the curves are the same as the curves for the ratio of the maximum horizontal to vertical displacement (Figure (8.3.19). Significant differences in the two sets of curves are apparent only for materials with large unequal amounts of absorption in bulk and shear.

8.3.2 Characteristics Versus Depth

Physical properties of a Rayleigh-Type surface wave specified as a function of depth in terms of fraction of a wavelength are specified by (8.3.7) through (8.3.14). Families of curves, computed as a function of depth for fixed non-negative values of σ, Q_K^{-1} and Q_{HS}^{-1}, provide useful insight into variations in the physical characteristics with distance from the free surface.

Normalized maximum horizontal and vertical amplitude is specified as a function of depth by (8.2.22). Corresponding families of curves computed as a function of

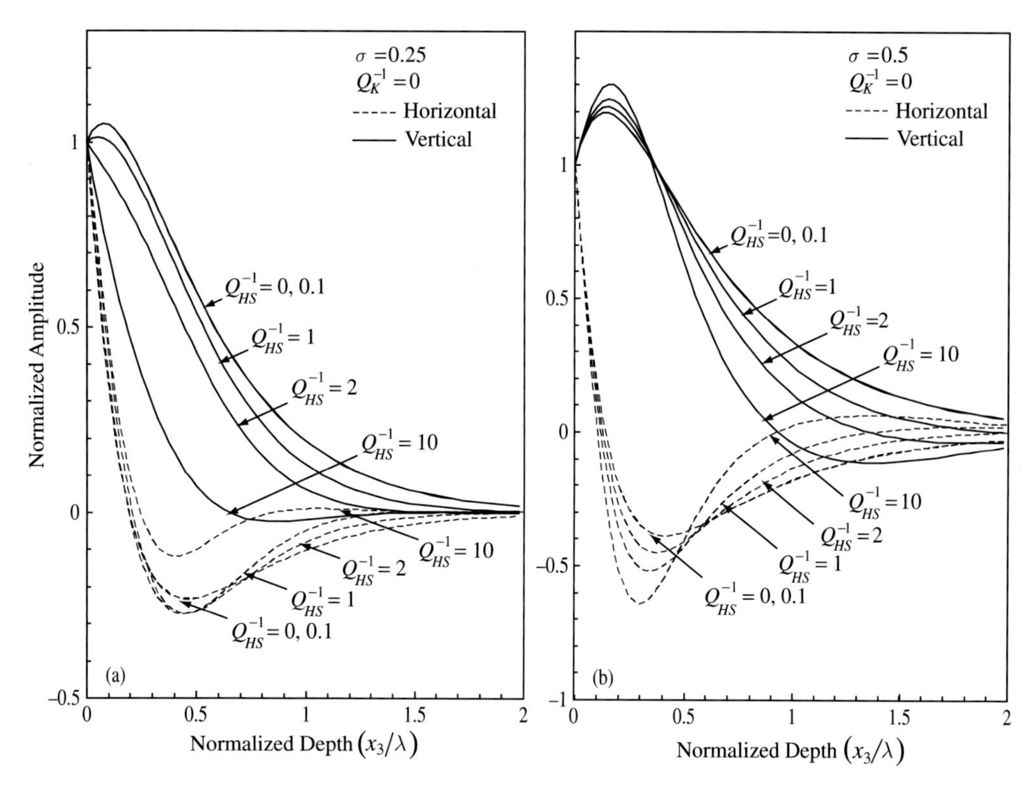

Figure (8.3.22). Normalized maximum horizontal and vertical amplitude for a Rayleigh-Type surface wave as a function of depth for media with no absorption in bulk ($Q_K^{-1} = 0$), various amounts of absorption in shear ($Q_{HS}^{-1} = 0,\ 0.1,\ 1,\ 2,\ 10$), $\sigma = 0.25$, and $\sigma = 0.5$.

normalized depth and material parameters are shown in Figures (8.3.21) through (8.3.23). The amplitude distribution curves show

(1) the majority of both the vertical and horizontal amplitude for a Rayleigh-Type surface wave is concentrated within one wavelength of the free surface,
(2) for fixed values of σ increasing the amounts of absorption tends to concentrate the normalized horizontal and vertical disturbances for a Rayleigh-Type surface wave toward the surface,
(3) the maximum vertical amplitude displacement increases to a maximum at a depth between 5 and 20 percent of a wavelength,
(4) the maximum horizontal displacements decrease rapidly away from the free surface assuming a zero value at depths between 10 and 20 percent of a wavelength, then assume minimum negative values ranging from about 30 to 50 percent of a wavelength,

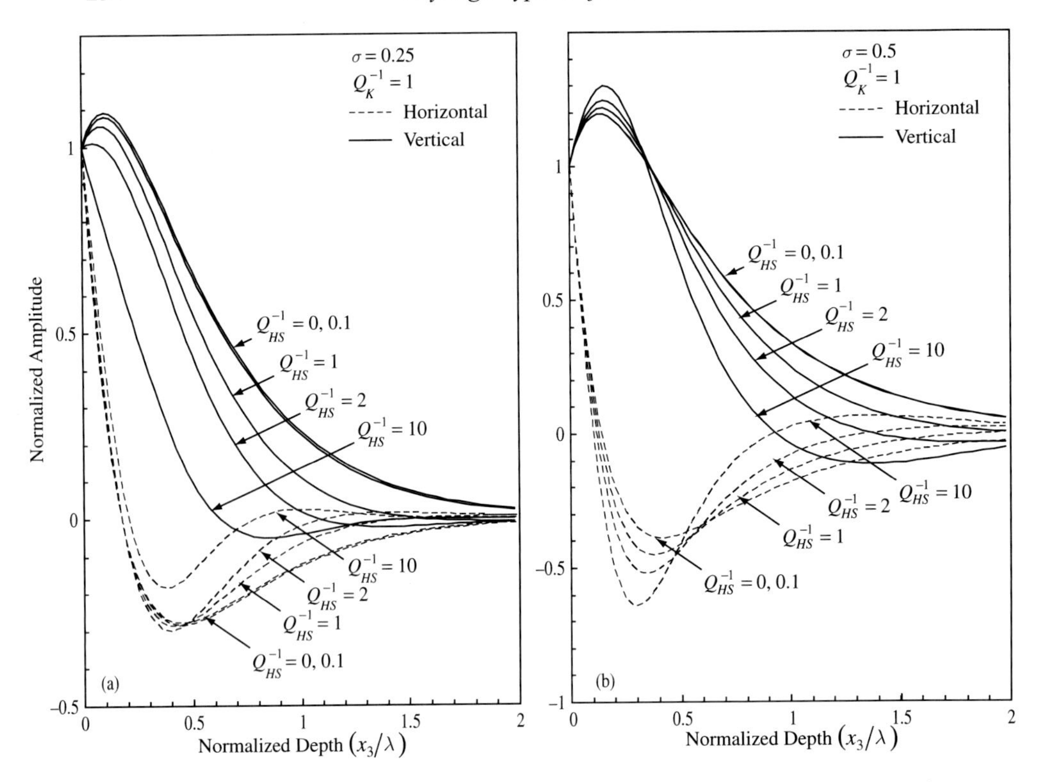

Figure (8.3.23). Normalized maximum horizontal and vertical amplitude for a Rayleigh-Type surface wave as a function of depth for media with a moderate amount of absorption in bulk ($Q_K^{-1} = 1$), various amounts of absorption in shear ($Q_{HS}^{-1} = 0,\ 0.1,\ 1,\ 2,\ 10$), $\sigma = 0.25$ and $\sigma = 0.5$.

(5) for fixed values of σ and large amounts of absorption ($Q_{HS}^{-1} > 1$), the horizontal amplitude distribution oscillates twice through zero within the distance of one wavelength of the free surface illustrating that the amplitude distributions for anelastic media exhibit a superimposed sinusoidal dependence on depth,

(6) the difference between corresponding amplitude distributions for low-loss anelastic media (i.e. $Q_{HS}^{-1} \ll 1;\ Q_K^{-1} \ll 1$) and those for elastic media increase with depth to a value less than 2 percent at a depth of two wavelengths,

(7) for elastic solids, the depth at which the normalized horizontal amplitude becomes negative is the depth at which the orbit of the particle motion changes from retrograde to prograde,

(8) for anelastic solids the particle motion changes from retrograde to prograde when the length of the minor axis of the particle motion ellipse vanishes.

The *Axis Ratio* of the particle motion ellipse is specified as a function of normalized depth by (8.2.34). Corresponding families of curves computed as a function of depth for various values of material parameters are shown in Figures (8.3.24) through (8.3.26). The curves show

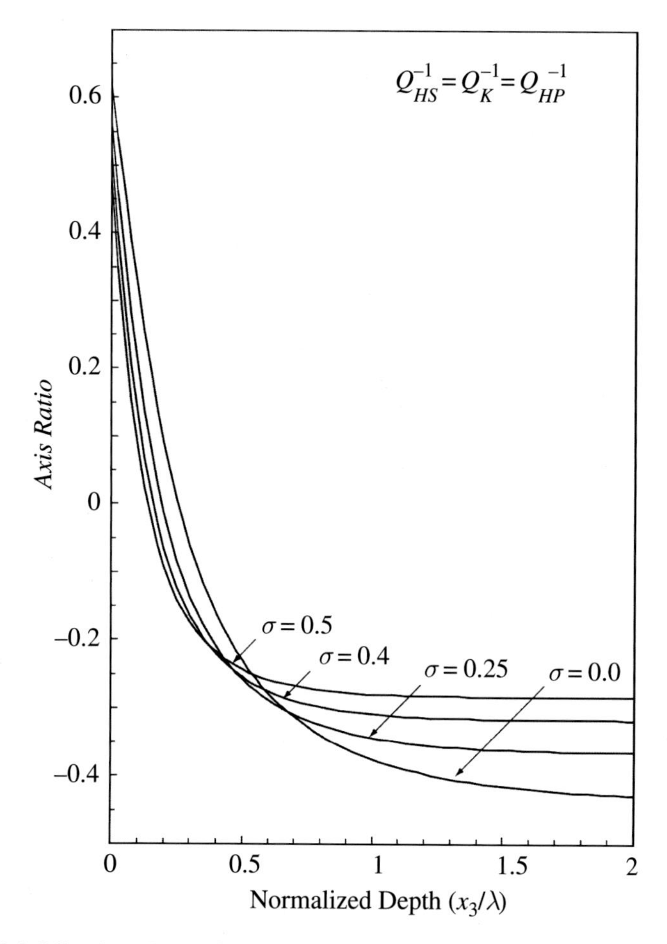

Figure (8.3.24). *Axis Ratio* for a Rayleigh-Type surface wave as a function of normalized depth for media with equal amounts of absorption in shear and bulk and values of $\sigma = 0.0,\ 0.25,\ 0.4,\ 0.5$. Positive values of the *Axis Ratio* indicate that the elliptical particle motion orbit is retrograde.

(1) the *Axis Ratio* decreases rapidly with depth below the free surface indicating that the orbit of a particle changes from retrograde at the surface to linear then prograde at depths ranging from 0.1λ to 0.4λ,

(2) for solids with equal amounts of absorption in shear and bulk $\left(Q_K^{-1} = Q_{HS}^{-1} = Q_{HP}^{-1}\right)$, which includes elastic media, the curves in Figure (8.3.24)) indicate that the *Axis Ratio* at each depth does not depend on the amount of absorption, only on σ,

(3) for low-loss solids ($Q_{HS}^{-1} \ll 1,\ Q_{HP}^{-1} \ll 1,\ Q_K^{-1} \ll 1$) the *Axis Ratio* is approximately equal to that for corresponding elastic media,

(4) for the range of parameters considered, the orbit of particles as a function of depth changes from retrograde to prograde only once in the depth interval of two wavelengths,

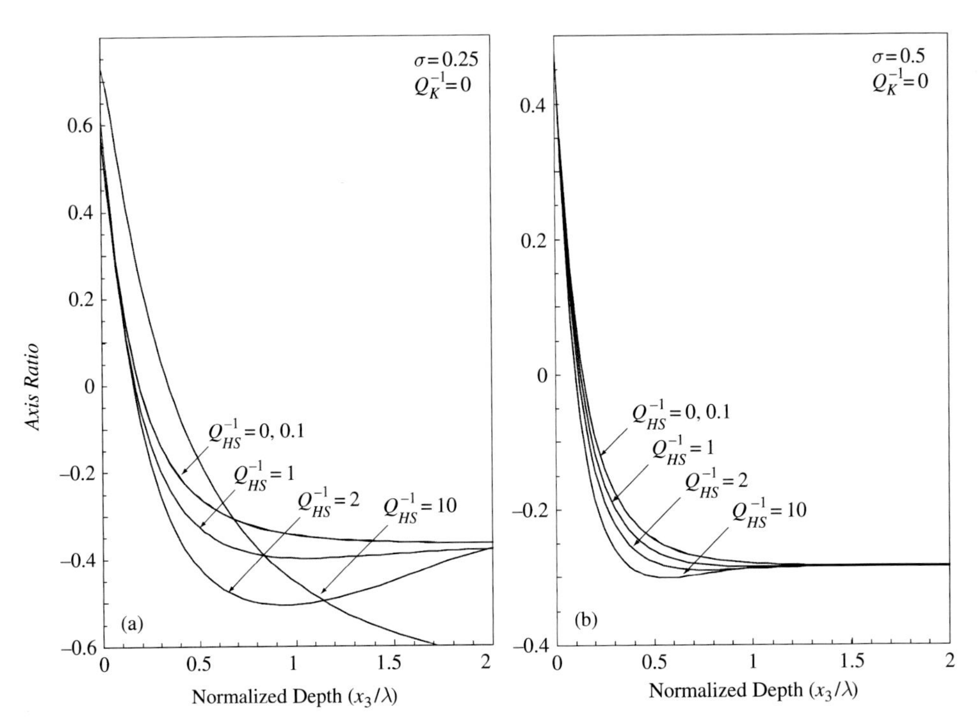

Figure (8.3.25). *Axis Ratio* as a function of normalized depth for a Rayleigh-Type surface wave in media with no absorption in bulk ($Q_K^{-1} = 0$), various amounts of absorption in shear ($Q_{HS}^{-1} = 0,\ 0.1,\ 1,\ 2,\ 10$), $\sigma = 0.25$ and $\sigma = 0.5$.

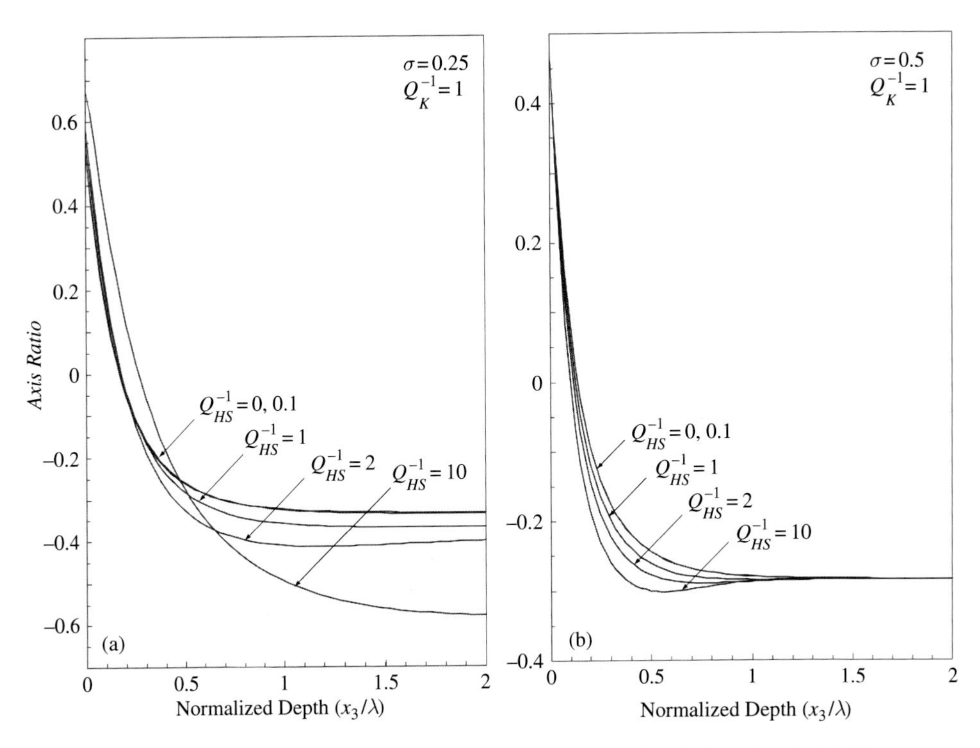

Figure (8.3.26). *Axis Ratio* as a function of normalized depth for a Rayleigh-Type surface wave in media with a moderate amount of absorption in bulk ($Q_K^{-1} = 1$), various amounts of absorption in shear ($Q_{HS}^{-1} = 0,\ 0.1,\ 1,\ 2,\ 10$), $\sigma = 0.25$, and $\sigma = 0.5$.

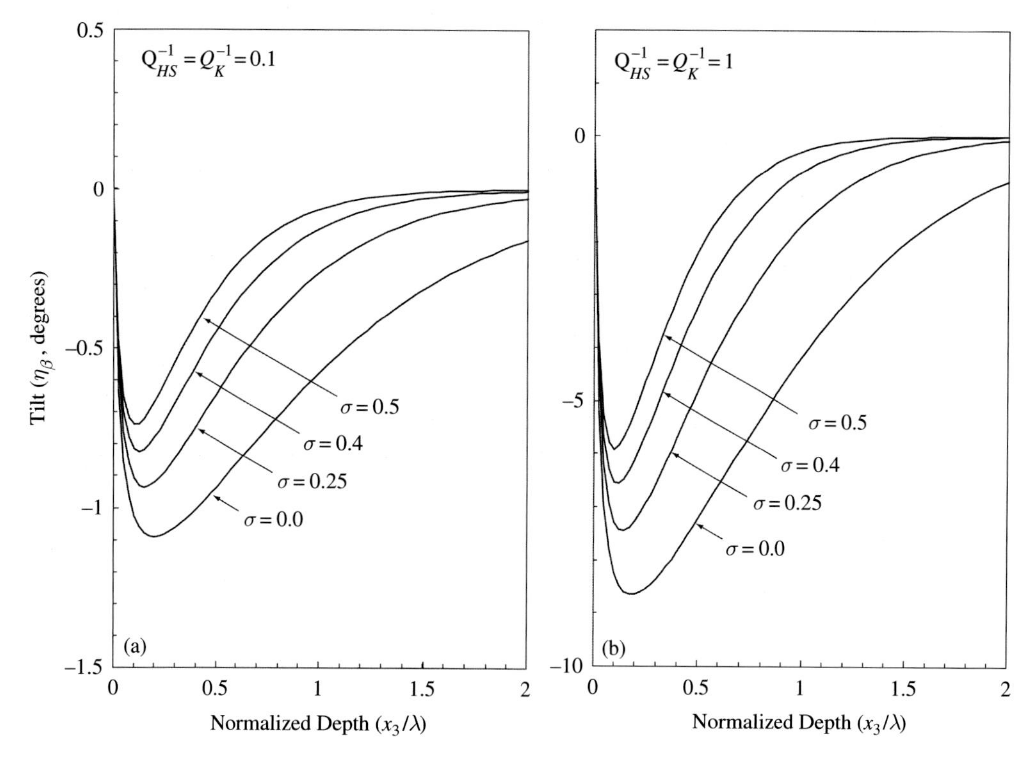

Figure (8.3.27). Tilt of the major axis of the particle motion ellipse for a Rayleigh-Type surface wave with respect to vertical versus normalized depth for media with equal small ($Q_K^{-1} = Q_{HS}^{-1} = 0.1$) and moderate ($Q_K^{-1} = Q_{HS}^{-1} = 1$) amounts of absorption in bulk and shear for the indicated values of $\sigma = 0.0,\ 0.25,\ 0.4,\ 0.5$.

(5) for solids with $\sigma = 0.5$ the *Axis Ratio* is independent of the amount of absorption in shear and bulk (compare Figures (8.3.25) and (8.3.26).

The tilt η_β of the particle motion ellipse with respect to the vertical is specified as a function of depth by (8.2.27). Corresponding families of curves computed as a function of normalized depth for various sets of material parameters are shown in Figures (8.3.27) through (8.3.29). The curves show

(1) for anelastic media with equal amounts of absorption in bulk and shear ($Q_K^{-1} = Q_{HS}^{-1} = 0.1$; $Q_K^{-1} = Q_{HS}^{-1} = 1$; see Figure (8.3.27)) the tilt of the particle motion ellipse exhibits a well-defined dependence on depth below the surface, σ, and the amount of absorption; the magnitude of the tilt rapidly increases to a maximum value near the surface at depths between about 0.05λ and 0.15λ for values of $0 \le \sigma \le 0.5$,
(2) for elastic media the tilt of the particle motion ellipse is zero at each depth regardless of the value of σ (see Figure (8.3.28)),

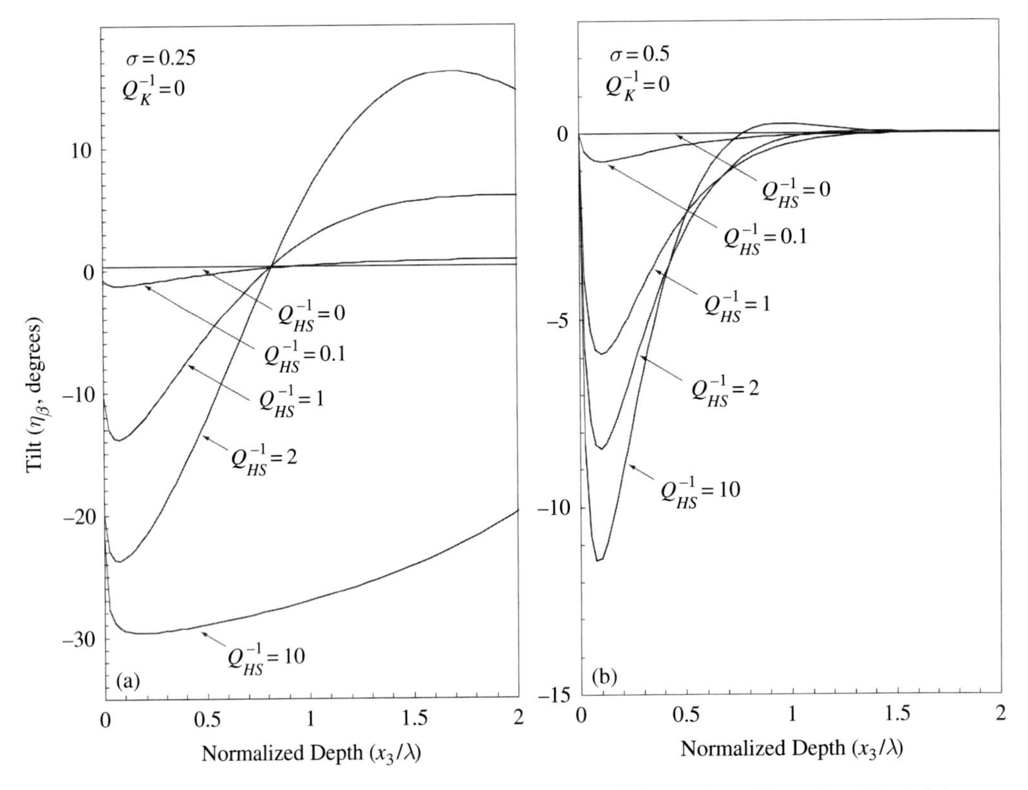

Figure (8.3.28). Tilt of the major axis of the particle motion ellipse for a Rayleigh-Type surface wave with respect to vertical versus normalized depth for materials with no absorption in bulk ($Q_K^{-1} = 0$), various amounts of absorption in shear ($Q_{HS}^{-1} = 0,\ 0.1,\ 1,\ 2,\ 10$), $\sigma = 0.25$ and $\sigma = 0.5$.

(3) for anelastic media the direction of the tilt of the major axis of the ellipse is counter clockwise with respect to the vertical if $Q_K^{-1} < Q_{HS}^{-1}$ and in the clockwise direction if $Q_{HS}^{-1} < Q_K^{-1}$, unless $\sigma = 0.5$, in which case the tilt vanishes at all depths (see Figure (8.3.29),
(4) for anelastic media in which $\sigma \neq 0.5$ and the amounts of absorption in bulk and shear are not equal, the tilt with respect to the vertical changes from counter clockwise to clockwise or vice versa often within a depth of one wavelength depending on the contrast in absorption between bulk and shear,
(5) for anelastic media with small amounts of absorption ($Q_K^{-1} \ll 1,\ Q_{HS}^{-1} \ll 1$) the tilt of the particle motion ellipse reaches its maximum of less than 1.3° at a depth near the surface of about 0.075λ.

The volumetric strain for a Rayleigh-Type surface wave for a fixed time as a function of depth as normalized by the corresponding maximum value at the surface is specified by (8.2.46).

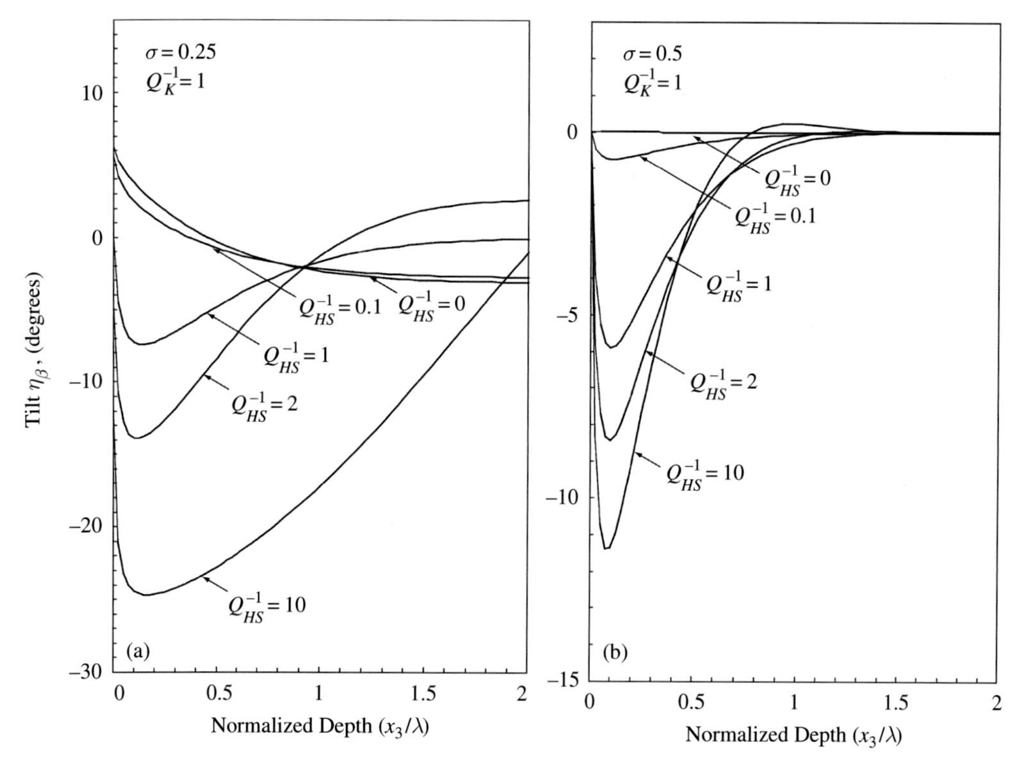

Figure (8.3.29). Tilt of the major axis of the particle motion ellipse for a Rayleigh-Type surface wave with respect to vertical versus normalized depth for materials with a significant amount of absorption in bulk ($Q_K^{-1} = 1$), various amounts of absorption in shear ($Q_{HS}^{-1} = 0,\ 0.1,\ 1,\ 2,\ 10$), $\sigma = 0.25$ and $\sigma = 0.5$

Corresponding families of curves showing normalized volumetric strain as a function of depth for various sets of material parameters are shown in Figures (8.3.30) through (8.3.32). The curves show

(1) volumetric strain associated with a Rayleigh-Type surface wave is concentrated near the free surface for both elastic and anelastic media; the maximum volumetric strain decreases by 50 percent within 0.1 wavelength (0.1λ) and by 90 percent within 0.5 wavelength (0.5λ) of the free surface,
(2) the concentration of volumetric strain towards the surface increases with increasing σ, and increasing absorption in shear (Q_{HS}^{-1}) and or absorption in bulk (Q_K^{-1}),
(3) for anelastic media with significant amounts of absorption the volumetric strain may change from dilation to compression or vice versa within the depth interval of about one-half of a wavelength (see Figures (8.3.31) and (8.3.32)); these changes are indicative of the sinusoidal dependence of volumetric strain on depth as indicated in (8.2.47).

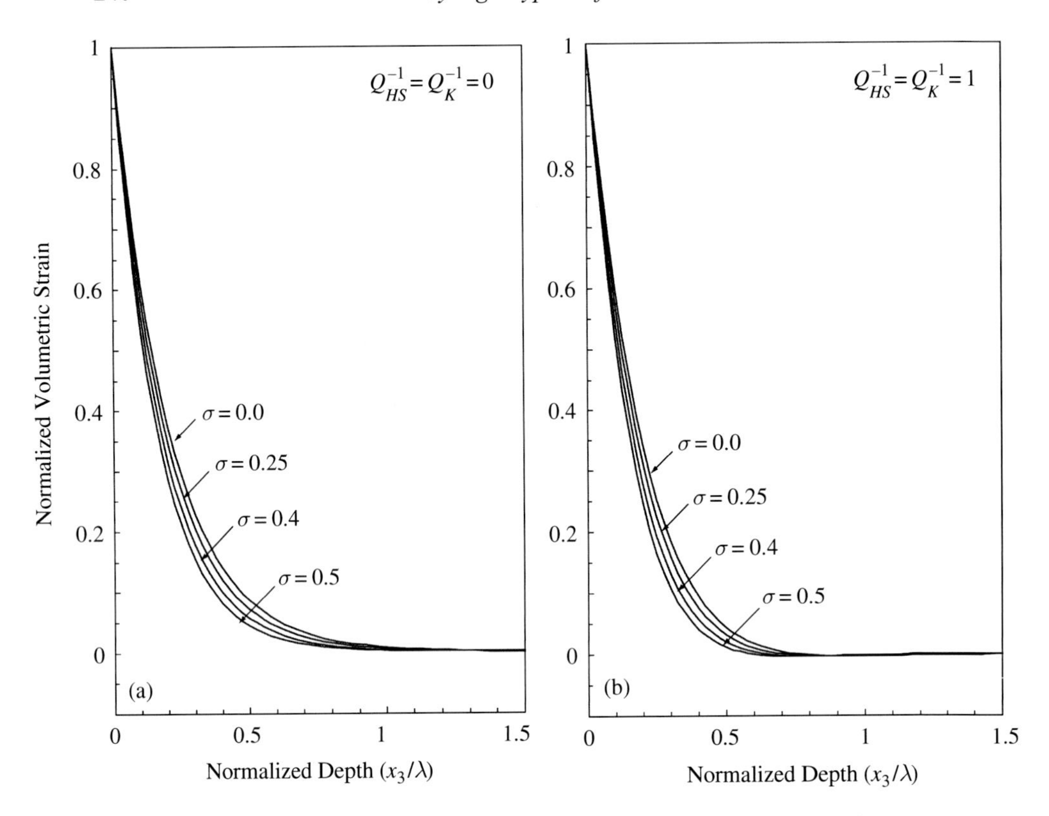

Figure (8.3.30). Normalized volumetric strain for a Rayleigh-Type surface wave as a function of normalized depth for elastic media and anelastic media with a moderate amount of equal absorption in bulk and shear ($Q_K^{-1} = Q_{HS}^{-1} = 1$) for values of $\sigma = 0.0,\ 0.25,\ 0.4,\ 0.5$.

Particle displacements at a fixed time are illustrated for a Rayleigh-Type surface wave propagating in the direction of the arrows in media with various amounts of intrinsic absorption (Figures (8.3.33) and (8.3.34)). Positions of the grid-line intersections illustrate particle positions computed from (8.2.23) and (8.2.24) as normalized by the maximum vertical amplitude at the surface and scaled for purposes of illustration by factors of 0.1 (Figures (8.3.33)a, b), 0.2 (Figure (8.3.34)a) and 0.3 (Figure (8.3.34)b).

For elastic media the particle positions (Figure (8.3.33)a) illustrate that the maximum amplitude of the surface wave does not decrease in the direction of phase propagation. For anelastic media, the particle positions (Figures (8.3.33) b and (8.3.34)a, b) illustrate that the maximum amplitude does decrease in the direction of phase propagation with the rate of decrease dependent on the amount of intrinsic material absorption. The particle positions also illustrate that particle motion amplitudes rapidly decay in amplitude with depth below

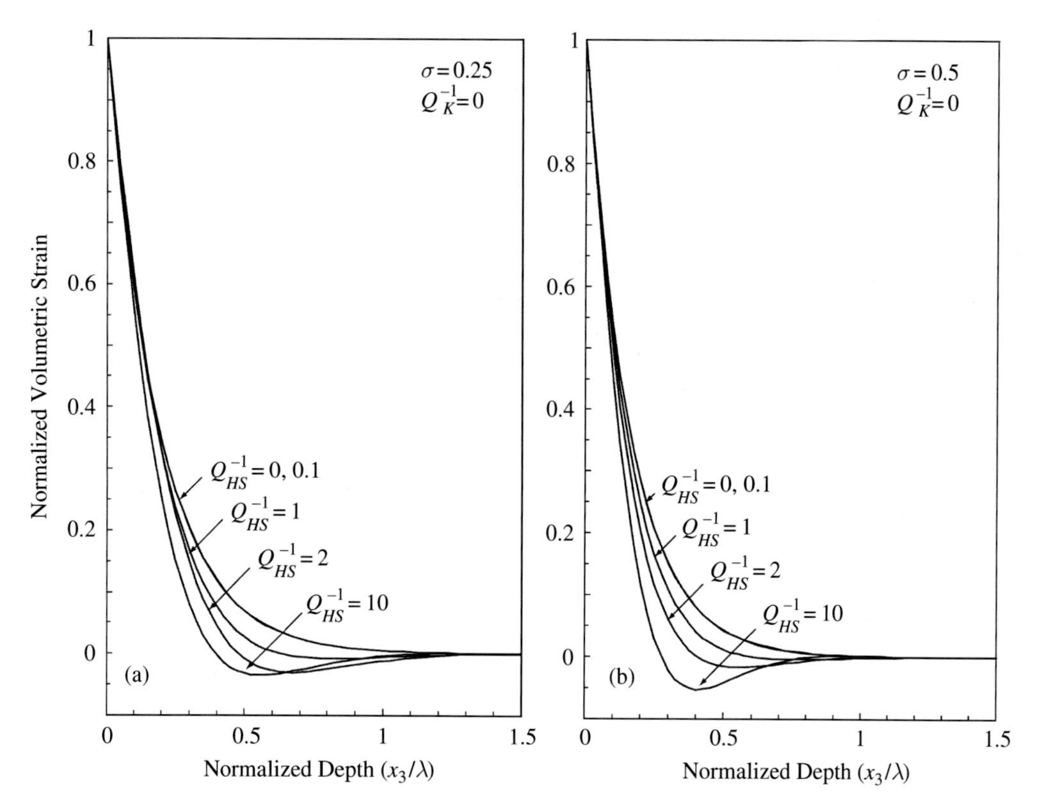

Figure (8.3.31). Normalized volumetric strain for a Rayleigh-Type surface wave as a function of normalized depth for media with no absorption in bulk ($Q_K^{-1} = 0$), various amounts of absorption in shear ($Q_{HS}^{-1} = 0,\ 0.1,\ 1,\ 2,\ 10$), $\sigma = 0.25$ and $\sigma = 0.5$.

the surface, which is consistent with quantitative results in Figures (8.3.21) through (8.3.23). Curvature of the vertical grid lines for media with large amounts of intrinsic absorption (Figures (8.3.34)a, b) is suggestive of the superimposed sinusoidal dependence of the amplitudes on depth that occurs for anelastic media, but not elastic media.

8.4 Problems

(1) Show that the propagation and attenuation vectors for the component P and Type-I S solutions of a Rayleigh-Type surface wave on a viscoelastic half space are not parallel and perpendicular to the free surface unless the half space is elastic.

(2) Compare characteristics of the particle motion for a Rayleigh-Type surface wave on a linear anelastic viscoelastic half space as stated in Theorem (8.2.26) with those of a Rayleigh wave on an elastic half space as stated in Corollary (8.2.37).

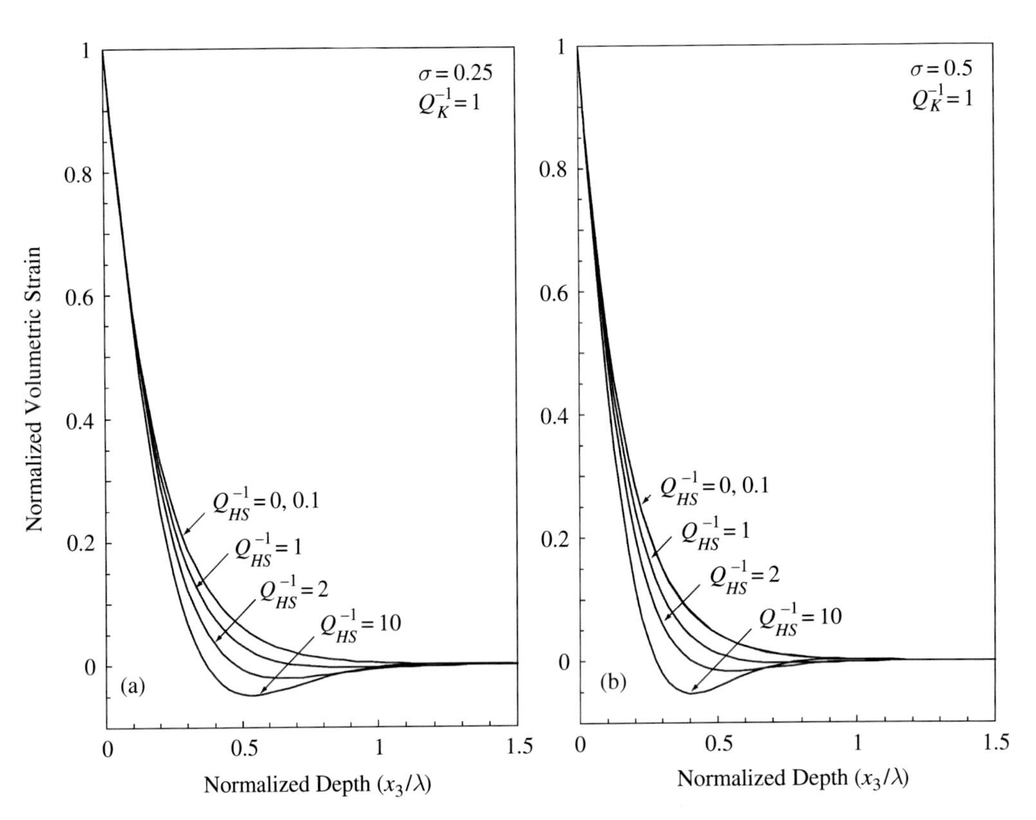

Figure (8.3.32). Normalized volumetric strain for a Rayleigh-Type surface wave as a function of normalized depth for media with a moderate amount of absorption in bulk ($Q_K^{-1} = 1$), various amounts of absorption in shear ($Q_{HS}^{-1} = 0,\ 0.1,\ 1,\ 2,\ 10$), $\sigma = 0.25$ and $\sigma = 0.5$.

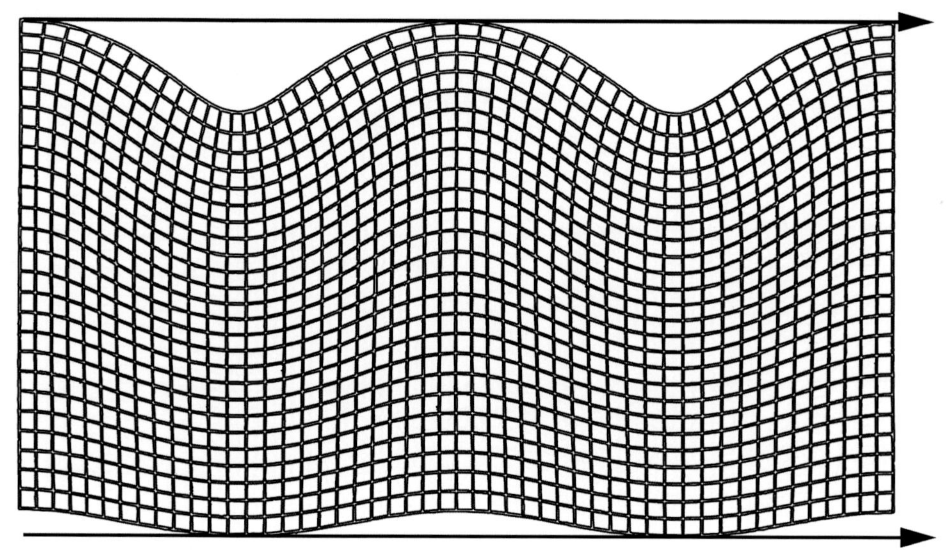

(a) Elastic media

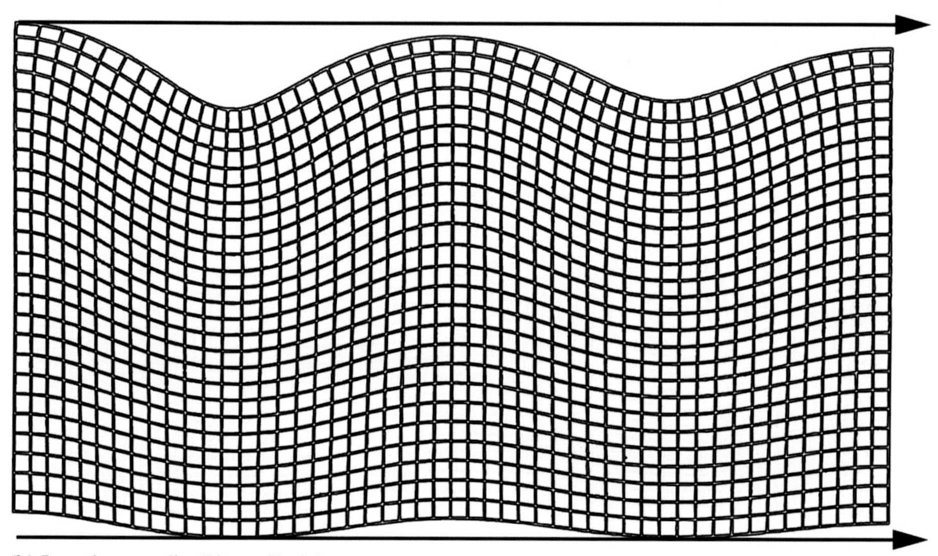

(b) Low-loss media (Pierre Shale)

Figure (8.3.33). Grid illustrating particle displacements at a fixed time for a Rayleigh-Type surface wave on a half space comprised of (a) elastic media with $\sigma = 0.25$ and (b) low-loss media (Pierre Shale, $\sigma = 0.491$, $Q_{HS}^{-1} = 0.104$, $Q_{HP}^{-1} = 0.031$, $Q_{K}^{-1} = 0.0147$; McDonal *et al.*, 1958).

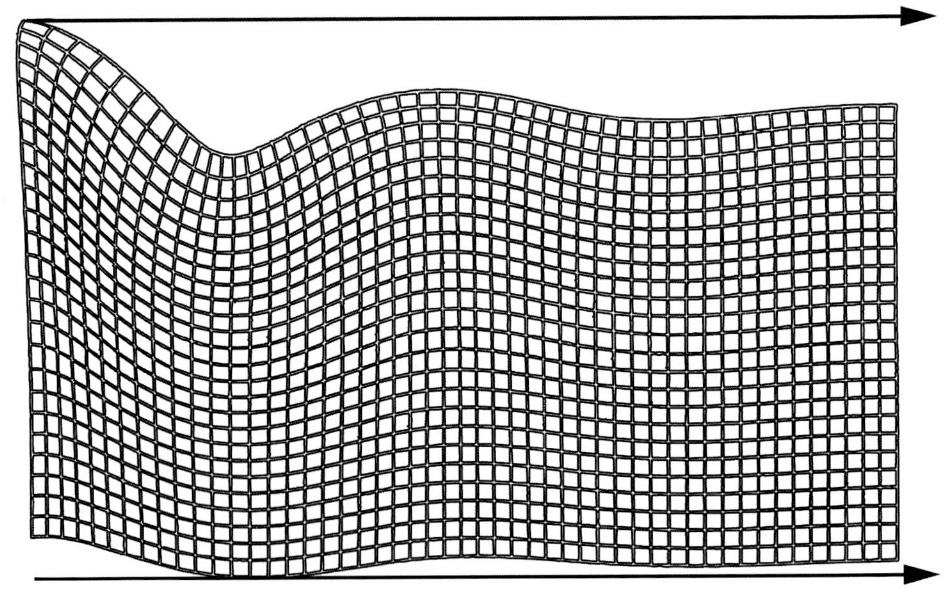

(a) Moderate-loss media (water-saturated sediments)

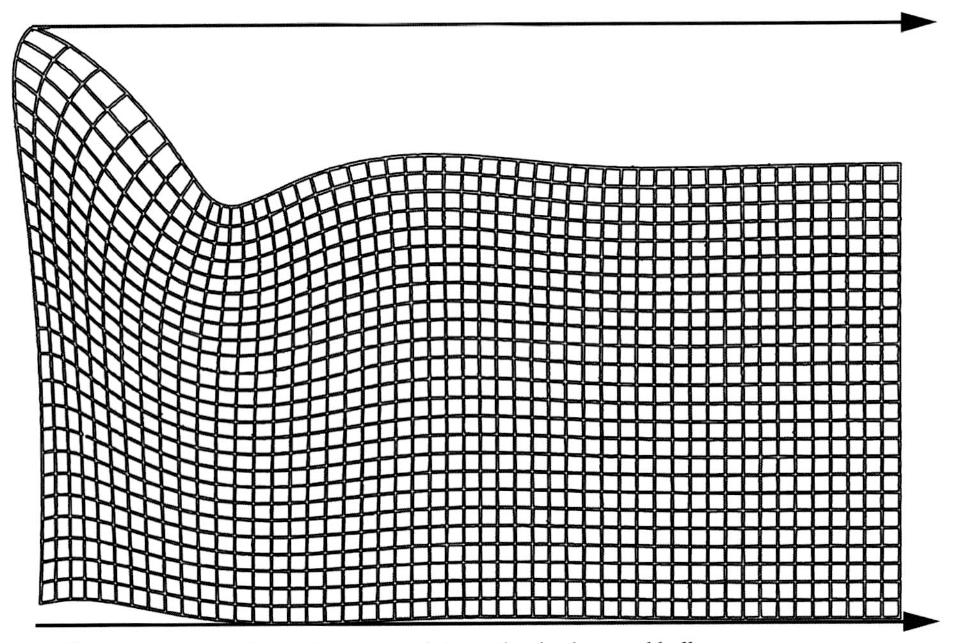

(b) High-loss media with equal amounts of absorption in shear and bulk

Figure (8.3.34). Grid illustrating particle displacements at a fixed time for a Rayleigh-Type surface wave on a half space comprised of (a) moderate-loss media (water-saturated sediments with $\sigma = 0.497$, $Q_{HS}^{-1} = 0.555$, $Q_{HP}^{-1} = 0.00441$, $Q_K^{-1} = 0$ (Hamilton *et al.*, 1970) and (b) high-loss media with $\sigma = 0.25$ and $Q_{HS}^{-1} = Q_K^{-1} = 1$.

(3) Derive the results of Theorem (8.2.26) pertaining to particle motion orbit and tilt of the particle motion ellipse as a function of depth for viscoelastic media with equal complex Lamé parameters, where $v_{HP} = \sqrt{3} v_{HS}$ and $Q_{HS}^{-1} = Q_{HP}^{-1}$.

(4) Show that the ratio of β^2/α^2 may be written in terms of the wave speeds (v_{HS}, v_{HP}) and reciprocal quality factors $(Q_{HS}^{-1}, Q_{HP}^{-1})$ for homogeneous S and P waves as

$$\frac{\beta^2}{\alpha^2} = \left(\frac{v_{HS}^2}{v_{HP}^2}\right) \frac{1 + \sqrt{1 + Q_{HS}^{-2}}}{1 + \sqrt{1 + Q_{HP}^{-2}}} \frac{1 - iQ_{HP}^{-1}}{1 - iQ_{HS}^{-1}}$$

and in terms σ and reciprocal quality factors for bulk and shear (Q_K^{-1}, Q_{HS}^{-1}) as

$$\frac{\beta^2}{\alpha^2} = \frac{1 + iQ_{HS}^{-1}}{\frac{2}{3}\left(\frac{1+\sigma}{1-2\sigma}\right)(1 + iQ_K^{-1}) + \frac{4}{3}(1 + iQ_{HS}^{-1})}.$$

(5) For a solid with $Q_K^{-1} = 1$ sketch the dependences of the normalized wave speed and absorption coefficient of a Rayleigh-Type surface wave as a function of σ and various amounts of intrinsic absorption in shear (Q_{HS}^{-1}). Note that the curves for normalized wave speed and absorption coefficient for elastic media are a special case of the curves computed for viscoelastic media with $Q_K^{-1} = Q_{HS}^{-1}$.

(6) Sketch curves for Rayleigh-Type surface waves as a function of normalized depth for solids with $Q_K^{-1} = 0$, $\sigma = 0.25$, and $Q_{HS}^{-1} = 0, 1, 10$ showing
   (a) normalized horizontal and vertical amplitude,
   (b) *Axis Ratio*,
   (c) tilt of particle motion ellipse, and
   (d) normalized volumetric strain.

   Describe variations in each characteristic with increasing amounts of intrinsic material absorption in shear.

9

General SII Waves Incident on Multiple Layers of Viscoelastic Media

The response of a stack of multiple layers of viscoelastic media to waves incident at the base of the stack is of special interest in seismology. Solutions of the problem for elastic media with incident homogeneous waves have proven useful in understanding the response of the Earth's crust and near-surface soil and rock layers to earthquake-induced ground shaking. Solutions are provided here for general (homogeneous or inhomogeneous) SII waves incident at the base of a stack of viscoelastic layers. The derivations of solutions for the problems of incident general P and SI waves are similar, but more cumbersome. The method for deducing solutions of the incident P and SI wave problems will be illustrated by those developed here. The results provided here for viscoelastic media include those derived for elastic media (Haskell, 1953, 1960). The method used here to derive the solutions for viscoelastic waves uses a matrix formulation similar to that initially used by Thompson (1950) and implemented with the correct boundary condition for elastic media by Haskell (1953).

To set up the mathematical framework for multilayered media consider a stack of $n-1$ parallel viscoelastic layers in welded contact underlain by a viscoelastic half space. Spatial reference for the layers is provided by a rectangular coordinate system designated by (x_1, x_2, x_3) or (x, y, z) as shown in Figure (4.1.3) with the plane $x_3 = z = 0$ chosen to correspond to the boundary at the free surface. The layers are indexed sequentially with the index of each layer corresponding to that of its lower boundary as indicated in Figure (9.1.1). Notation for material parameters and various wave-field parameters as introduced for single-boundary reflection–refraction problems in previous chapters is extended to the multilayer problem upon introduction of an additional subscript corresponding to that of the layer as illustrated for the parameters of the incident waves in Figure (9.1.1). This extension of the notation convention used for single-layer problems is simple to recall, because the first subscript indicates an upgoing or downgoing solution ($j = 1, 2$) and the second subscript indicates the layer ($m = 1, \ldots, n$).

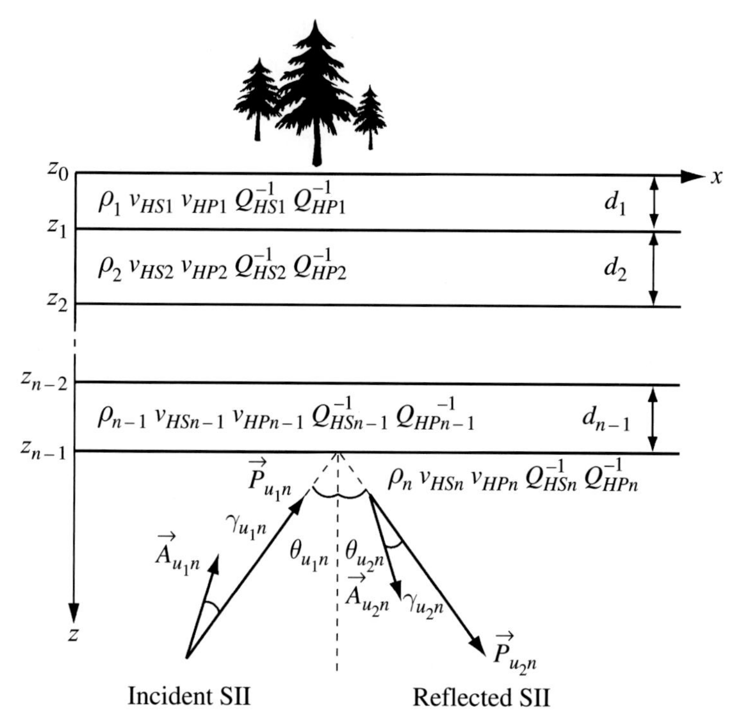

Figure (9.1.1). Diagram illustrating notation for the problem of a general Type-II S wave incident at the base of a stack of viscoelastic layers.

9.1 Analytic Solution (Multiple Layers)

The problem of a general SII wave incident at an arbitrary, but fixed, angle of incidence at the base of a stack of viscoelastic layers is formulated by assuming solutions for SII waves in each layer with directions of phase propagation toward and away from the boundary as specified by (4.2.26) with wave parameters defined by (4.2.28), (4.2.30), and (4.2.32). These solutions are distinguished for each layer by attaching the corresponding layer index. Specification of the general SII wave incident on the boundary of the viscoelastic half space is similar to that for the problem of a general SII wave on a single boundary. The incident general SII wave is specified by (4.2.26) through (4.2.32), where the layer index for the half space "n" is attached to each of the wave parameters except the complex wave number k. As for the single-boundary problem, application of the boundary conditions implies the complex wave numbers k for the solutions in each of the layers are the same.

Solution of the multilayer problem is accomplished by using the boundary conditions of continuity of displacement and stress at each of the welded boundaries and vanishing stress at the free surface to infer the displacement fields at the free surface and for the reflected wave in terms of the given amplitude, phase, angle of incidence, degree of inhomogeneity of the incident general SII wave, and a given set of material parameters.

The assumed complex steady-state displacement solution in the mth layer is, from (4.2.26), given by

$$\vec{u}_m(z) = u_m(z)\,\hat{x}_2 = \vec{u}_{1m} + \vec{u}_{2m} = \sum_{j=1}^{2} D_{jm} \exp\left[i\left(\omega t - \vec{K}_{u_j m} \cdot \vec{r}\right)\right] \hat{x}_2, \qquad (9.1.2)$$

where

$$\vec{K}_{u_j m} = \vec{P}_{u_j m} - i\vec{A}_{u_j m} = k\hat{x}_1 + (-1)^j d_{\beta m}\hat{x}_3, \qquad (9.1.3)$$

$$\vec{P}_{u_j m} = k_R\hat{x}_1 + (-1)^j d_{\beta m_R}\hat{x}_3, \qquad (9.1.4)$$

$$\vec{A}_{u_j m} = -k_I\hat{x}_1 + (-1)^{j+1} d_{\beta m_I}\hat{x}_3, \qquad (9.1.5)$$

$$d_{\beta m} = principal\ value\sqrt{k^2 - k_{Sm}^2} \qquad (9.1.6)$$

and the scalar components of the displacement solutions expressed in terms of the depth variable, z, are defined as

$$u_m(z) \equiv u_{1m}(z) + u_{2m}(z) \equiv \sum_{j=1}^{2} D_{jm} \exp\left[-\vec{A}_{u_j m} \cdot \vec{r}\right] \exp\left[i\left(\omega t - \vec{P}_{u_j m} \cdot \vec{r}\right)\right]$$

$$= \sum_{j=1}^{2} D_{jm} \exp[i(\omega t - kx - (-1)^j d_{\beta m} z)], \qquad (9.1.7)$$

where u_{1m} (z) represents the upgoing solution and u_{2m} (z) represents the downgoing solution and the physical displacement is given by the real part of the corresponding complex displacement. The complex wave number k, expressed in terms of parameters of the solutions and the material parameters of the mth layer is, from (4.2.26), given by

$$k = \frac{\omega}{v_{HSm}}\left(\sqrt{\frac{1+\chi_{Su_j m}}{1+\chi_{HSm}}}\sin\theta_{u_j m} - i\sqrt{\frac{-1+\chi_{Su_j m}}{1+\chi_{HSm}}}\sin\left[\theta_{u_j m} - \gamma_{u_j m}\right]\right), \qquad (9.1.8)$$

where $\chi_{Su_j m} \equiv \sqrt{1 + Q_{HSm}^{-2}\sec^2\gamma_{u_j m}}$ and $\chi_{HSm} \equiv \sqrt{1 + Q_{HSm}^{-2}}$ for $j = 1, 2$.

Shearing stress acting on planes parallel to the layering in the $\hat{x}_2$ direction in the mth layer may be written, from (9.1.7), using complex notation for steady state in terms of the assumed displacement solutions as

$$\begin{aligned} p_{32m}(z) = p_{zym}(z) &= M_m \frac{\partial u_m}{\partial z} \\ &= iM_m d_{\beta m} \sum_{j=1}^{2} D_{jm}(-1)^{j+1} \exp[i(\omega t - kx - (-1)^j d_{\beta m} z)] \\ &= iM_m d_{\beta m}(u_{1m}(z) - u_{2m}(z)). \end{aligned} \tag{9.1.9}$$

These expressions imply that the complex displacement and shearing stress in the mth layer at the $m-1$ boundary, $z = z_{m-1}$, are given by

$$u_m(z_{m-1}) = u_{1m}(z_{m-1}) + u_{2m}(z_{m-1}) \tag{9.1.10}$$

and

$$p_{zym}(z_{m-1}) = iM_m d_{\beta m}(u_{1m}(z_{m-1}) - u_{2m}(z_{m-1})), \tag{9.1.11}$$

and at the m boundary with $z = z_m = z_{m-1} + d_m$, they are given by

$$u_m(z_m) = \sum_{j=1}^{2} u_{jm}(z_{m-1}) \exp\left[i(-1)^{j+1} d_{\beta m} d_m\right] \tag{9.1.12}$$

and

$$p_{zym}(z_m) = -iM_m d_{\beta m} \sum_{j=1}^{2} (-1)^j u_{jm}(z_{m-1}) \exp\left[i(-1)^{j+1} d_{\beta m} d_m\right]. \tag{9.1.13}$$

For subsequent steps in the derivation it is expedient to recall that the cosine and sine functions of a complex number $c = a + ib$ are given in terms of the real and imaginary parts of the complex number by

$$\begin{aligned} \cos\, c &= \frac{1}{2}\left(e^{ic} + e^{-ic}\right) = \cos\, a\, \cosh\, b\, - i \sin\, a\, \sinh\, b, \\ \sin\, c &= \frac{1}{2i}\left(e^{ic} - e^{-ic}\right) = \sin\, a\, \cosh\, b\, + i \cos\, a\, \sinh\, b, \end{aligned} \tag{9.1.14}$$

the complex hyperbolic functions are given by

$$\begin{aligned} \cosh\, c &= \cos[ic], \\ \sinh\, c &= -i \sin[ic], \end{aligned} \tag{9.1.15}$$

and Euler's formula for an arbitrary complex argument c (Brand, 1960, p. 454) is valid as specified by

$$e^{ic} = \cos c + i \sin c. \tag{9.1.16}$$

Euler's formula implies that the complex stress and displacement at the m boundary may be written in terms of complex trigonometric functions with a complex argument as

$$u_m(z_m) = \sum_{j=1}^{2} u_{jm}(z_{m-1}) \left(\cos[d_{\beta m} d_m] + i(-1)^{j+1} \sin[d_{\beta m}\, d_m] \right) \tag{9.1.17}$$

and

$$p_{zym}(z_m) = -iM_m d_{\beta m} \sum_{j=1}^{2} (-1)^j u_{jm}(z_{m-1}) \Big(\cos[d_{\beta m} d_m] + i(-1)^{j+1} \sin[d_{\beta m} d_m] \Big). \tag{9.1.18}$$

Rearrangement of the terms in (9.1.17) and (9.1.18) shows that the complex displacement and stress in layer m on the m boundary is related to the corresponding quantities on the $m-1$ boundary by

$$u_m(z_m) = \cos[d_{\beta m}\, d_m] u_m(z_{m-1}) + \sin[d_{\beta m}\, d_m] \frac{p_{zym}(z_{m-1})}{M_m d_{\beta m}} \tag{9.1.19}$$

and

$$p_{zym}(z_m) = -M_m d_{\beta m} \sin[d_{\beta m}\, d_m] u_m(z_{m-1}) + \cos[d_{\beta m}\, d_m]\, p_{zym}(z_{m-1}). \tag{9.1.20}$$

In matrix notation these equations may be written as

$$\begin{pmatrix} u_m(z_m) \\ p_{zym}(z_m) \end{pmatrix} = \mathbf{f}_m \begin{pmatrix} u_m(z_{m-1}) \\ p_{zym}(z_{m-1}) \end{pmatrix}, \tag{9.1.21}$$

where $\mathbf{f}_m$ is defined by

$$\mathbf{f}_m \equiv \begin{pmatrix} \cos[d_{\beta m}\, d_m] & \dfrac{\sin[d_{\beta m}\, d_m]}{M_m d_{\beta m}} \\ -M_m d_{\beta m} \sin[d_{\beta m} d_m] & \cos[d_{\beta m} d_m] \end{pmatrix}. \tag{9.1.22}$$

In layer $m-1$ the displacement and stress on the $m-1$ boundary may be related to those on the $m-2$ boundary by an analogous equation, where the index m is replaced by the index $m-1$. The boundary conditions at the $m-1$ boundary imply that the displacement and stress at the m boundary are related to those at the $m-2$ boundary by

$$\begin{aligned} \begin{pmatrix} u_m(z_m) \\ p_{zym}(z_m) \end{pmatrix} &= \mathbf{f}_m \begin{pmatrix} u_m(z_{m-1}) \\ p_{zym}(z_{m-1}) \end{pmatrix} = \mathbf{f}_m \begin{pmatrix} u_{m-1}(z_{m-1}) \\ p_{zym-1}(z_{m-1}) \end{pmatrix} \\ &= \mathbf{f}_m \mathbf{f}_{m-1} \begin{pmatrix} u_{m-1}(z_{m-2}) \\ p_{zym-1}(z_{m-2}) \end{pmatrix}. \end{aligned} \tag{9.1.23}$$

Continuing this procedure shows that the displacement and stress on the $m-1$ boundary are related to those at the free surface by

$$\begin{pmatrix} u_m(z_{m-1}) \\ p_{zym}(z_{m-1}) \end{pmatrix} = \mathbf{f}_{m-1} \mathbf{f}_{m-2} \ldots \mathbf{f}_1 \begin{pmatrix} u_1(z_0) \\ p_{zy1}(z_0) \end{pmatrix}. \tag{9.1.24}$$

Defining

$$\mathbf{F}_{m-1} = \mathbf{f}_{m-1} \mathbf{f}_{m-2} \ldots \mathbf{f}_1 \tag{9.1.25}$$

the elements of (9.1.24) may be written as

$$\begin{aligned} u_m(z_{m-1}) &= \mathbf{F}_{m-1_{11}} u_1(z_0) + \mathbf{F}_{m-1_{12}}\, p_{zy1}(z_0), \\ p_{zym}(z_{m-1}) &= \mathbf{F}_{m-1_{21}} u_1(z_0) + \mathbf{F}_{m-1_{22}}\, p_{zy1}(z_0). \end{aligned} \tag{9.1.26}$$

Writing the complex displacement and stress at the $m-1$ boundary in terms of the displacements at the $m-1$ boundary corresponding to the upgoing and downgoing solutions (9.1.10) and (9.1.11) implies

$$\begin{aligned} u_{1m}(z_{m-1}) + u_{2m}(z_{m-1}) &= \mathbf{F}_{m-1_{11}} u_1(z_0) + \mathbf{F}_{m-1_{12}}\, p_{zy1}(z_0), \\ u_{1m}(z_{m-1}) - u_{2m}(z_{m-1}) &= \frac{-i}{M_m\, d_{\beta m}} \left(\mathbf{F}_{m-1_{21}} u_1(z_0) + \mathbf{F}_{m-1_{22}}\, p_{zy1}(z_0)\right). \end{aligned} \tag{9.1.27}$$

The boundary condition of vanishing stress at the free surface, namely $p_{zx1}\,(z_0)=0$, implies that (9.1.27) may be written in terms of the complex amplitudes of the solutions in the mth layer as

$$\begin{aligned} D_{1m} \exp[i d_{\beta m}\, z_{m-1}] + D_{2m} \exp[-i d_{\beta m}\, z_{m-1}] &= \mathbf{F}_{m-1_{11}} (D_{11} + D_{21}), \\ D_{1m} \exp[i d_{\beta m}\, z_{m-1}] - D_{2m} \exp[-i d_{\beta m}\, z_{m-1}] &= \frac{-i\mathbf{F}_{m-1_{21}}}{M_m\, d_{\beta m}} (D_{11} + D_{21}). \end{aligned} \tag{9.1.28}$$

Considering the case that the mth layer corresponds to the half space, that is $m=n$, the preceding two equations readily permit the sum of the complex amplitudes of the upgoing and downgoing waves at the free surface, $D_{11}+D_{21}$, and the complex amplitude of the reflected wave D_{2n} to be expressed in terms of the complex amplitude of the incident general SII wave, namely,

$$\frac{D_{11} + D_{21}}{D_{1n}} = \frac{2 M_n\, d_{\beta n}}{\mathbf{F}_{n-1_{11}} M_n\, d_{\beta n} - i\mathbf{F}_{n-1_{21}}} \exp[i d_{\beta n}\, z_{n-1}] \tag{9.1.29}$$

and

$$\frac{D_{2n}}{D_{1n}} = \frac{\mathbf{F}_{n-1_{11}} M_n\, d_{\beta n} + i\mathbf{F}_{n-1_{21}}}{\mathbf{F}_{n-1_{11}} M_n\, d_{\beta n} - i\mathbf{F}_{n-1_{21}}} \exp[i\, 2 d_{\beta n}\, z_{n-1}], \tag{9.1.30}$$

where (9.1.11) and the condition of vanishing stress at the free surface imply the amplitudes of the upgoing and downgoing solutions in the top layer are equal, that is

$$D_{11} = D_{21}. \tag{9.1.31}$$

To show explicitly that (9.1.29) and (9.1.30) provide the desired solutions recall that

(1) k for the general SII wave incident in viscoelastic medium n is given from (9.1.8) by

$$k = \frac{\omega}{v_{HS_n}} \left(\sqrt{\frac{1 + \chi_{Su_1 n}}{1 + \chi_{HSn}}} \sin \theta_{u_1 n} - i \sqrt{\frac{-1 + \chi_{Su_1 n}}{1 + \chi_{HSn}}} \sin \left[\theta_{u_1 n} - \gamma_{u_1 n} \right] \right) \quad (9.1.32)$$

with $\chi_{Su_1 n} = \sqrt{1 + Q_{HSn}^{-2} \sec^2 \gamma_{u_1 n}}$ and $\chi_{HSn} = \sqrt{1 + Q_{HSn}^{-2}}$, which simplifies for an elastic medium n to

$$k = \frac{\omega}{v_{HSn}} \sin \theta_{u_1 n}, \quad (9.1.33)$$

(2) k_{Sm} from (3.6.13) and (3.6.14) is given by

$$k_{Sm} = \frac{\omega}{v_{HSm}} \left(1 - i \frac{Q_{HSm}^{-1}}{1 + \chi_{HSm}} \right), \quad (9.1.34)$$

(3) $d_{\beta m}$ from (4.2.10) is given by

$$d_{\beta m} = \sqrt{k_{Sm}^2 - k^2}, \quad (9.1.35)$$

and

(4) the complex modulus M_m from (3.6.24) is given by

$$M_m = \frac{\rho_m v_{HSm}^2}{2} \frac{1 + \chi_{HSm}}{\chi_{HSm}^2} \left(1 + i Q_{HSm}^{-1} \right) \quad (9.1.36)$$

for $m = 1, \ldots, n$.

Substitution of the pertinent material parameters given for each layer, namely v_{HSm}, Q_{HSm}^{-1}, ρ_m, d_m, and the given parameters for the incident general SII wave, namely $\theta_{u_1 n}$, $\gamma_{u_1 n}$, and ω, into (9.1.32) through (9.1.36) implies that each of the corresponding parameters, $\mathbf{f}_m$, as defined by (9.1.22) for each layer, and $\mathbf{F}_{n-1} = \mathbf{f}_{n-1} \mathbf{f}_{n-2} \ldots \mathbf{f}_1$ are determined. Hence, (9.1.29) and (9.1.30) represent the desired solutions for the complex amplitude at the free surface and the complex amplitude of the general SII wave reflected at the base in terms of the complex amplitude (D_{1n}) of the incident general SII wave, its angle of incidence ($\theta_{u_1 n}$), its degree of inhomogeneity ($\gamma_{u_1 n}$), and the material parameters v_{HSm}, Q_{HSm}^{-1}, ρ_m, d_m for each of the layers.

Equations (9.1.29) through (9.1.36) show that the amplitude and phase response of the stack of viscoelastic layers depends explicitly on the intrinsic absorption Q_{HSm}^{-1} in each of the layers. In addition, they show that the response varies with angle of incidence $\theta_{u_1 n}$ and the degree of inhomogeneity $\gamma_{u_1 n}$ of the general SII wave incident at the base of the stack of layers.

The solution as derived here for multilayered media is valid for any linear viscoelastic solid. In particular, the solution is valid with the layers chosen to be any solid whose constitutive law may be expressed as a combination of springs and dashpots in parallel and or in series with the type of viscoelastic solid not necessarily being the same for each layer. A special case of the general solution derived here that has been used in geotechnical engineering is that in which the incident SII wave is assumed to be homogeneous and the material in each layer is modeled as a Voight solid (Kanai, 1950; Kramer, 1996, pp. 268–269). The general solution derived here may be written readily in terms of the parameters μ_m and η_m for the mth layer of a stack of Voight solids (see Tables (1.3.29) and (1.3.30)) upon writing material parameters Q_{HSm}^{-1} and v_{HSm} using (3.5.2) and (3.5.5) as

$$Q_{HSm}^{-1} = \frac{\omega \eta_m}{\mu_m} \tag{9.1.37}$$

and

$$v_{HSm} = \sqrt{\frac{\mu_m}{\rho_m} \frac{2\left(1 + \left(\frac{\omega \eta_m}{\mu_m}\right)^2\right)}{1 + \sqrt{1 + \left(\frac{\omega \eta_m}{\mu_m}\right)^2}}}. \tag{9.1.38}$$

Substitution of these expressions for the material parameters into the general solutions (9.1.29) and (9.1.30) yields the solution for the special case of a stack of viscoelastic Voight layers. With the additional assumption that the incident SII wave is homogeneous the result is easily shown to agree with that derived by Kanai (1950) and Kramer (1996).

For the case that the incident wave is assumed to be homogeneous, that is $\gamma_{u_1 n} = 0$, expression (9.1.32) for k simplifies to

$$k = \frac{\omega}{v_{HSn}} \left(1 - i \frac{Q_{HSn}^{-1}}{1 + \chi_{HSn}}\right) \sin \theta_{u_1 n} = k_{Sn} \sin \theta_{u_1 n} \tag{9.1.39}$$

so (9.1.35) for $d_{\beta m}$ may be written as

$$d_{\beta m} = \sqrt{k_{Sm}^2 - k^2} = \sqrt{k_{Sm}^2 - k_{Sn}^2 \sin^2 \theta_{u_1 n}}. \tag{9.1.40}$$

With the additional assumption that the media are elastic with $Q_{HSm}^{-1} = 0$, (9.1.39) simplifies to the familiar expression for a homogeneous SII (SH) wave incident in elastic media, namely

$$k = k_{Sn} \sin \theta_{u_1 n} = \frac{\omega}{v_{HSn}} \sin \theta_{u_1 n}, \tag{9.1.41}$$

with

$$M_m = \rho_m v_{HSm}^2. \tag{9.1.42}$$

Substitution of (9.1.41) and (9.1.42) into (9.1.29) and (9.1.30) shows that thc equations derived for multilayered viscoelastic media simplify to the corresponding equations derived for elastic media. (See e.g. equations 4 and 5 in Haskell (1960) and equations 3.165 and 3.166 in Ben-Menehem and Singh (1981). Their equations derived for an incident homogeneous SH wave incident on a stack of elastic layers are easily shown to agree with those derived here in terms of displacement for elastic media upon realizing that their matrix coefficients were defined for velocity and are related by $\mathbf{F}_{n-1_{11}} = A_{11}$ and $\mathbf{F}_{n-1_{21}} = iA_{21}k$ in notation used by Haskell.)

9.2 Analytic Solution (One Layer)

Consideration of a single viscoelastic layer with $n=2$ and definitions (9.1.25) and (9.1.22) imply

$$\mathbf{F}_1 = \mathbf{f}_1 = \begin{pmatrix} \cos[d_{\beta 1} d_1] & \sin[d_{\beta 1} d_1]/M_1 d_{\beta 1} \\ -M_1 d_{\beta 1} \sin[d_{\beta 1} d_1] & \cos[d_{\beta 1} d_1] \end{pmatrix} \tag{9.2.1}$$

from which it follows that (9.1.29) and (9.1.30) simplify to

$$\frac{D_{11} + D_{21}}{D_{12}} = \frac{2\exp\left[i d_{\beta 2}\, d_1\right]}{\left[\cos d_{\beta 1} d_1\right] + i\dfrac{M_1 d_{\beta 1}}{M_2 d_{\beta 2}} \sin\left[d_{\beta 1} d_1\right]} \tag{9.2.2}$$

and

$$\frac{D_{22}}{D_{12}} = \frac{\cos\left[d_{\beta 1} d_1\right] - i\dfrac{M_1 d_{\beta 1}}{M_2 d_{\beta 2}} \sin\left[d_{\beta 1} d_1\right]}{\cos\left[d_{\beta 1} d_1\right] + i\dfrac{M_1 d_{\beta 1}}{M_2 d_{\beta 2}} \sin\left[d_{\beta 1} d_1\right]} \exp\left[i\, 2 d_{\beta 2}\, d_1\right], \tag{9.2.3}$$

where

$$\begin{aligned}
k &= \frac{\omega}{v_{HS2}} \left(\sqrt{\frac{1 + \chi_{Su_1 2}}{1 + \chi_{HS2}}} \sin\theta_{u_1 2} - i\sqrt{\frac{-1 + \chi_{Su_1 2}}{1 + \chi_{HS2}}} \sin\left[\theta_{u_1 2} - \gamma_{u_1 2}\right] \right), \\
k_{Sm} &= \frac{\omega}{v_{HSm}} \left(1 - i\frac{Q_{HSm}^{-1}}{1 + \chi_{HSm}} \right), \\
d_{\beta m} &= \sqrt{k_{Sm}^2 - k^2}, \\
M_m &= \frac{\rho_m v_{HSm}^2}{2} \frac{1 + \chi_{HSm}}{\chi^2{}_{HSm}} \left(1 + i Q_{HSm}^{-1}\right), \\
\chi_{Su_j m} &= \sqrt{1 + Q_{HSm}^{-2} \sec^2 \gamma_{u_j m}}, \\
\chi_{HSm} &= \sqrt{1 + Q_{HSm}^{-2}} \qquad \text{for } m = 1, 2 \text{ and } j = 1, 2.
\end{aligned} \tag{9.2.4}$$

Substitution of (9.2.4) into (9.2.2) and (9.2.3) shows explicitly that the surface response of the single layer and the complex amplitude of the reflected wave depend on the angle of phase incidence ($\theta_{u_1 2}$) and degree of inhomogeneity ($\gamma_{u_1 2}$) of the incident wave and the material parameters as specified by quality factors and wave speeds for homogeneous waves and densities, namely Q_{HSm}^{-1}, v_{HSm}, and ρ_m, $m = 1, 2$.

9.3 Numerical Response of Viscoelastic Layers (Elastic, Earth's Crust, Rock, Soil)

Equations (9.2.2) through (9.2.4) readily permit computation of the response of the layer as a function of viscoelastic material parameters and the given parameters of the incident general Type-II S wave. For purposes of computation it is convenient to write $d_1 d_{\beta m}$ in terms of a frequency f_0 as

$$d_1 d_{\beta m} = d_1 k_{S1_R} \sqrt{\frac{k_{Sm}^2}{k_{S1_R}^2} - \frac{k^2}{k_{S1_R}^2}} = \frac{\pi f}{2 f_0} \sqrt{\frac{k_{Sm}^2}{k_{S1_R}^2} - \frac{k^2}{k_{S1_R}^2}}, \tag{9.3.1}$$

where f_0 is defined as

$$f_0 \equiv \frac{v_{HS1}}{4 d_1} \tag{9.3.2}$$

and the components of the complex wave numbers k and $k_{Sm} (m = 1, 2)$ as normalized by the real part of that for a homogeneous S wave are given by

$$\frac{k_{Sm}}{k_{S1_R}} = \frac{v_{HS1}}{v_{HSm}} \left(1 - i \frac{Q_{HSm}^{-1}}{1 + \chi_{HSm}} \right) \quad \text{for } m = 1, 2, \tag{9.3.3}$$

$$\frac{k_R}{k_{S1_R}} = \frac{v_{HS1}}{v_{HS2}} \left(\frac{\sqrt{1 + \chi_{Su12}}}{\sqrt{1 + \chi_{HS2}}} \right) \sin \theta_{u_1 2}, \tag{9.3.4}$$

and

$$\frac{k_I}{k_{S1_R}} = - \frac{v_{HS1}}{v_{HS2}} \left(\frac{\sqrt{-1 + \chi_{Su_1 2}}}{\sqrt{1 + \chi_{HS2}}} \right) \sin \left[\theta_{u_1 2} - \gamma_{u_1 2} \right]. \tag{9.3.5}$$

and the ratio of the complex moduli by

$$\frac{M_1}{M_2} = \frac{\rho_1}{\rho_2} \frac{v_{HS1}^2}{v_{HS2}^2} \frac{1 + \chi_{HS1}}{1 + \chi_{HS2}} \frac{1 - i Q_{HS2}^{-1}}{1 - i Q_{HS1}^{-1}}. \tag{9.3.6}$$

Expressions (9.3.1) through (9.3.6) when substituted into (9.2.2) and (9.2.3) provide explicit expressions for the normalized response in terms of the parameters of the incident wave, namely the angle of phase incidence $\theta_{u_1 2}$, the degree of inhomogeneity $\gamma_{u_1 2}$, the normalized frequency f/f_0, where f_0 depends on the thickness d_1 and material wave speed of the first layer v_{HS1}, and material parameters of the viscoelastic media, namely the ratio of densities ρ_1/ρ_2, ratio of material wave speeds for homogeneous waves v_{HS1}/v_{HS2}, and the amounts of material intrinsic absorption for homogeneous waves as expressed by Q_{HS1}^{-1} and Q_{HS2}^{-1}.

For elastic media (9.3.3) through (9.3.6) reduce to

$$\frac{k_{Sm}}{k_{S1_R}} = \frac{v_{HS1}}{v_{HSm}} \quad \text{for } m = 1, 2, \tag{9.3.7}$$

$$\frac{k_R}{k_{S1_R}} = \frac{v_{HS1}}{v_{HS2}} \sin\theta_{u_1 2}, \tag{9.3.8}$$

$$\frac{k_I}{k_{S1_R}} = 0, \tag{9.3.9}$$

and

$$\frac{M_1}{M_2} = \frac{\rho_1}{\rho_2}\frac{v_{HS1}^2}{v_{HS2}^2}, \tag{9.3.10}$$

where $f_0 = v_{HS1}/4d_1$ represents the lowest frequency of maximum response or the fundamental frequency of the elastic layer.

To illustrate the influence of various parameters on the response of the layer, the normalized response to incident SII waves (9.2.2) is calculated as a function of normalized frequency and one other chosen parameter with the other parameters held fixed. As a first example, the normalized response of the layer to a normally incident homogeneous SII wave is shown as a function of the ratio of the material wave speeds (v_{HS1}/v_{HS2}) and specified amounts of intrinsic absorption (Q_{HS1}^{-1}, Q_{HS2}^{-1}) that range in the layer from near-elastic (Figure (9.3.11)a) to low-loss amounts appropriate for a model of the Earth's crust and near-surface soil–rock layer (Figures (9.3.11)b, c) to non-low-loss amounts appropriate for soft soils (Figure (9.3.11)d). Considering that variations in density are in part accounted for by variations in material wave speeds, the ratio of densities for the models is chosen fixed, $\rho_1/\rho_2 = 0.9$. The calculations illustrate that the amplitude response of the layer to a normally incident homogeneous wave is frequency-dependent, with local maxima occurring approximately at odd multiples of $f_0 = v_{HS1}/(4d_1)$ for materials with $v_{HS1}/v_{HS2} \leq 1$ and approximately at even multiples for $v_{HS1}/v_{HS2} \geq 1$. The figures indicate that the amplitudes of the local maxima are approximately constant as a function of frequency for media that are nearly elastic (Figure (9.3.11)a) and that they decrease with frequency for increasing amounts of intrinsic absorption or damping in the layer. They indicate that the response of the layer to a normally

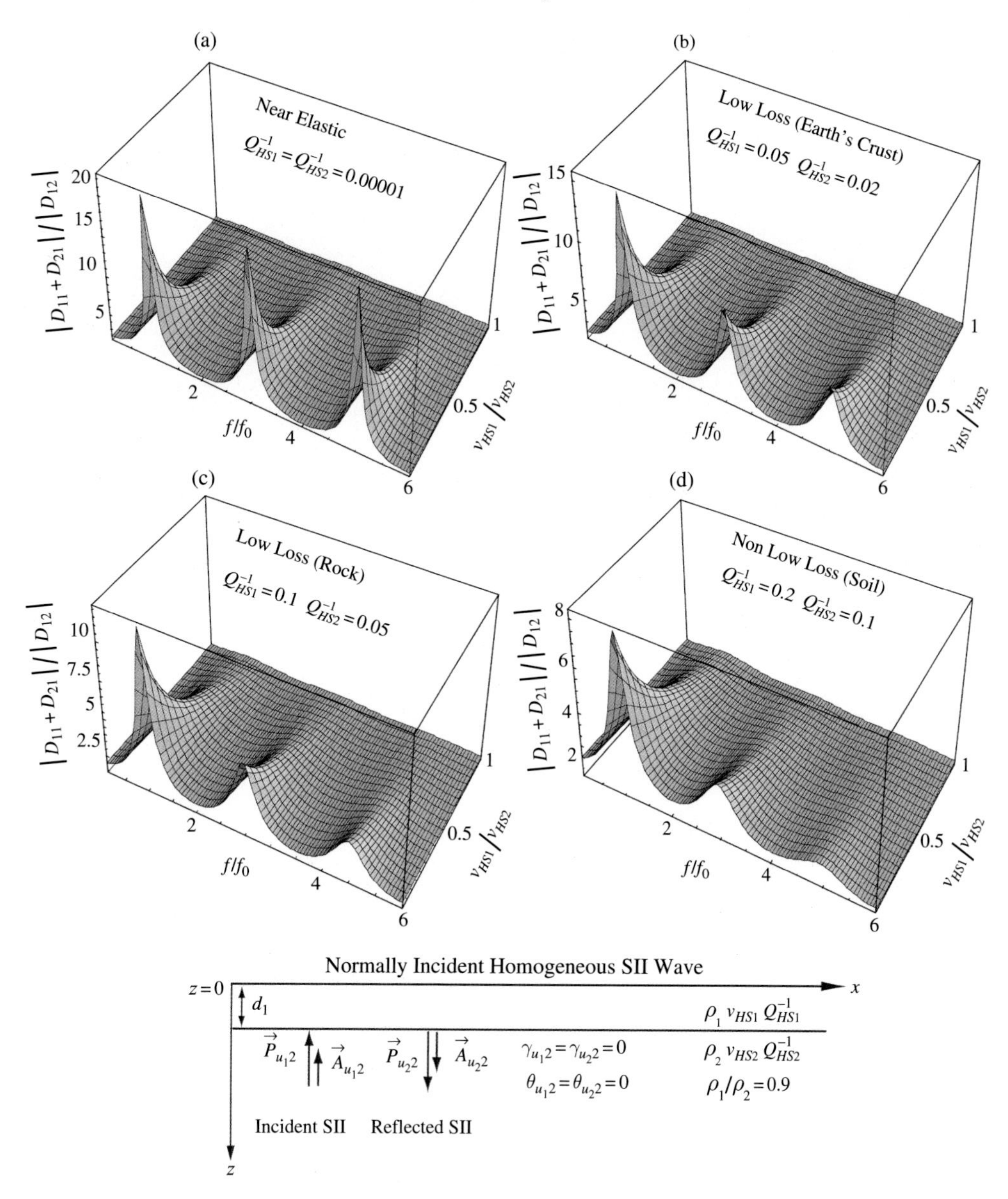

Figure (9.3.11). Amplitude response of viscoelastic layers to normally incident homogeneous Type-II S waves as a function of normalized frequency and the ratio of material velocities for layers with increasing amounts of intrinsic absorption.

incident homogeneous wave decreases as the material wave-speed ratio v_{HS1}/v_{HS2} tends toward unity and the amount of intrinsic absorption increases.

To consider the influence of inhomogeneity of the incident SII wave, the response of the layer to a normally incident inhomogeneous SII wave is calculated for material

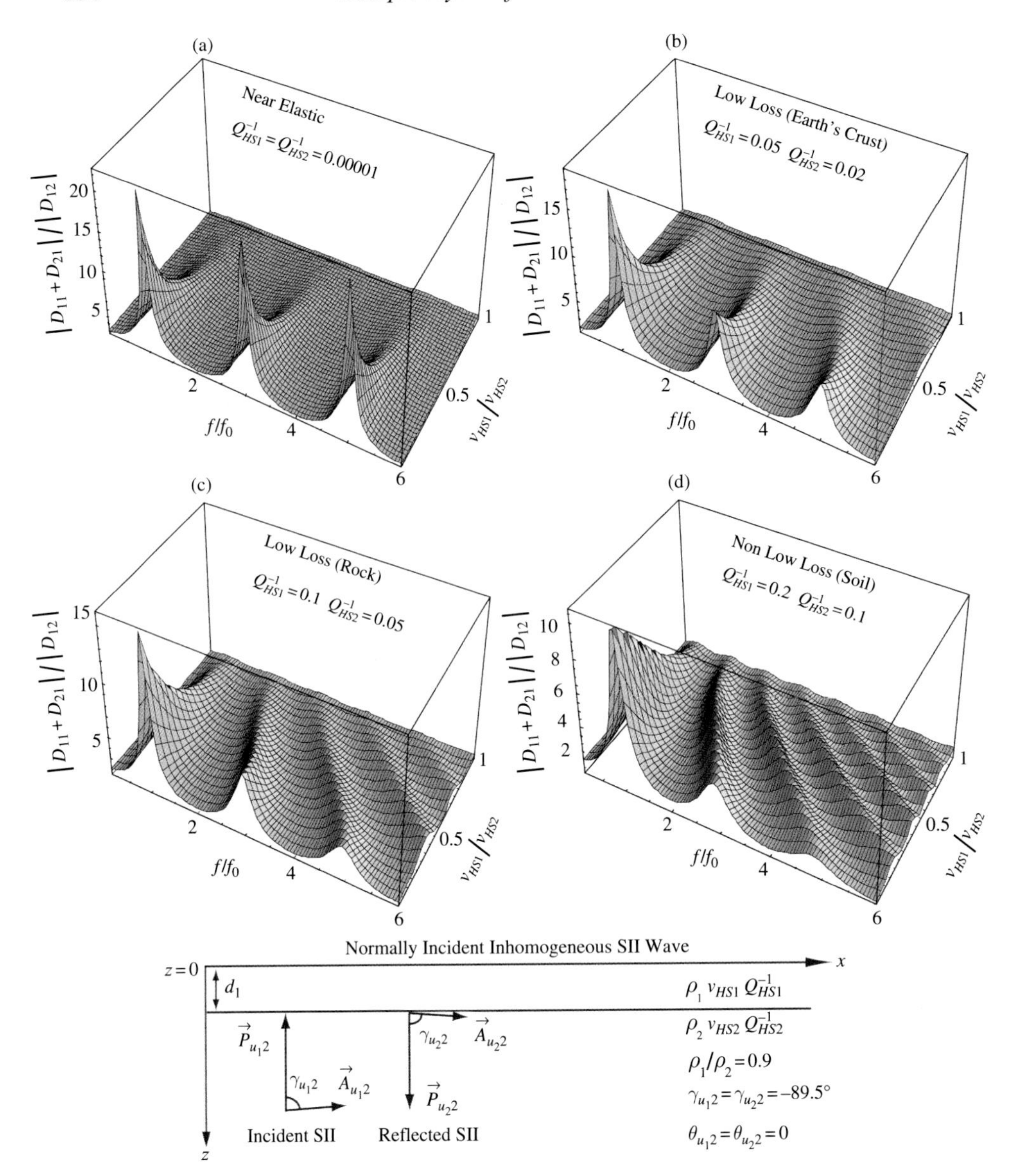

Figure (9.3.12). Amplitude response of viscoelastic layers to normally incident inhomogeneous Type-II S waves as a function of normalized frequency and the ratio of material velocities with increasing amounts of intrinsic absorption.

parameters corresponding to near-elastic media (Figure (9.3.12)a), to low-loss media (Figures (9.3.12)b, c), and to non-low-loss media (Figure (9.3.12)d). The plots indicate that for a chosen large degree of inhomogeneity of $\gamma_{u_1 2} = 89.5°$ for the normally incident wave, the response of the layer varies with material wave speed and intrinsic absorption of the layer. The local maxima associated with the fundamental mode and higher modes vary as the ratios of material wave speeds v_{HS1}/v_{HS2} vary in a manner not evident for the case of a normally incident homogeneous wave (compare Figures (9.3.12) and (9.3.11)). However, similar plots (not shown) for smaller amounts of inhomogeneity, say $\gamma_{u_1 2} = 80°$, show that the response for the material parameters considered is much more similar to that for a normally incident homogeneous wave. Hence, for these material parameters the influence of inhomogeneity on the response decreases rapidly with a decrease in the degree of inhomogeneity of the incident wave. For the range of material parameters considered, the response of the viscoelastic surface layer to a normally incident inhomogeneous wave shows a significant dependence on inhomogeneity of the incident wave only for large values of inhomogeneity with the dependence increasing with the amount of intrinsic absorption.

To further consider the influence of inhomogeneity as well as angle of incidence, the response of the layer is calculated as a function of angle of incidence for low-loss material parameters appropriate for a firm soil layer overlying rock for chosen amounts of inhomogeneity of the incident wave of $\gamma_{u_1 2} = 0°,\ -60°,\ -80°,\ -86°$ (Figures (9.3.13)a, b, c, d). Calculations of the angle between the direction of phase propagation and energy flux for the incident waves indicate the angles are $\angle(\vec{P}_{u_1 2}, \langle\vec{\mathcal{J}}_{u_1 2}\rangle) = 0°,\ 0.12°,\ 0.40°,\ 0.92°$, respectively, for the incident waves with $\gamma_{u_1 2} = 0°,\ -60°,\ -80°,\ -86°$. Hence, to ensure that the energy flux associated with the incident general SII waves is toward the boundary, as discussed in Section 6.1.2, model results are calculated only for angles for which $0 \leq \theta_{u_1 2} < 90°$, 89.88°, 89.60°, and 89.08°, respectively.

The response of the layer for each incident general SII wave (Figures (9.3.13)a, b, c, d) reveals local maxima corresponding to the fundamental and higher modes with the maximum response for each mode decreasing with increasing frequency. For the example in which the incident wave is homogeneous (Figure (9.3.13)a), the response of the layer for each normalized frequency decreases monotonically with increasing angles of incidence. The response indicates that the vertical energy flux across the boundary associated with the incident wave decreases as the angle of incidence approaches grazing incidence for the energy flux of the incident wave.

For the examples in which the incident wave is inhomogeneous (Figures (9.3.13) b, c, d), the response of the layer for each normalized frequency shows a distinct difference from that for incident homogeneous waves. The response is similar to that for an incident homogeneous wave (Figure (9.3.13)a) for a lower range in angles of incidence, but for larger angles of incidence the response increases with increasing angle of incidence. This important difference in the nature of the response for incident inhomogeneous waves is explained by an increase in the vertical energy flux across the boundary as the angle of incidence increases due to energy flow associated with interaction of the velocity and stress fields of the incident and reflected waves at the base of the layer (see Section 6.1.2 for additional discussion).

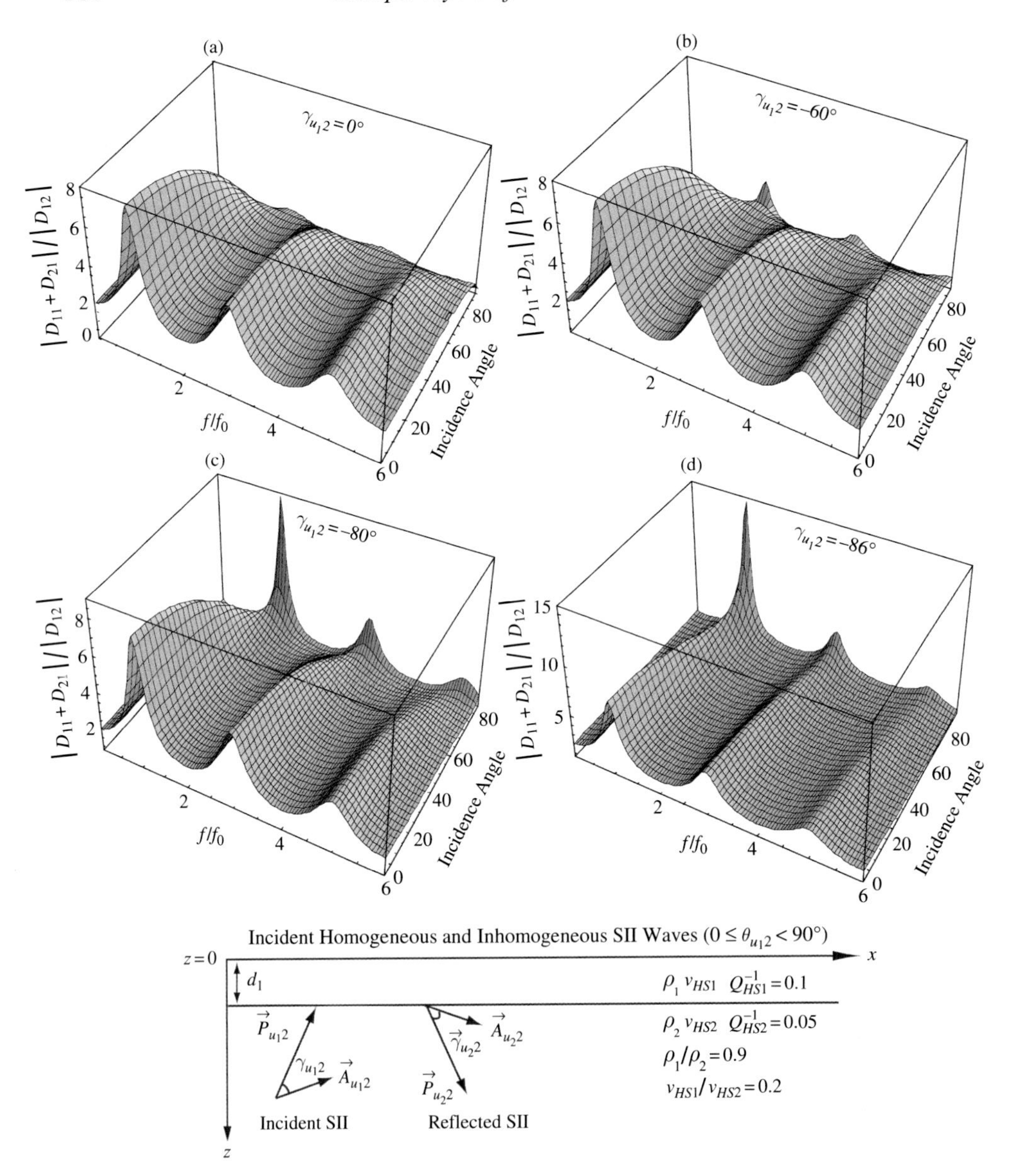

Figure (9.3.13). Amplitude response of a viscoelastic layer as a function of normalized frequency and angle of incidence for incident Type-II S waves with degrees of inhomogeneity of $\gamma_{u_1 2} = 0°,\ -60°,\ -80°,\ -86°$ (a, b, c, d).

Calculations (not shown) indicate that the influence of inhomogeneity of the incident wave begins to be apparent for the chosen material parameters for degrees of inhomogeneity of −60° and larger in magnitude (Figures (9.3.13)b, c, d). The figures indicate that the influence of inhomogeneity increases as the magnitude of inhomogeneity of the incident wave increases with a principal result being an increase in the response of the layer for each of the various modes as grazing incidence for energy flux is approached. The range of angles of incidence over which the increase in response occurs increases with the degree of inhomogeneity of the incident wave. Additional calculations (not shown) also indicate that this range of angles over which the increase occurs increases as the amount of intrinsic absorption in the layer increases.

9.4 Problems

(1) Use the solution for the response of an arbitrary viscoelastic layer to an incident inhomogeneous SII wave as specified by (9.2.2) and (9.2.3) to derive the expressions for the corresponding response if the media are standard linear viscoelastic solids.

(2) Find the classic solution for a stack of n elastic layers to an incident homogeneous SII wave using the general solutions (9.1.29) and (9.1.30) derived for an arbitrary stack of n viscoelastic layers.

(3) Sketch curves from Figure (9.3.11) showing the response of a single layer as a function of normalized frequency for $v_{HS1}/v_{HS2} \approx 0.1$ to a vertically incident homogeneous SII wave for near-elastic ($Q_{HS1}^{-1} = Q_{HS2}^{-1} = 0.000\,01$), low-loss ($Q_{HS1}^{-1} = 0.05; Q_{HS2}^{-1} = 0.02$) and non-low-loss media ($Q_{HS1}^{-1} = 0.2; Q_{HS2}^{-1} = 0.1$). Describe the dependence of the response on the amount of intrinsic absorption in the layer.

(4) Sketch curves from Figure (9.3.13) showing the response of a single low-loss layer as a function of normalized frequency for an incident SII wave at angles of incidence of $\theta_{u_1 2} = 0°$ and $\theta_{u_1 2} \approx 85°$ with degrees of inhomogeneity of $\gamma_{u_1 2} = 0°$, $\gamma_{u_1 2} = -60°$, and $\gamma_{u_1 2} = -80°$. Describe the dependences of the response on angle of incidence and degree of inhomogeneity of the incident wave.

10
Love-Type Surface Waves in Multilayered Viscoelastic Media

In 1911 Love first established the solution for surface waves with horizontal particle motions in elastic media comprised of a layer over a half space. He showed that solutions for SII (SH) waves with attenuation perpendicular to the boundary could be found that satisfy the boundary conditions. A solution is presented here for the problem of multiple viscoclastic layers overlying a viscoelastic half space. The solution for the postulated surface waves is found by showing that a complex wave number k can be found such that assumed plane-wave solutions in each of the layers and in the half space satisfy the boundary conditions at the free surface and at the interfaces between the layers and the half space.

10.1 Analytic Solution (Multiple Layers)

To specify the problem consider $n-1$ layers over a half space with notation specifying the media and boundaries as shown in Figure (9.1.1) for the problem of an incident SII wave on a stack of $n-1$ layers. Without loss of generality, solutions in each of the layers as specified by (9.1.2) through (9.1.7) are rewritten in terms of

$$b_\beta \equiv principal\ value\sqrt{k^2 - k_S^2} = id_\beta, \tag{10.1.1}$$

where "*principal value* $\sqrt{\ }$ " denotes the principal value (4.2.13) of the square root. Assuming that the solutions in the mth layer are of this form implies that $b_{\beta m_R} \geq 0$ and hence the direction of attenuation for the chosen solutions is unambiguously specified with respect to the $+\hat{x}_3$ direction.

The plane-wave solution assumed for the mth layer $(1 < m \leq n)$ written in terms of $b_{\beta m}$ is

$$\vec{u}_m(z) = u_m(z)\hat{x}_2 = \sum_{j=1}^{2} u_{jm}\hat{x}_2 = \sum_{j=1}^{2} D_{jm} \exp\left[i\left(\omega t - \vec{K}_{u_{jm}} \bullet \vec{r}\right)\right]\hat{x}_2$$
$$= \sum_{j=1}^{2} D_{jm} \exp\left[i\left(\omega t - kx - (-1)^{j+1} i b_{\beta m} z\right)\right]\hat{x}_2, \tag{10.1.2}$$

where

$$\vec{K}_{u_{jm}} = \vec{P}_{u_{jm}} - i\vec{A}_{u_{jm}} = k\hat{x}_1 + i(-1)^{j+1} b_{\beta m}\hat{x}_3, \tag{10.1.3}$$

$$\vec{P}_{u_{jm}} = k_R\hat{x}_1 + (-1)^j b_{\beta m_I}\hat{x}_3, \tag{10.1.4}$$

$$\vec{A}_{u_{jm}} = -k_I\hat{x}_1 + (-1)^j b_{\beta m_R}\hat{x}_3, \tag{10.1.5}$$

with the solution for the half space specified by $m = n$ and $D_{1n} = 0$. Choice of these solutions assures that the attenuation vector for the solution in the half space is directed in the $+\hat{x}_3$ direction into the half space away from the free surface.

Application of the boundary conditions of vanishing stress at the free surface and continuity of stress and displacement at the boundaries allows the derivation used to derive (9.1.28) to be repeated using the solutions assumed here. Hence, the complex displacement at the boundary of the half space is related to that at the free surface by

$$D_{2n} \exp[-b_{\beta n} z_{n-1}] = \mathbf{F}_{n-1_{11}}(D_{11} + D_{21}) \tag{10.1.6}$$

and

$$D_{2n} \exp[-b_{\beta n} z_{n-1}] = \frac{-\mathbf{F}_{n-1_{21}}}{M_n b_{\beta n}}(D_{11} + D_{21}), \tag{10.1.7}$$

where the condition of vanishing stress at the free surface implies from (9.1.9), (10.1.1), and (10.1.2) that

$$D_{11} = D_{21}. \tag{10.1.8}$$

Hence, for non-trivial plane-wave solutions

$$\mathbf{F}_{n-1_{11}} = \frac{-\mathbf{F}_{n-1_{21}}}{M_n b_{\beta n}}, \tag{10.1.9}$$

where $\mathbf{F}_{n-1} \equiv \mathbf{f}_{n-1}\,\mathbf{f}_{n-2} \ldots \mathbf{f}_1$ and

$$\mathbf{f}_m \equiv \begin{pmatrix} \cos[-i\, b_{\beta m}\, d_m] & \dfrac{\sin[-i\, b_{\beta m}\, d_m]}{-iM_m\, b_{\beta m}} \\ iM_m\, b_{\beta m} \sin[-i\, b_{\beta m}\, d_m] & \cos[-i\, b_{\beta m}\, d_m] \end{pmatrix}. \tag{10.1.10}$$

Hence, if the postulated solutions are to satisfy the boundary conditions for the posed problem, then the complex wave number k for the solutions must satisfy (10.1.9). It turns out that an infinite number of values for the complex wave number k exist that satisfy (10.1.9) with each value dependent on the given frequency of the wave and the given set of layer thicknesses and viscoelastic material parameters. The existence of these values establishes that the postulated solutions for a Love-Type surface wave as specified by (10.1.2) through (10.1.5) satisfy the boundary conditions as specified for a stack of $n-1$ viscoelastic layers overlying a viscoelastic half space. Hence, the solution for a Love-Type surface wave on a stack of $n-1$ viscoelastic layers is established upon determining the real and imaginary components of the complex wave number k that satisfies (10.1.9). A procedure for finding the components of k will be presented in a subsequent section.

The magnitude of the phase velocity v_L and the absorption coefficient a_L for a Love-Type surface wave of circular frequency ω are given in terms of the real and imaginary parts of the complex wave number k by

$$v_L = \frac{\omega}{k_R} \qquad (10.1.11)$$

and

$$a_L = -k_I. \qquad (10.1.12)$$

These expressions indicate that the problem of finding the solution for a Love-Type surface wave may also be stated as the problem of finding pairs of values for the wave speed and absorption coefficient (v_L, a_L) that satisfy (10.1.9) for a given set of viscoelastic material parameters characterizing each layer and the half space.

Definitions given for the propagation and attenuation vectors of the solutions in each layer (10.1.4) and (10.1.5) imply that the angles the vectors make with respect to the horizontal $(e_{u_j m})$ and the vertical $(d_{u_j m})$ are given by

$$\tan e_{u_j m} = (-1)^j \, b_{\beta m_I}/k_R = (-1)^j \, b_{\beta m_I} \frac{v_L}{\omega} \qquad (10.1.13)$$

and

$$\tan d_{u_j m} = \frac{(-1)^{j+1} \, k_I}{b_{\beta m_R}} = \frac{(-1)^j \, a_L}{b_{\beta m_R}}. \qquad (10.1.14)$$

Hence, the existence of non-vanishing pairs of values (v_L, a_L) for a Love-Type surface wave on anelastic media indicates the propagation and attenuation vectors for the component solutions in each layer are inclined with respect to the horizontal and vertical, respectively. A vanishing value of a_L for elastic media indicates that the attenuation vectors for the component solutions in each layer are perpendicular to the free surface and hence the corresponding propagation vectors are parallel to the free surface, because the only type of inhomogeneous component solution that can propagate in elastic media is one for which the vectors are perpendicular (see Theorem (3.1.17)). The inclinations of these vectors for anelastic media imply the characteristics of the physical displacement field in each layer are different from those for elastic media.

10.2 Displacement (Multiple Layers)

Assuming that complex values of k can be found such that (10.1.9) can be satisfied, solution of (9.1.28) rewritten for the problem of a Love-Type surface wave allows the complex amplitudes of the upgoing and downgoing inhomogeneous solutions in the mth layer to be written in terms of the complex amplitudes of the solutions at the free surface, specifically

$$D_{1m} = \left(\mathbf{F}_{m-1_{11}} + \frac{\mathbf{F}_{m-1_{21}}}{M_m\, b_{\beta m}} \right) \exp[-b_{\beta m}\, z_{m-1}] D_{11} \tag{10.2.1}$$

and

$$D_{2m} = \left(\mathbf{F}_{m-1_{11}} - \frac{\mathbf{F}_{m-1_{21}}}{M_m\, b_{\beta m}} \right) \exp[b_{\beta m}\, z_{m-1}] D_{11}, \tag{10.2.2}$$

where $m > 1$. Hence, the complex displacement in the mth layer expressed in terms of the complex amplitudes of the solutions at the free surface is given from (10.1.2) through (10.1.5) by

$$\vec{u}_m(z) = D_{11} \sum_{j=1}^{2} \left(\mathbf{F}_{m-1_{11}} - \frac{(-1)^j \mathbf{F}_{m-1_{21}}}{2 M_m\, b_{\beta m}} \right) \exp[(-1)^{j+1} b_{\beta m}(z - z_{m-1})] \\ \exp[i(\omega t - kx)]\hat{x}_2. \tag{10.2.3}$$

This expression with $m = n$ and (10.1.9) implies that the complex displacement in the half space is

$$\vec{u}_n(z) = 2 D_{11} \mathbf{F}_{n-1_{11}} \exp[k_I x - b_{\beta n}(z - z_{n-1})] \exp[i(\omega t - k_R x)]\hat{x}_2. \tag{10.2.4}$$

The complex displacement in the first layer, from (10.1.2) and (10.1.8) is

$$\vec{u}_1(z) = 2 D_{11} \cos[i b_{\beta 1} z] \exp[k_I x] \exp[i(\omega t - k_R x)]\hat{x}_2. \tag{10.2.5}$$

Hence, the normalized physical displacement in the half space as implied by the real part of (10.2.4) is

$$\frac{\vec{u}_{n_R}(z)}{|D_{11}|} = 2\left|\mathbf{F}_{n-1_{11}}\right| \exp[k_I x - b_{\beta n_R}(z - z_{n-1})]$$
$$\sin\left[\omega t - k_R x + \pi/2 - b_{\beta n_I}(z - z_{n-1}) + \arg D_{11} + \arg \mathbf{F}_{n-1_{11}}\right]\hat{x}_2 \tag{10.2.6}$$

and that at the free surface from (10.2.5), is

$$\frac{\vec{u}_{1_R}(z = z_0)}{|D_{11}|} = 2\exp[k_I x]\sin[\omega t - k_R x + \pi/2 + \arg D_{11}]\hat{x}_2. \tag{10.2.7}$$

Expression (10.2.6) indicates that the physical displacement in the viscoelastic half space shows superimposed a sinusoidal dependence on depth that decays exponentially with distance from the interface. It indicates that the displacement in the half space is out of phase with that at the surface (10.2.7) by an amount that depends on the thickness of intervening layers and their viscoelastic material parameters. It indicates that the physical displacement attenuates with absorption coefficient $a_L = -k_I$ along both the interface at depth and the free surface.

The physical displacement field in the mth layer as inferred from the real part of (10.1.2) in terms of the amplitudes of the upgoing and downgoing solutions is

$$\vec{u}_{m_R}(z) = \sum_{j=1}^{2} u_{jm_R}\hat{x}_2 = \sum_{j=1}^{2} \left|D_{jm}\right| \exp\left[-\left(-k_I x + (-1)^j b_{\beta m_R} z\right)\right]$$
$$\sin\left[\omega t - \left(k_R x + (-1)^j b_{\beta m_I} z\right) + \arg D_{jm} + \pi/2\right]\hat{x}_2. \tag{10.2.8}$$

Upon introducing notation

$$\begin{aligned} E_{jm} &\equiv \left|D_{jm}\right| \exp\left[(-1)^{j+1} b_{\beta m_R} z\right], \\ g_{jm} &\equiv (-1)^{j+1} b_{\beta m_I} z + \arg D_{jm}, \end{aligned} \tag{10.2.9}$$

for $j = 1, 2$, (10.2.8) may be written as

$$\vec{u}_{m_R}(z) = \exp[k_I x] \sum_{j=1}^{2} \left|E_{jm}\right| \sin\left[\omega t - k_R x + \pi/2 + g_{jm}\right]\hat{x}_2 \tag{10.2.10}$$

and further simplified to yield the desired expression for the physical displacement in the mth layer, namely,

$$\vec{u}_{m_R}(z) = \exp[k_I x] H_m \sin[\omega t - k_R x + \pi/2 + \delta_m]\hat{x}_2, \tag{10.2.11}$$

where

$$
\begin{aligned}
H_m &\equiv \sqrt{E_{1m}^2 + E_{2m}^2 + 2E_{1m}E_{2m}\cos[g_{1m} - g_{2m}]}, \\
\delta_m &\equiv \arctan\left[\frac{E_{1m}\sin[g_{1m}] + E_{2m}\sin[g_{2m}]}{E_{1m}\cos[g_{1m}] + E_{2m}\cos[g_{2m}]}\right].
\end{aligned}
\tag{10.2.12}
$$

10.3 Analytic Solution and Displacement (One Layer)

Solutions of (10.1.9) for a Love-Type surface wave in viscoelastic media are most readily considered for the case of a single layer overlying a half space. For the case that $n = 2$ (10.1.9) simplifies using (10.1.10) to

$$
\cos[-ib_{\beta 1}d_1] = \frac{-iM_1 b_{\beta 1}\sin[-ib_{\beta 1}d_1]}{M_2 b_{\beta 2}}, \tag{10.3.1}
$$

or

$$
\tan[-ib_{\beta 1}d_1] = -\tan[ib_{\beta 1}d_1 \pm n\pi] = \frac{iM_2 b_{\beta 2}}{M_1 b_{\beta 1}}, \tag{10.3.2}
$$

where for anelastic media the argument of the tangent and the right-hand side of (10.3.2) are in general complex variables dependent on the intrinsic absorption of the media. This dependence causes the equations to be more cumbersome to solve than for elastic media. Using the inverse of the complex tangent function and an identity in terms of the complex natural logarithm (Kreysig, 1967, p. 559), (10.3.2) may be expressed in terms of circular frequency in two alternate forms as

$$
\begin{aligned}
\omega\frac{d_1}{v_L} &= -i\frac{k_R}{b_{\beta 1}}\left(\arctan\left[-\frac{iM_2 b_{\beta 2}}{M_1 b_{\beta 1}}\right] \mp n\pi\right) \\
&= -i\frac{k_R}{b_{\beta 1}}\left(\frac{i}{2}\ln\left[\left(1 - \frac{M_2 b_{\beta 2}}{M_1 b_{\beta 1}}\right)\Big/\left(1 + \frac{M_2 b_{\beta 2}}{M_1 b_{\beta 1}}\right)\right] \mp n\pi\right).
\end{aligned}
\tag{10.3.3}
$$

Towards deriving solutions of (10.3.3) the right-hand side of the equation can be expressed in terms of the ratios of wave speeds for a Love-Type surface wave and a homogeneous shear wave in the first layer, namely v_L/v_{HS1} and corresponding ratios for the absorption coefficient, namely a_L/a_{HS1} using (10.1.11) and (10.1.12). To derive this result, k/k_R may be written using (3.6.14) as

$$
\frac{k}{k_R} = 1 - i\frac{a_L v_L}{\omega} = 1 - i\frac{a_L}{a_{HS1}}\frac{v_L}{v_{HS1}}\frac{Q_{HS1}^{-1}}{1 + \chi_{HS1}} \tag{10.3.4}
$$

and k_{Sj}/k_R as

$$\frac{k_{Sj}}{k_R} = \frac{v_L}{v_{HS1}}\frac{v_{HS1}}{v_{HSj}}\left(1 - i\frac{a_{HSj}v_{HSj}}{\omega}\right) = \frac{v_L}{v_{HS1}}\frac{v_{HS1}}{v_{HSj}}\left(1 - i\frac{Q_{HSj}^{-1}}{1+\chi_{HSj}}\right) \tag{10.3.5}$$

from which it follows that $b_{\beta j}/k_R$ for $j = 1, 2$ may be written as

$$\frac{b_{\beta j}}{k_R} = \sqrt{\frac{k^2}{k_R^2} - \frac{k_{Sj}^2}{k_R^2}} = \sqrt{\left(1 - i\frac{a_L}{a_{HS1}}\frac{v_L}{v_{HS1}}\frac{Q_{HS1}^{-1}}{1+\chi_{HS1}}\right)^2 - \left(\frac{v_L}{v_{HS1}}\frac{v_{HS1}}{v_{HSj}}\left(1 - i\frac{Q_{HSj}^{-1}}{1+\chi_{HSj}}\right)\right)^2}. \tag{10.3.6}$$

Identities (3.5.11) and (3.5.9) show that the ratio of the complex shear moduli for the two media may be written as

$$\frac{M_2}{M_1} = \frac{\rho_2}{\rho_1}\frac{v_{HS2}^2}{v_{HS1}^2}\frac{1+\chi_{HS2}}{1+\chi_{HS1}}\frac{1 - iQ_{HS1}^{-1}}{1 - iQ_{HS2}^{-1}}. \tag{10.3.7}$$

Hence, substituting (10.3.6) and (10.3.7) into (10.3.3) allows (10.3.3) to be written as a function of the given material parameters and the ratio for the velocity and absorption coefficient of a Love-Type surface wave as

$$\omega\frac{d_1}{v_L} = F\left[\frac{\rho_1}{\rho_2}, Q_{HS1}^{-1}, Q_{HS2}^{-1}, \frac{a_L}{a_{HS1}}, \frac{v_L}{v_{HS1}}\right], \tag{10.3.8}$$

where the function $F[\,]$ is defined by

$$F\left[\frac{\rho_1}{\rho_2}, Q_{HS1}^{-1}, Q_{HS2}^{-1}, \frac{a_L}{a_{HS1}}, \frac{v_L}{v_{HS1}}\right] \equiv -i\left(\left(1 - i\frac{a_L}{a_{HS1}}\frac{v_L}{v_{HS1}}\frac{Q_{HS1}^{-1}}{1+\chi_{HS1}}\right)^2 - \left(\frac{v_L}{v_{HS1}}\left(1 - i\frac{Q_{HS1}^{-1}}{1+\chi_{HS1}}\right)\right)^2\right)^{-1/2}$$
$$\left(\arctan\left[\frac{\left(-\frac{i\rho_2}{\rho_1}\frac{{v_{HS2}}^2}{{v_{HS1}}^2}\frac{1+\chi_{HS2}}{1+\chi_{HS1}}\frac{1 - iQ_{HS1}^{-1}}{1 - iQ_{HS2}^{-1}}\right)\left(\sqrt{\left(1 - i\frac{a_L}{a_{HS1}}\frac{v_L}{v_{HS1}}\frac{Q_{HS1}^{-1}}{1+\chi_{HS1}}\right)^2 - \left(\frac{v_L}{v_{HS1}}\frac{v_{HS1}}{v_{HS2}}\left(1 - \frac{iQ_{HS2}^{-1}}{1+\chi_{HS2}}\right)\right)^2}\right)}{\left(\sqrt{\left(1 - i\frac{a_L}{a_{HS1}}\frac{v_L}{v_{HS1}}\frac{Q_{HS1}^{-1}}{1+\chi_{HS1}}\right)^2 - \left(\frac{v_L}{v_{HS1}}\left(1 - \frac{iQ_{HS1}^{-1}}{1+\chi_{HS1}}\right)\right)^2}\right)}\right] \mp n\pi\right). \tag{10.3.9}$$

Hence, for a given set of material parameters ρ_2/ρ_1, v_{HS2}/v_{HS1}, Q_{HS1}^{-1}, and Q_{HS2}^{-1} and a given value of n, (10.3.9) shows explicitly that pairs of values of v_L/v_{HS1}, and a_L/a_{HS1} that yield a non-negative real number for the right-hand side of (10.3.9) represent the solution of (10.3.8) for a Love-Type surface wave of circular frequency ω and wave speed v_L in viscoelastic media with a layer of thickness d_1. For a given set of material parameters the pair of values for v_L/v_{HS1} and a_L/a_{HS_1} that satisfy (10.3.8) depend on frequency. This dependence of the wave speed on frequency is termed dispersion.

For elastic media with $Q_{HS1}^{-1} = Q_{HS2}^{-1} = 0$, (10.3.7), (10.3.6), and (10.3.5) imply

$$b_{\beta 1} = i\frac{\omega}{v_L}\sqrt{\frac{v_L^2}{v_{HS1}^2} - 1} = ib_{\beta 1_I}, \qquad b_{\beta 1_R} = 0,$$
$$b_{\beta 2} = \frac{\omega}{v_L}\sqrt{1 - \frac{v_L^2}{v_{HS2}^2}} = b_{\beta 2_R}, \qquad b_{\beta 2_I} = 0. \tag{10.3.10}$$

Hence, the period equations (10.3.3) and (10.3.8) for viscoelastic media simplify to the familiar period equation for elastic media, namely

$$\frac{\omega}{v_L}d_1 = \left(\sqrt{\frac{v_L^2}{v_{HS1}^2} - 1}\right)^{-1}\left(\arctan\left[\frac{\rho_2}{\rho_1}\frac{v_{HS2}^2}{v_{HS1}^2}\sqrt{1 - \frac{v_L^2}{v_{HS2}^2}}\Big/\sqrt{\frac{v_L^2}{v_{HS1}^2} - 1}\right] \mp n\pi\right), \tag{10.3.11}$$

showing that for a given value of n, values of the wave speed $v_L = \omega/k$ exist such that the period equation (10.3.11) is satisfied for values of v_L such that $v_{HS1} < v_L < v_{HS2}$.

The normalized physical displacement field in the viscoelastic half space is given by (10.2.6) with $n = 2$ as

$$\frac{\vec{u}_{2_R}(z)}{|D_{11}|} = 2|\mathbf{F}_{1_{11}}|\exp[k_I x - b_{\beta 2_R}(z - d_1)]$$
$$\sin\left[\omega t - k_R x + \pi/2 - b_{\beta 2_I}(z - d_1) + \arg D_{11} + \arg \mathbf{F}_{1_{11}}\right]\hat{x}_2, \tag{10.3.12}$$

where $\mathbf{F}_{1_{11}} = \cos\left[i\, b_{\beta 1} d_1\right]$. The corresponding normalized displacement in the viscoelastic layer is given by (10.2.11) with $m = 1$, as

$$\frac{\vec{u}_{1_R}(z)}{|D_{11}|} = \exp[k_I x]\frac{H_1}{|D_{11}|}\sin[\omega t - k_R x + \pi/2 + \delta_1]\hat{x}_2, \tag{10.3.13}$$

where

$$\frac{H_1}{|D_{11}|} = \sqrt{2\cosh\left[2b_{\beta 1_R}z\right] + 2\cos\left[2b_{\beta 1_I}z\right]}, \tag{10.3.14}$$

$$\delta_1 = \tan^{-1}\left[\frac{\exp\left[b_{\beta 1_R}z\right]\sin[g_{11}] + \exp\left[-b_{\beta 1_R}z\right]\sin[g_{21}]}{\exp\left[b_{\beta 1_R}z\right]\cos[g_{11}] + \exp\left[-b_{\beta 1_R}z\right]\cos[g_{21}]}\right], \tag{10.3.15}$$

and

$$g_{j1} \equiv (-1)^{j+1} b_{\beta 1_I} z + \arg D_{11} \tag{10.3.16}$$

for $j = 1, 2$.

For elastic media with $Q_{HS1}^{-1} = Q_{HS2}^{-1} = 0$, (10.3.10) implies $\mathbf{F}_{1_{11}} = \cos\left[b_{\beta 1_I} d_1\right]$, so (10.3.12) yields the familiar expression for the displacement in the elastic half space as

$$\frac{\vec{u}_{2_R}(z)}{|D_{11}|} = 2\cos\left[\frac{\omega}{v_L}\sqrt{\frac{v_L^2}{v_{HS1}^2} - 1}\, d_1\right]\exp\left[-\frac{\omega}{v_L}\sqrt{1 - \frac{v_L^2}{v_{HS2}^2}}(z - d_1)\right] \sin[\omega t - k_R x + \pi/2 + \arg D_{11}]\hat{x}_2, \tag{10.3.17}$$

where $v_{HS_1} < v_L < v_{HS2}$. Similarly, (10.3.10) and (10.3.13) yield the familiar expression for the displacement in the elastic layer as

$$\vec{u}_{1_R}(z) = 2|D_{11}|\cos\left[b_{\beta 1_I} z\right]\sin[\omega t - k_R x + \pi/2 + \arg D_{11}]\hat{x}_2, \tag{10.3.18}$$

where it can be readily shown that

$$H_1 = |D_{11}|\sqrt{2 + 2\cos\left[2 b_{\beta 1_I} z\right]} = |D_{11}| 2\cos\left[b_{\beta 1_I} z\right] \tag{10.3.19}$$

and

$$\delta_1 = \arctan\left[\frac{\sin[g_{11}] + \sin[g_{21}]}{\cos[g_{11}] + \cos[g_{21}]}\right] = \arg D_{11}. \tag{10.3.20}$$

Comparison of (10.3.17) with (10.3.18) and (10.3.19) together with (10.3.10) shows that the displacement is continuous across the boundary. In contrast to the situation for anelastic media, (10.3.17) shows that the displacement in the elastic half space shows no superimposed sinusoidal dependence with depth and that the displacement at any depth is in phase with that at the surface.

10.4 Numerical Characteristics of Love-Type Surface Waves

Quantitative descriptions of the physical characteristics of a Love-Type surface wave can be derived from solutions for pairs of values for wave speed v_L and absorption coefficient a_L that satisfy (10.3.8) for a given set of viscoelastic material parameters and chosen frequency for the wave. If pairs of values of v_L and a_L are solutions of (10.3.8) for a Love-Type surface wave, then the imaginary part of $F(\,)$ as defined in (10.3.9) must vanish and the real part must be positive, that is

$$\operatorname{Im}\left[\mathrm{F}\left(\frac{\rho_1}{\rho_2}, Q_{HS1}^{-1}, Q_{HS2}^{-1}, \frac{a_L}{a_{HS1}}, \frac{v_L}{v_{HS1}}\right)\right] = 0 \tag{10.4.1}$$

and

$$\mathrm{Re}\left[F\left(\frac{\rho_1}{\rho_2}, Q_{HS1}^{-1}, Q_{HS2}^{-1}, \frac{a_L}{a_{HS1}}, \frac{v_L}{v_{HS1}}\right)\right] > 0. \tag{10.4.2}$$

These conditions suggest an approach towards determining the desired pairs of v_L and a_L for a given set of material parameters. They indicate that upon specification of an appropriate wave-speed ratio v_L/v_{HS1} the corresponding absorption-coefficient ratio a_L/a_{HS1} can be determined by finding the root of (10.4.1) that satisfies (10.4.2). Numerical calculations that use the secant method in Mathematica (Wolfram, 1999) to locate the roots of (10.4.1) show that this approach yields the desired pairs of v_L and a_L values for a wide range of viscoelastic materials as a function of frequency and thickness of the layer.

Quantitative estimates of the absorption coefficient ratio a_L/a_{HS1} as a function of the wave-speed ratio v_L/v_{HS1} for values of the ratio that satisfy $1 < v_L/v_{HS1} < v_{HS2}/v_{HS1}$ are shown for a given set of viscoelastic material parameters for the fundamental mode of a Love-Type surface wave in Figure (10.4.6). Values chosen for the material parameters needed to solve (10.4.1) are $\rho_2/\rho_1 = 1.283$, $v_{HS2}/v_{HS1} = 1.297$, $Q_{HS2}^{-1} = 0.01$, and a set of values for intrinsic absorption in the layer as indicated in the figure ranging from low loss $(Q_{HS1}^{-1} = 0.01)$ to significant loss $\left(Q_{HS1}^{-1} = 0.5\right)$. The curves indicate that the dependence of the absorption-coefficient ratio on the wave-speed ratio varies significantly as the amount of intrinsic absorption in the layer increases.

Dispersion curves corresponding to the absorption curves in Figure (10.4.6) are shown in Figure (10.4.7) for the fundamental mode of Love-Type surface waves. The plots indicate that for the chosen material parameters, variations in the curves due to variations in the amount of intrinsic absorption in the layer are most evident for amounts of intrinsic absorption $Q_{HS1}^{-1} \geq 0.1$. For low-loss amounts of intrinsic absorption in the layer with $Q_{HS1}^{-1} < 0.1$ the deviations in the curves from those for an elastic solid are small and less discernible at the scale plotted. The curve computed for the special case of elastic media agrees with that previously published (see e.g. Ewing, Jardetsky, and Press, 1957, p. 213).

The distribution of physical displacement with depth as specified by (10.3.12) and (10.3.13) can be computed efficiently as a function of depth in units of fractions of a wavelength along the surface. With the wavelength along the surface defined by $\lambda \equiv 2\pi/k_R$ the normalized maximum amplitude of a Love-Type surface wave, expressed as a function of depth in terms of fractions of a wavelength, z/λ, is given in the half space by

$$\max\left[\frac{\vec{u}_{2R}(z)}{|D_{11}|}\right] = 2\left|\cos\left[i\frac{b_{\beta 1}}{k_R}2\pi\frac{d_1}{\lambda}\right]\right|\exp\left[-a_L x - 2\pi\frac{b_{\beta 2R}}{k_R}\left(\frac{z}{\lambda} - \frac{d_1}{\lambda}\right)\right] \tag{10.4.3}$$

and in the layer from (10.3.13) as

$$\frac{\max[\vec{u}_{1_R}(z)]}{|D_{11}|} = \exp[-a_L x]\sqrt{2\cosh\left[4\pi\frac{b_{\beta 1_R}}{k_R}\frac{z}{\lambda}\right] + 2\cos\left[4\pi\frac{b_{\beta 1_I}}{k_R}\frac{z}{\lambda}\right]}. \quad (10.4.4)$$

For purposes of calculation, fractions of a wavelength, z/λ, may be expressed in terms of $\omega d_1/v_L$ as

$$\frac{z}{\lambda} = z\frac{k_R}{2\pi} = \frac{1}{2\pi}\frac{z}{d_1}\frac{\omega\, d_1}{v_L}. \quad (10.4.5)$$

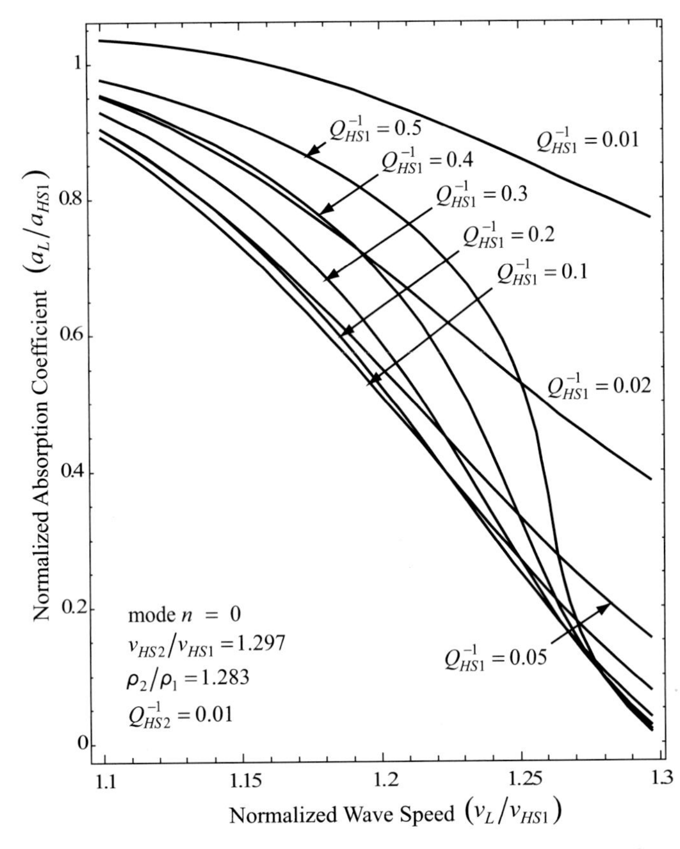

Figure (10.4.6). Normalized absorption coefficient as a function of normalized wave speed for the fundamental mode of a Love-Type surface wave in viscoelastic media with material parameters as indicated.

Hence, a given pair of values of v_L/v_{HS1} and a_L/a_{HS1} that satisfy (10.4.1) and (10.4.2) when substituted into (10.3.8) yield estimates of $\omega d_1/v_L$ and in turn estimates of $d_1/\lambda = \omega d_1/v_L$. As a result the amplitude distribution in the half space (10.4.3) when specified in units of fractions of a wavelength may be computed without an explicit specification of the thickness d_1.

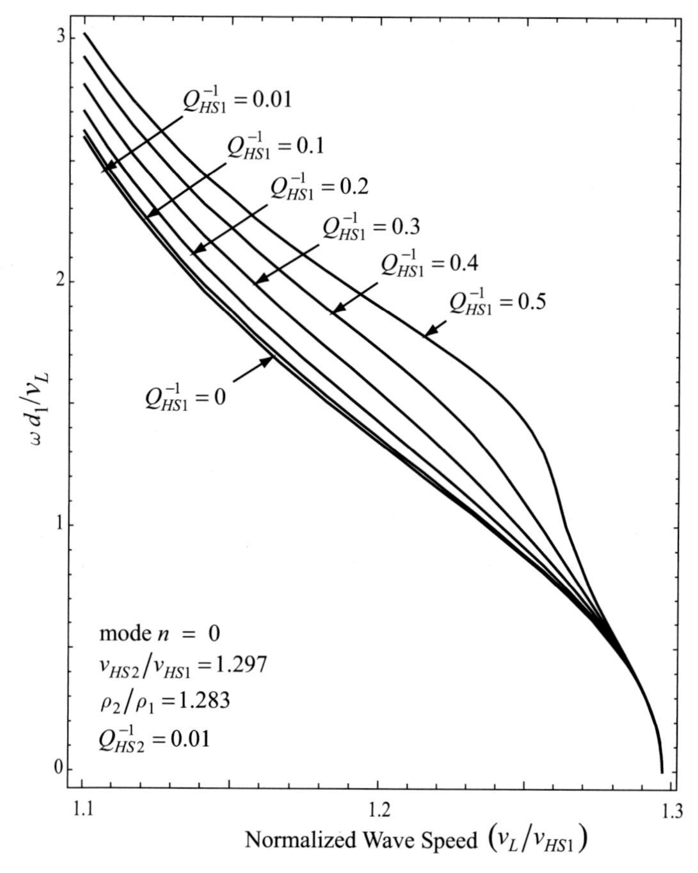

Figure (10.4.7). Dispersion curves showing the correspondence between normalized wave speed and $\omega d_1/v_L$ for the fundamental mode of a Love-Type surface wave in viscoelastic media with material parameters as indicated.

The normalized maximum physical amplitude of a Love-Type surface wave is shown as a function of depth for various amounts of intrinsic absorption in the layer in Figure (10.4.8). The amplitude distribution corresponding to each value of Q_{HS1}^{-1} is that for which the wave-speed ratio $v_L/v_{HS1} = 1.25$ and the viscoelastic material parameters are as indicated.

The curves in Figure (10.4.8) indicate that for the chosen material parameters, the amplitudes decrease by about 25 percent within a depth of about 17 percent of a wavelength below the surface. The curves indicate that deviations in the amplitude distributions due to increases in intrinsic absorption in the layer begin to become apparent for depths greater than about 20 percent of a wavelength and values of $Q_{HS1}^{-1} > 0.1$. For depths greater than about 20 percent, the decrease in amplitude with depth increases with an increase in intrinsic absorption. For the material parameters chosen, the dependence of the normalized amplitude distribution on depth for low-loss media ($Q_{HS1}^{-1} < 0.1$) is nearly indistinguishable at the scale plotted from that for corresponding elastic media.

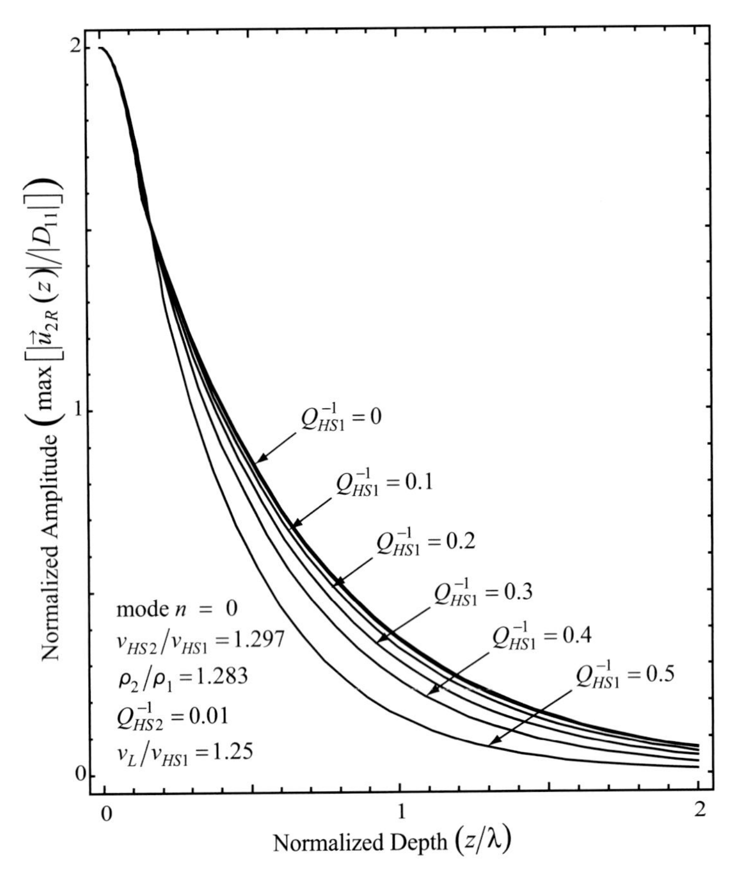

Figure (10.4.8). Normalized amplitude distribution for the fundamental mode of a Love-Type surface wave in viscoelastic media with material parameters as indicated.

Normalized absorption, dispersion, and normalized amplitude distribution curves are shown for the first higher mode ($n=1$) of a Love-Type surface wave in Figures (10.4.9), (10.4.10), and (10.4.15). These curves for the first higher mode are computed for the same set of viscoelastic material parameters as in corresponding Figures (10.4.6), (10.4.7), and (10.4.8) for the fundamental mode.

The curves for the first higher mode in Figures (10.4.9) and (10.4.10) indicate that for the material parameters chosen, the range of wave-speed ratios v_L/v_{HS1} for which valid roots of (10.4.1) can be determined, that is valid absorption coefficients ratios can be determined, is restricted to large amounts of intrinsic absorption ($Q_{HS1}^{-1} \geq 0.4$) compared with the corresponding range for the fundamental mode.

The normalized absorption coefficient for the first higher mode shows a strong dependence on Q_{HS1}^{-1} in the layer (Figure (10.4.9)). The corresponding dispersion curve (Figure (10.4.10)) shows a distinguishable dependence on Q_{HS1}^{-1} for media with $Q_{HS1}^{-1} \geq 0.1$. The dispersion curve for an equal amount of absorption in the layer and the half space ($Q_{HS1}^{-1} = Q_{HS2}^{-1} = 0.01$) is indistinguishable from the corresponding curve for elastic media.

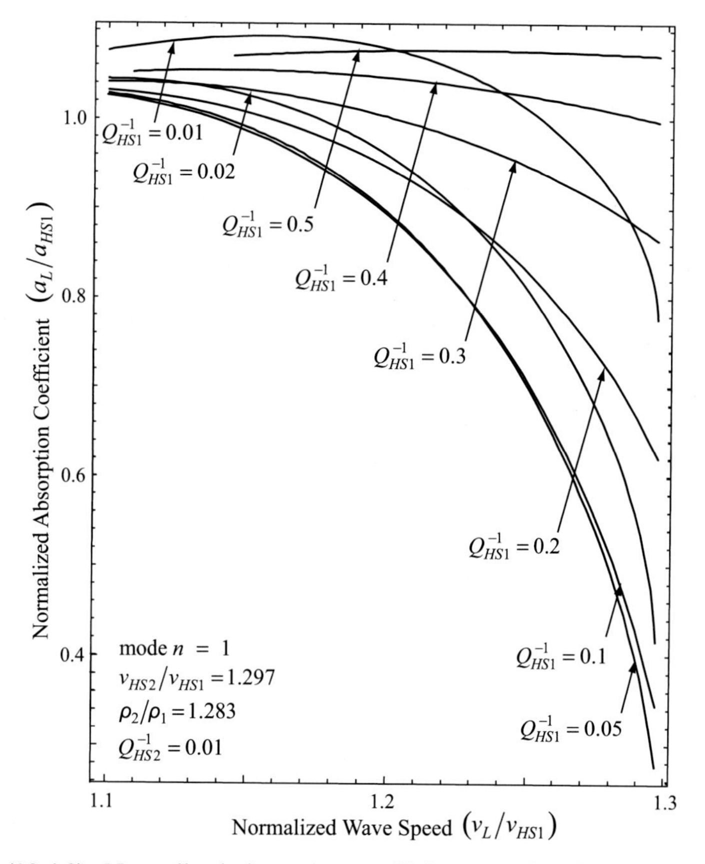

Figure (10.4.9). Normalized absorption coefficient as a function of normalized wave speed for the first higher mode ($n = 1$) of a Love-Type surface wave in viscoelastic media with material parameters as indicated.

The normalized amplitude distribution for the first higher mode (Figure (10.4.15)) indicates that at a depth of about 33 percent of a wavelength the amplitude of a Love-Type surface wave vanishes for elastic media but does not for anelastic media. For the parameters chosen, distinguishable dependences on Q_{HS1}^{-1} are apparent for depths near the minimum and maximum values in the distribution at depths near 33 and 66 percent of a wavelength. The largest dependences on Q_{HS1}^{-1} are apparent at depths near and greater than a wavelength.

The procedure used here to derive quantitative estimates of the physical characteristics of Love-Type surface waves on an arbitrary viscoelastic media with general material parameters $v_{HSj}, Q_{HSj}^{-1}, \rho_j$ for $j = 1, 2$ can be readily extended to any particular viscoelastic solid as might be characterized by combinations of springs and dashpots. As an example, if the viscoelastic model for the layer is chosen as a Standard Linear solid with material parameters τ_p, τ_e, M_r and ρ_1 and the model for the half space is

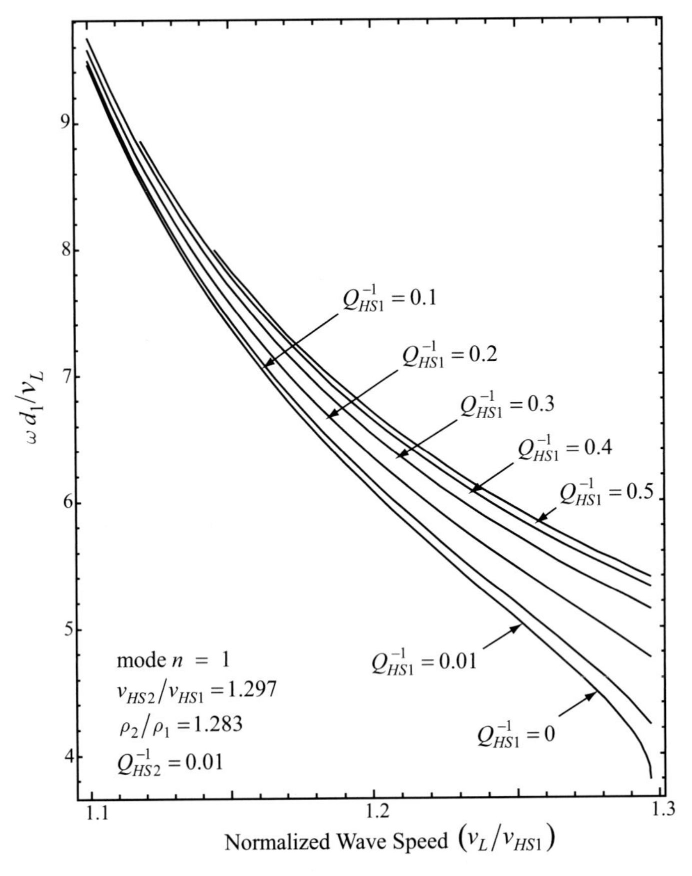

Figure (10.4.10). Dispersion curves showing the correspondence between normalized wave speed and $\omega d_1/v_L$ for the first higher mode ($n=1$) of a Love-Type surface wave in viscoelastic media with material parameters as indicated.

chosen as a Maxwell solid with material parameters μ, η, and ρ_2, then specification of circular frequency ω and these material parameters implies Q_{HSj}^{-1} and v_{HSj} for $j = 1, 2$ are given in the layer from Table (1.3.30) using (3.5.2) and (3.5.5) by

$$Q_{HS1}^{-1} = \frac{\omega(\tau_e - \tau_p)}{1 + \omega^2 \tau_p \tau_e} \tag{10.4.11}$$

and

$$v_{HS1} = \sqrt{\frac{\frac{M_r}{\rho_1} \frac{1 + \omega^2 \tau_e \tau_p}{1 + \omega^2 \tau_p^2}\, 2\left(1 + \left(\frac{\omega(\tau_e - \tau_p)}{(1 + \omega^2 \tau_e \tau_p)}\right)^2\right)}{1 + \sqrt{1 + \left(\frac{\omega(\tau_e - \tau_p)}{(1 + \omega^2 \tau_e \tau_p)}\right)^2}}} \tag{10.4.12}$$

and in the half space by

$$Q_{HS2}^{-1} = \frac{\mu}{\omega\eta} \tag{10.4.13}$$

and

$$v_{HS2} = \sqrt{\frac{\mu}{\rho_2} \frac{\omega^2\eta^2}{\mu^2 + \omega^2\eta^2} 2\left(1 + \left(\frac{\mu}{\omega\eta}\right)^2\right) \Big/ \left(1 + \sqrt{1 + \left(\frac{\mu}{\omega\eta}\right)^2}\right)}. \tag{10.4.14}$$

Hence, specification of the material parameters for these particular viscoelastic models together with frequency ω allows the general material parameters v_{HS2}/v_{HS1}, ρ_2/ρ_1, and Q_{HSj}^{-1} for $j = 1, 2$ to be determined, from which curves similar to those in Figures (10.4.6) through (10.4.15) for a Love-Type surface wave can be readily derived.

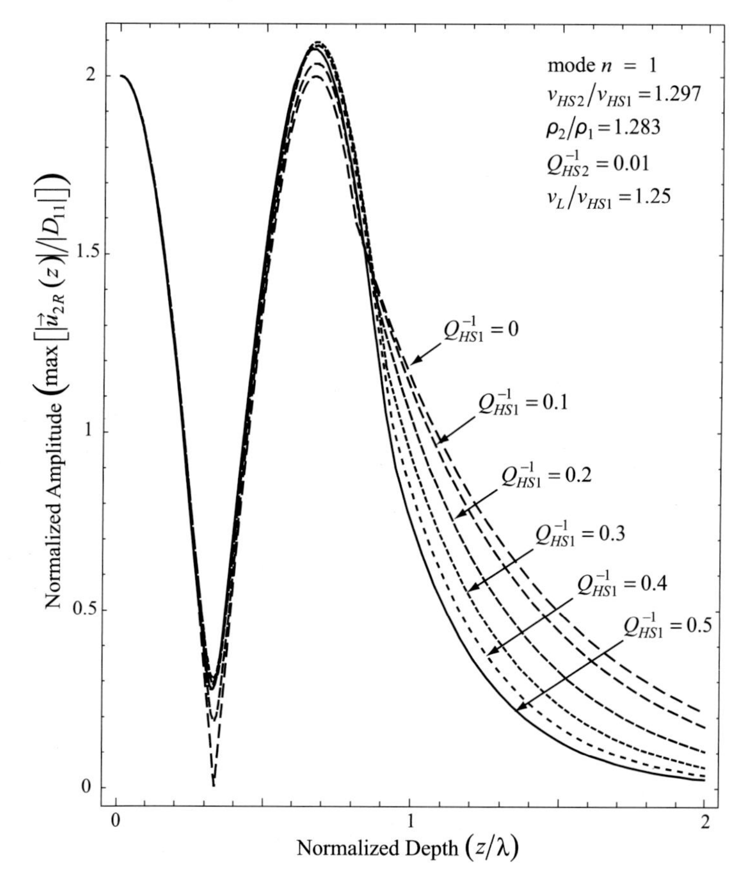

Figure (10.4.15). Normalized amplitude distribution for the first higher mode ($n = 1$) of a Love-Type surface wave in viscoelastic media with material parameters as indicated.

10.5 Problems

(1) Use the period equation for a Love-Type surface wave propagating in a layer over a half space of arbitrary viscoelastic media as specified by (10.3.8) and (10.3.9) to derive the period equation for elastic media (10.3.11). Similarly, find the period equation if the media are Voight solids.

(2) Use the period equation for a Love-Type surface wave specified in terms of the complex shear moduli by (10.3.3) to derive the period equation as expressed in terms of the wave-speed and absorption-coefficient ratios v_L/v_{HS1} and a_L/a_{HS1} and material parameters ρ_2/ρ_1, v_{HS2}/v_{HS1}, Q_{HS1}^{-1}, and Q_{HS2}^{-1} as specified by (10.3.8) and (10.3.9).

(3) Describe each of the steps used to solve the complex period equation as specified by (10.3.8) and (10.3.9) for a Love-Type surface wave in a viscoelastic layer over a half space.

(4) Sketch curves showing pairs of values for the absorption coefficient ratio and wave-speed ratio $(a_L/a_{HS1}, v_L/v_{HS1})$ that satisfy the complex period equation as specified by (10.3.8) and (10.3.9) for viscoelastic layers with increasing amounts of intrinsic absorption, that is $Q_{HS1}^{-1} = 0.01$, $Q_{HS1}^{-1} = 0.02$, $Q_{HS1}^{-1} = 0.1$, and $Q_{HS1}^{-1} = 0.5$ for the fundamental mode of a Love-Type surface wave using the curves in Figure (10.4.6).

(5) Plot values for the absorption-coefficient ratio versus values of $Q_{HS1}^{-1} = 0.01$, 0.02, 0.05, 0.1, 0.2, 0.3, 0.4, 0.5 for a fixed value of the wave-speed ratio of $v_L/v_{HS1} = 1.2$ for the fundamental mode of a Love-Type surface wave using Figure (10.4.6). Describe how the absorption coefficient ratio as plotted in Figure (10.4.6) varies with increasing amounts of intrinsic absorption in shear of the layer.

(6) Sketch dispersion curves for the fundamental mode of a Love-Type surface wave showing the correspondence between normalized wave speed v_L/v_{HS1} and $\omega d_1/v_L$ for viscoelastic layers with $Q_{HS1}^{-1} = 0.01$, $Q_{HS1}^{-1} = 0.02$, $Q_{HS1}^{-1} = 0.1$, and $Q_{HS1}^{-1} = 0.5$ using Figure (10.4.7). Describe the influence of increasing intrinsic absorption in the layer on the dispersion curve.

(7) Sketch curves showing the normalized amplitude distribution versus normalized depth for the fundamental mode of a Love-Type surface wave for viscoelastic layers with $Q_{HS1}^{-1} = 0$, $Q_{HS1}^{-1} = 0.1$, and $Q_{HS1}^{-1} = 0.5$, using Figure (10.4.8). Describe the influence of increasing intrinsic absorption in the layer on the normalized amplitude distribution in the half space.

(8) Compare curves for the fundamental mode of a Love-Type surface wave in Figures (10.4.6), (10.4.7), and (10.4.8) with corresponding curves for the first higher mode, namely (10.4.9), (10.4.10), and (10.4.15). Comment on differences observed in the curves.

11

Appendices

11.1 Appendix 1 – Properties of Riemann–Stieltjes Convolution Integral

Useful properties of the Riemann–Stieltjes convolution integral as proved by Gurtin and Sternberg (1962, pp. 298–299) are as follows.

Theorem (11.1.1) If $f(t)$ *is continuous on* $[0, \infty]$, *the first derivatives of* $g(t)$ and $h(t)$ *exist on* $[0, \infty]$, *and* $f(t) = g(t) = h(t) = 0$ *for* $t < 0$, *then*

$$f * dg = g * df, \tag{11.1.2}$$

$$f * d(g * h) = (f * dg) * dh = f * dg * dh, \tag{11.1.3}$$

$$f * d(g + h) = f * dg + f * dh, \tag{11.1.4}$$

$$f * d(g) = 0 \qquad \text{implies } f = 0 \text{ or } g = 0, \tag{11.1.5}$$

$$f * d(g) = g(0)f(t) + \int_0^\infty f(t - \tau)\frac{dg(\tau)}{d\tau}d\tau \qquad \text{for } 0 \le t < \infty. \tag{11.1.6}$$

11.2 Appendix 2 – Vector and Displacement-Potential Identities

This appendix provides identities for scalar and vector products and identities for displacement potentials.

11.2.1 Vector Identities

Useful vector identities are

(1) scalar triple product,

$$\vec{A} \bullet \left(\vec{B} \times \vec{C}\right) = \vec{B} \bullet \left(\vec{C} \times \vec{A}\right) = \vec{C} \bullet \left(\vec{A} \times \vec{B}\right), \tag{11.2.1}$$

(2) vector triple product,

$$\vec{A} \times \left(\vec{B} \times \vec{C}\right) = \vec{B}\left(\vec{A} \cdot \vec{C}\right) - \vec{C}\left(\vec{A} \cdot \vec{B}\right), \tag{11.2.2}$$

(3) vector quadruple product,

$$\left(\vec{A} \times \vec{B}\right) \cdot \left(\vec{C} \times \vec{D}\right) = \left(\vec{A} \cdot \vec{C}\right)\left(\vec{B} \cdot \vec{D}\right) - \left(\vec{A} \cdot \vec{D}\right)\left(\vec{B} \cdot \vec{C}\right), \tag{11.2.3}$$

$$\left(\vec{A} \times \vec{B}\right)^2 = \left(\vec{A} \cdot \vec{A}\right)\left(\vec{B} \cdot \vec{B}\right) - \left(\vec{A} \cdot \vec{B}\right)\left(\vec{B} \cdot \vec{A}\right) = \vec{A}^2\vec{B}^2 - \left(\vec{A} \cdot \vec{B}\right)^2. \tag{11.2.4}$$

11.2.2 Displacement-Potential Identities

Useful displacement-potential identities for the scalar potential

$$\phi = G_0 \exp\left[-\vec{A} \cdot \vec{r}\right] \exp\left[i\left(\omega t - \vec{P} \cdot \vec{r}\right)\right] = G_0 \exp\left[i\left(\omega t - \vec{K} \cdot \vec{r}\right)\right] \tag{11.2.5}$$

are

$$\nabla\phi = -(\vec{A} + i\vec{P})\phi = -i\vec{K}\phi \tag{11.2.6}$$

and

$$\nabla^2\phi = (\vec{A} + i\vec{P}) \cdot (\vec{A} + i\vec{P})\phi = -\vec{K} \cdot \vec{K}\phi. \tag{11.2.7}$$

Useful identities for the vector potential (Cowan, 1968, p. 226)

$$\vec{\psi} = \vec{G}_0 \exp\left[-\vec{A} \cdot \vec{r}\right] \exp\left[i\left(\omega t - \vec{P} \cdot \vec{r}\right)\right] = \vec{G}_0 \exp\left[i\left(\omega t - \vec{K} \cdot \vec{r}\right)\right] \tag{11.2.8}$$

are

$$\nabla \cdot \vec{\psi} = -i(\vec{P} - i\vec{A}) \cdot \vec{\psi} = -i\vec{K} \cdot \vec{\psi}, \tag{11.2.9}$$

$$\nabla \times \vec{\psi} = -i(\vec{P} - i\vec{A}) \times \vec{\psi} = -i\vec{K} \times \vec{\psi}, \tag{11.2.10}$$

and

$$\nabla^2\vec{\psi} = -(\vec{P} - i\vec{A}) \cdot (\vec{P} - i\vec{A})\vec{\psi} = -\vec{K} \cdot \vec{K}\vec{\psi}. \tag{11.2.11}$$

11.3 Appendix 3 – Solution of the Helmholtz Equation

This appendix uses the method of separation of variables to derive solutions of the scalar Helmholtz equation. The derivation provides additional insight regarding the nature of the solutions used to describe P, S, and surface waves.

For two-dimensional wave-propagation problems in layered media it is sufficient to assume that particle motions along any line parallel to one of the coordinates are equal. This assumption, applied to the x_2 axis of an assumed Cartesian coordinate system (x_1, x_2, x_3), is shown in Chapter 4 to imply that the Helmholtz equations governing the propagation of P, S, and surface waves can be reduced to three scalar equations involving the scalar potential Φ, the scalar potential Ψ_2 as the x_2 component of the vector potential $\vec{\Psi}$, and the scalar displacement U_2 as the x_2 component of the displacement field $\vec{U}_2$. The resulting scalar Helmholtz equations are

$$\nabla^2 \Phi + k_P^2 \Phi = 0, \tag{11.3.1}$$

$$\nabla^2 \Psi_2 + k_S^2 \Psi_2 = 0, \tag{11.3.2}$$

and

$$\nabla^2 U_2 + k_S^2 U_2 = 0. \tag{11.3.3}$$

The solution to each of these equations may be found using the method known as separation of variables. To apply the method, consider (11.3.1) governing P wave solutions and suppose the solution for Φ may be written so that the dependences on the two spatial variables may be separated into the product of two functions, each dependent on only one spatial coordinate as

$$\Phi = X_1(x_1) X_3(x_3). \tag{11.3.4}$$

Substitution of this expression into (11.3.1) yields

$$\frac{X_{1,11}(x_1)}{X_1(x_1)} = -\left(\frac{X_{3,33}(x_3)}{X_3(x_3)} + k_P^2 \right), \tag{11.3.5}$$

where $X_{j,ii}$ denotes $\partial^2 X_j / \partial x_i^2$. Each side of (11.3.5) depends on a different spatial coordinate. Hence, each side of the equation must be independent of the spatial coordinates and equal to some arbitrary, but fixed constant, say $-b$. As a consequence, two second-order linear differential equations are to be solved, namely,

$$X_{1,11} + b X_1 = 0 \tag{11.3.6}$$

and

$$X_{3,33} + \left(k_P^2 - b\right) X_3 = 0. \tag{11.3.7}$$

The auxiliary equation for (11.3.6) namely,

$$m^2 + b = 0, \tag{11.3.8}$$

implies a solution of (11.3.6) involving the x_1 spatial variable is

$$X_1 = D_1 \exp[-ik_\phi x_1] + D_2 \exp[ik_\phi x_1], \tag{11.3.9}$$

where $\pm k_\phi$ are roots of the auxiliary equation (11.3.8) defined as

$$k_\phi \equiv \textit{principal value}\ \sqrt{b}. \tag{11.3.10}$$

Use of the principal value of the square root as defined in Chapter 4 to define k_ϕ implies $k_{\phi_R} \geq 0$. Writing the root k_ϕ of the auxiliary equation in this form ensures a known direction of propagation with respect to the x_1 axis for each term in (11.3.9) with the term needed to consider propagation in the positive x_1 direction being

$$X_1 = D_1 \exp[-ik_\phi x_1]. \tag{11.3.11}$$

Substitution of the solution (11.3.9) into (11.3.5) reconfirms that the complex amplitudes for the terms in the solution, namely D_1 and D_2, are constants independent of the spatial coordinates.

Similarly, a solution to (11.3.7) involving the x_3 spatial coordinate is

$$X_3 = E_1 \exp[id_\alpha x_3] + E_2 \exp[-id_\alpha x_3], \tag{11.3.12}$$

where the auxiliary equation is

$$m^2 = k_\phi^2 - k_P^2. \tag{11.3.13}$$

Writing roots of the auxiliary equation in the form of

$$d_\alpha \equiv \textit{principal value}\ \sqrt{k_P^2 - k_\phi^2} \tag{11.3.14}$$

implies $d_{\alpha_R} \geq 0$, which will ensure a known direction of propagation with respect to the x_3 axis for each term in (11.3.12). Writing roots of the auxiliary equation in the form of

$$b_\alpha \equiv \textit{principal value}\sqrt{k_\phi^2 - k_P^2} = -i\, d_\alpha \tag{11.3.15}$$

implies $b_{\alpha_R} \geq 0$ and hence $d_{\alpha_I} \geq 0$, which ensures a known direction of attenuation with respect to the x_3 axis for each term in (11.3.12). Choice of the expression for the root to ensure a known direction of propagation along the x_3 axis is convenient for consideration of reflection–refraction problems. Choice of the expression for the root to ensure a known direction for attenuation along the x_3 axis is convenient for consideration of the problem of a Rayleigh-Type surface wave.

Combining the solutions (11.3.11) and (11.3.12) implies the desired solution for the Helmholtz equation for P wave solutions may be written as

$$\Phi = X_1(x_1)X_3(x_3) = B_1 \exp[-ik_\phi x_1 + id_\alpha x_3] + B_2 \exp[-ik_\phi x_1 - id_\alpha x_3], \tag{11.3.16}$$

which upon separation of the real and imaginary parts of the exponents allows the solution of the corresponding wave equation to be written in terms of propagation and attenuation vectors as

$$\phi = B_1 \exp\left[i\left(\omega t - \left(\vec{P}_{\phi_1} - i\vec{A}_{\phi_1}\right)\bullet\vec{r}\right)\right] + B_2 \exp\left[i\left(\omega t - \left(\vec{P}_{\phi_2} - i\vec{A}_{\phi_2}\right)\bullet\vec{r}\right)\right], \tag{11.3.17}$$

where

$$\vec{P}_{\phi_j} \equiv k_{\phi_R}\hat{x}_1 + (-1)^j d_{\alpha_R}\hat{x}_3 \qquad \text{for } j = 1, 2 \tag{11.3.18}$$

and

$$\vec{A}_{\phi_j} \equiv -k_{\phi_I}\hat{x}_1 + (-1)^{j+1} d_{\alpha_I}\hat{x}_3 \qquad \text{for } j = 1, 2. \tag{11.3.19}$$

With the expression for d_α chosen as in (11.3.14) $\vec{P}_{\phi_1}$ and $\vec{P}_{\phi_2}$ are directed in the $-x_3$ and $+x_3$ directions, respectively. With the expression for d_α chosen as in (11.3.15) $\vec{A}_{\phi_1}$ and $\vec{A}_{\phi_2}$ are directed in the $+x_3$ and $-x_3$ directions, respectively.

Similarly, solutions of equations (11.3.2) and (11.3.3) for propagation in the $+x_1$ direction with $\Psi \equiv \Psi_2$ and $U = U_2$ are given by

$$\begin{aligned} \Psi &= C_{12} \exp\left[-ik_\psi x_1 + id_\beta x_3\right] + C_{22} \exp\left[-ik_\psi x_1 - id_\beta x_3\right], \\ \psi(t) &= C_{12} \exp\left[i\left(\omega t - \left(\vec{P}_{\psi_1} - i\vec{A}_{\psi_1}\right)\bullet\vec{r}\right)\right] + C_{22} \exp\left[i\left(\omega t - \left(\vec{P}_{\psi_2} - i\vec{A}_{\psi_2}\right)\bullet\vec{r}\right)\right], \end{aligned} \tag{11.3.20}$$

and

$$\begin{aligned} U &= D_1 \exp\left[-ik_u x_1 + id_\beta x_3\right] + D_2 \exp\left[-ik_u x_1 - id_\beta x_3\right], \\ u(t) &= C_{12} \exp\left[i\left(\omega t - \left(\vec{P}_{u_1} - i\vec{A}_{u_1}\right)\bullet\vec{r}\right)\right] + C_{22} \exp\left[i\left(\omega t - \left(\vec{P}_{u_2} - i\vec{A}_{u_2}\right)\bullet\vec{r}\right)\right], \end{aligned} \tag{11.3.21}$$

where

(1) k_ψ and k_u are chosen such that $k_{\psi_R} \geq 0$ and $k_{u_R} \geq 0$,
(2) d_β and d_u are defined either by

$$d_\beta \equiv \textit{principal value}\sqrt{k_S^2 - k_\psi^2} \tag{11.3.22}$$

and

$$d_u \equiv \textit{principal value}\sqrt{k_S^2 - k_u^2} \tag{11.3.23}$$

for known directions of propagation with respect to the x_3 axis or by

$$d_\beta \equiv i\ \textit{principal value}\sqrt{k_\psi^2 - k_S^2} \tag{11.3.24}$$

and

$$d_u \equiv i \; principal\, value \sqrt{k_u^2 - k_S^2} \tag{11.3.25}$$

for known directions of maximum attenuation with respect to the x_3 axis,

(3) the propagation and attenuation vectors are defined by

$$\vec{P}_{\psi_j} \equiv k_{\psi_R}\hat{x}_1 + (-1)^j d_{\beta_R}\hat{x}_3 \qquad \text{for } j = 1, 2, \tag{11.3.26}$$

$$\vec{P}_{u_j} \equiv k_{u_R}\hat{x}_1 + (-1)^j d_{\beta_R}\hat{x}_3 \qquad \text{for } j = 1, 2, \tag{11.3.27}$$

and

$$\vec{A}_{\psi_j} \equiv -k_{\psi_I}\hat{x}_1 + (-1)^{j+1} d_{\beta_I}\hat{x}_3 \qquad \text{for } j = 1, 2 \tag{11.3.28}$$

$$\vec{A}_{u_j} \equiv -k_{u_I}\hat{x}_1 + (-1)^{j+1} d_{\beta_I}\hat{x}_3 \qquad \text{for } j = 1, 2, \tag{11.3.29}$$

(4) and the complex amplitudes C_{j2} and D_j for $j = 1, 2$ are constants independent of the spatial coordinates.

For purposes of generality, distinct complex wave numbers are assumed for the solutions to each scaler Helmholtz equation, that is, k_ϕ, k_ψ, and k_u are not necessarily assumed to be the same arbitrary, but fixed, constant. However, consideration of various reflection–refraction and surface-wave problems shows that upon application of the boundary conditions to the solutions derived here, the complex wave number used in the various solutions must be the same. Hence, for purposes of brevity, the distinction between the complex wave numbers as specified by the subscript is often neglected.

11.4 Appendix 4 – Roots of Squared Complex Rayleigh Equation

This appendix derives explicit expressions for the roots of the squared complex Rayleigh equation (8.1.20) for a HILV half space.

The square of the complex Rayleigh equation, namely

$$\left(\frac{c^2}{\beta^2}\right)^3 - 8\left(\frac{c^2}{\beta^2}\right)^2 + \left(24 - 16\frac{\beta^2}{\alpha^2}\right)\left(\frac{c^2}{\beta^2}\right) - 16\left(1 - \frac{\beta^2}{\alpha^2}\right) = 0, \tag{11.4.1}$$

is a cubic polynomial with coefficients in the complex field. The Fundamental Theorem of Algebra implies the equation has three roots not all of which may be distinct. The technique for finding the roots of such an equation is well known in mathematics (see for example, Birkoff and MacLane, 1953, p. 112). To derive explicit expressions for the roots, define

$$a \equiv -8, \qquad b \equiv 24 - 16\frac{\beta^2}{\alpha^2}, \text{ and } \qquad c \equiv -16\left(1 - \frac{\beta^2}{\alpha^2}\right), \tag{11.4.2}$$

then the equation may be written as

$$y^3 + ay^2 + by + c = 0, \tag{11.4.3}$$

where

$$y \equiv \frac{c^2}{\beta^2}. \tag{11.4.4}$$

Equation (11.4.3) may be put in reduced form with no second-order term upon introduction of the transformation $y = z - a/3$, resulting in

$$z^3 + pz + q = 0, \tag{11.4.5}$$

where

$$p \equiv -\frac{a^2}{3} + b; \qquad q \equiv \frac{2}{27}a^3 - \frac{ab}{3} + c. \tag{11.4.6}$$

Introduction of a second transformation allows the equation to be reduced to a quadratic equation, namely, let

$$z = w - \frac{p}{3w}, \tag{11.4.7}$$

then substitution of (11.4.7) into (11.4.5) yields the desired quadratic equation

$$\left(w^3\right)^2 + q\left(w^3\right) - \left(\frac{p}{3}\right)^3 = 0. \tag{11.4.8}$$

With $v \equiv w^3$ roots of the quadratic equation are

$$v_+ \equiv \frac{-q}{2} + \sqrt{\left(\frac{q}{2}\right)^2 + \left(\frac{p}{3}\right)^3} \tag{11.4.9}$$

and

$$v_- \equiv \frac{-q}{2} - \sqrt{\left(\frac{q}{2}\right)^2 + \left(\frac{p}{3}\right)^3}. \tag{11.4.10}$$

Roots of the original equation (11.4.3) are found by tracing the roots (11.4.9) and (11.4.10) through the three successive transformations which yields a total of six roots. Defining the principal value of the cube root of a complex number as

$$\text{p.v.}\left(z^{1/3}\right) \equiv \sqrt[3]{|z|}\ \exp[i \arg[z]/3], \tag{11.4.11}$$

where $-\pi < \arg[z] \leq \pi$ allows the six roots to be written as

$$z_n^+ \equiv w_0^+ u^{n-1} - \frac{p}{3w_0{}^+ u^{n-1}} \qquad (n = 1,\ 2,\ 3) \tag{11.4.12}$$

and

$$z_n^- \equiv w_0^- u^{n-1} - \frac{p}{3w_0^- u^{n-1}} \qquad (n = 1,\ 2,\ 3), \tag{11.4.13}$$

where

$$w_0^+ \equiv \text{p.v.}(v_+)^{1/3}, \tag{11.4.14}$$

$$w_0^- \equiv \text{p.v.}(v_-)^{1/3}, \tag{11.4.15}$$

and

$$u \equiv \exp[2\pi i/3]. \tag{11.4.16}$$

Each of the roots z_n^+ can be quickly shown to be equal to one of the corresponding roots z_n^- for $n = 1,\ 2,\ 3$. Hence, the three roots of the original equation are

$$y_n \equiv z_n^- - \frac{a}{3} \qquad \text{for } n = 1,\ 2,\ 3, \tag{11.4.17}$$

which when expressed in terms of β^2/α^2 yield the desired roots as specified in (8.1.21)

11.5 Appendix 5 – Complex Root for a Rayleigh-Type Surface Wave

This appendix provides the proof of Lemma (8.1.25) restated here for convenience.

Lemma (8.1.25). *If* $y_j = c^2/\beta^2$ *as given by* (8.1.21) *is a root of the squared complex Rayleigh equation* (8.1.20) *and* $|y_j| < 1$, *then* $y_j = c^2/\beta^2$ *is a root of the complex Rayleigh equation* (8.1.18) *and in turn a solution for a Rayleigh-Type surface wave on a HILV half space.*

To initiate the proof, (8.1.18) is rewritten here in terms of the complex velocities as

$$4\sqrt{1 - \frac{c^2}{\beta^2}}\sqrt{1 - \frac{c^2}{\alpha^2}} = \left(2 - \frac{c^2}{\beta^2}\right)^2, \tag{11.5.1}$$

where each of the square roots is interpreted as the principal value. Denoting the left-hand side of (11.5.1) by "f" and the right-hand side by "g", (11.5.1) may be expressed as

$$f = g. \tag{11.5.2}$$

Hence, in this notation if by assumption c^2/β^2 is a root of (8.1.20), then c^2/β^2 is a root of

$$f^2 = g^2, \tag{11.5.3}$$

which implies either

$$f - g = 0, \tag{11.5.4}$$

in which case c^2/β^2 is the desired root of the complex Rayleigh equation (11.5.1) and the desired result is obtained, or

$$f + g = 0. \tag{11.5.5}$$

The condition that $0 < |c^2/\beta^2| < 1$ will be used to show that this second option as stated in (11.5.5) is not possible, because the algebraic signs of the real parts of f and g and their imaginary parts are the same. To show this result, define

$$e_\alpha \equiv \sqrt{1 - \frac{c^2}{\alpha^2}} \qquad \text{and} \qquad e_\beta \equiv \sqrt{1 - \frac{c^2}{\beta^2}}, \tag{11.5.6}$$

then the real part of f may be written as

$$f_R = \text{Re}\left[4e_\alpha e_\beta\right] = 4\left(e_{\alpha_R} e_{\beta_R} - e_{\alpha_I} e_{\beta_I}\right) \tag{11.5.7}$$

and the real part of g as

$$g_R = \left(1 - \left(\text{Im}\left[e_\beta{}^2\right]\right)^2\right) + 2\ \text{Re}\left[e_\beta^2\right] + \left(\text{Re}\left[e_\beta^2\right]\right)^2. \tag{11.5.8}$$

The condition that $0 < |c^2/\beta^2| < 1$ implies

$$\text{Re}\left[e_\beta^2\right] = \text{Re}\left[1 - \frac{c^2}{\beta^2}\right] > 0 \tag{11.5.9}$$

and

$$\left(\text{Im}\left[e_\beta^2\right]\right)^2 = \left(\text{Im}\left[-\frac{c^2}{\beta^2}\right]\right)^2 < 1. \tag{11.5.10}$$

These inequalities together with (11.5.8) show that the algebraic sign of g_R is positive. Definitions (3.1.7) and (3.1.8) for α and β imply

$$\frac{|\beta^2|}{|\alpha^2|} < \frac{3}{4}. \tag{11.5.11}$$

Hence, it follows that

$$\text{Re}\left[e_\alpha^2\right] = \text{Re}\left[1 - \frac{\beta^2}{\alpha^2}\frac{c^2}{\beta^2}\right] > \frac{1}{4} > 0. \tag{11.5.12}$$

In addition, definition (11.4.11) for principal values and (4.2.13) imply

$$e_{\alpha_R} = \sqrt{\frac{\left|e_\alpha^2\right| + \mathrm{Re}\left[e_\alpha^2\right]}{2}},$$
$$\left|e_{\alpha_I}\right| = \sqrt{\frac{\left|e_\alpha^2\right| - \mathrm{Re}\left[e_\alpha^2\right]}{2}} \tag{11.5.13}$$

and

$$e_{\beta_R} = \sqrt{\frac{\left|e_\beta^2\right| + \mathrm{Re}\left[e_\beta^2\right]}{2}},$$
$$\left|e_{\beta_I}\right| = \sqrt{\frac{\left|e_\beta^2\right| - \mathrm{Re}\left[e_\beta^2\right]}{2}}, \tag{11.5.14}$$

from which inequalities (11.5.9) and (11.5.12) imply

$$e_{\alpha_R} \geq \left|e_{\alpha_I}\right| \qquad \text{and} \qquad e_{\beta_R} \geq \left|e_{\beta_I}\right|. \tag{11.5.15}$$

These inequalities together with (11.5.7) imply that the algebraic sign of f_R is positive and the same as that of g_R. By assumption (11.5.3) is valid, hence the imaginary parts of each side of the equation imply

$$f_R f_I = g_R g_I, \tag{11.5.16}$$

from which it follows that positive algebraic signs for f_R and g_R imply the algebraic signs of f_I and g_I are the same, which completes the proof.

11.6 Appendix 6 – Particle-Motion Characteristics for a Rayleigh-Type Surface Wave

This appendix provides an explicit derivation of the characteristics of the particle motion for a Rayleigh-Type surface wave on a HILV half space as stated in Theorem (8.2.26).

Parametric equations (8.2.23) and (8.2.24) describing the motion of a particle are restated here for convenience. They are

$$u_{Rx_1} = D\ \exp[-a_B x_1] H_1(x_3) \sin[\vartheta_B(t)] \tag{11.6.1}$$

and

$$u_{Rx_3} = D\ \exp[-a_B x_1] H_3(x_3) \cos[\vartheta_B(t) + S(x_3)], \tag{11.6.2}$$

where

$$\vartheta_B(t) \equiv \omega t + f_1(x_1) + \varsigma_1(x_3) \quad \text{and} \quad S(x_3) \equiv \varsigma_3(x_3) - \varsigma_1(x_3) \tag{11.6.3}$$

and the additional parameters are defined in (8.2.16), (8.2.18), and (8.2.19).

To show that these parametric equations represent the equation of an ellipse as a function of time consider the trigonometric identity for the cosine of the sum of two angles. It implies that (11.6.2) may be written as

$$u_{Rx_3} = D\ \exp[-a_B x_1] H_3(x_3)(\cos[\vartheta_B(t)] \cos[S(x_3)] - \sin[\vartheta_B(t)] \sin[S(x_3)]). \tag{11.6.4}$$

Substitution of (11.6.1) into (11.6.4) and simplification yields

$$\cos[\vartheta_B(t)] = (D\ \exp[-a_B x_1]\ \cos[S(x_3)])^{-1} \left(\frac{u_{Rx_3}}{H_3(x_3)} + \frac{u_{Rx_1}}{H_1(x_3)} \sin[S(x_3)] \right). \tag{11.6.5}$$

Substitution of (11.6.1) and (11.6.5) into the trigonometric identity

$$\cos^2[\vartheta_B(t)] + \sin^2[\vartheta_B(t)] = 1 \tag{11.6.6}$$

yields the desired second-degree equation

$$A\left(u_{Rx_1}\right)^2 + B u_{Rx_1} u_{Rx_3} + C\left(u_{Rx_3}\right)^2 - 1 = 0, \tag{11.6.7}$$

where

$$A \equiv \frac{1}{(D \exp[-a_B x_1] H_1(x_3) \cos[S(x_3)])^2}, \tag{11.6.8}$$

$$B \equiv \frac{2}{(D\ \exp[-a_B x_1])^2 H_1(x_3) H_3(x_3)} \frac{\sin[S(x_3)]}{(\cos[S(x_3)])^2}, \tag{11.6.9}$$

and

$$C \equiv \frac{1}{(D \exp[-a_B x_1] H_3(x_3) \cos[S(x_3)])^2}. \tag{11.6.10}$$

For a given particle the coefficients A, B, and C are constant with respect to time. The discriminant of (11.6.7) is

$$B^2 - 4AC = 4 \frac{(\sin[S(x_3)])^2 - 1}{\left((D\ \exp[-a_B x_1])^2 H_1(x_3) H_3(x_3) (\cos[S(x_3)])^2\right)^2}. \tag{11.6.11}$$

For $S(x_3) \neq \pi/2$, the discriminant is negative, indicating that (11.6.7) describes an ellipse as a function of time.

To determine the direction in which a given particle describes an ellipse with time consider the polar angle θ defined as

$$\theta \equiv \arctan\left[u_{Rx_3}/u_{Rx_1}\right], \tag{11.6.12}$$

whose partial derivative with respect to time is given by

$$\frac{\partial \theta}{\partial t} = \frac{-H_3(x_3)}{H_1(x_3)} \frac{u_{Rx_1}^2}{u_{Rx_1}^2 + u_{Rx_3}^2} \frac{\cos[S(x_3)]}{(\sin[\vartheta_B(t)])^2}. \tag{11.6.13}$$

This equation implies that if $\cos[S(x_3)] < 0$, then θ increases with time and the elliptical motion of the particle is clockwise or prograde. If $\cos[S(x_3)] > 0$, θ decreases with time and the elliptical motion of the particle is counter clockwise or retrograde. If $\cos[S(x_3)] = 0$, then θ is not changing with respect to time and the motion of the particle is linear as a degenerate case of elliptical motion.

Additional properties of the elliptical orbit of a particle can be derived by defining a suitable rotation of coordinates, such that (11.6.7) reduces to standard form. Such a rotation is given by

$$u_{Rx_1} = u'_{x_1} \cos \eta_B - u'_{x_3} \sin \eta_B \tag{11.6.14}$$

and

$$u_{Rx_3} = u'_{x_1} \sin \eta_B + u'_{x_3} \cos \eta_B, \tag{11.6.15}$$

where the angle of rotation η_β is given by

$$\tan[2\eta_B] = \begin{Bmatrix} B/(A-C) & \text{for } A \neq C \\ \pi/4 & \text{for } A = C \end{Bmatrix}. \tag{11.6.16}$$

Substitution of the rotation defined by (11.6.14) and (11.6.15) into (11.6.7) yields the desired standard form of the equation

$$\frac{u_{x_1}'^2}{A'^{-1}} + \frac{u_{x_3}'^2}{C'^{-1}} = 1, \tag{11.6.17}$$

where

$$A' = A\cos^2 \eta_B + B \sin \eta_B \cos \eta_B + C \sin^2 \eta_B \tag{11.6.18}$$

and

$$C' = A\sin^2 \eta_B - B \sin \eta_B \cos \eta_B + C \cos^2 \eta_B. \tag{11.6.19}$$

Equation (11.6.17) represents an ellipse whose ellipticity is given by

$$\frac{\sqrt{A'}}{\sqrt{C'}}. \tag{11.6.20}$$

The angle of rotation η_β specifies the tilt of the particle motion ellipse with respect to the vertical. The tilt as specified by (11.6.16) may be rewritten using (11.6.8) through (11.6.10) as

$$\tan[2\eta_B] = \left\{ \begin{array}{ll} \dfrac{2\sin[S(x_3)]}{\dfrac{H_3(x_3)}{H_1(x_3)} - \dfrac{H_1(x_3)}{H_3(x_3)}} & \text{for } H_1(x_3) \neq H_3(x_3) \\ \pi/4 & \text{for } H_1(x_3) = H_3(x_3) \end{array} \right\}. \quad (11.6.21)$$

References

Aki, K., and P. G. Richards (1980). *Quantitative Seismology, Theory and Methods*, San Francisco, CA, W. H. Freeman and Company, p. 20.

Becker, F. L. (1971). Phase measurements of reflected ultrasonic waves near the Rayleigh critical angle, *J. Appl. Phys.*, **42**, 199–202.

Becker, F. L., and R. L. Richardson (1969). Critical angle reflectivity, *J. Acoust. Soc. Am.*, **45**, 793–794.

Becker, F. L., and R. L. Richardson (1970). Ultrasonic critical angle reflectivity, in *Research Techniques in Nondestructive Testing*, ed. R. S. Sharpe, Orlando, FL, Academic, pp. 91–131.

Benioff, H. (1935). A linear strain seismograph, *Bull. Seism. Soc. Am.*, **25**, 283–309.

Benioff, H., and B. Gutenberg. (1952). The response of strain and pendulum seismographs to surface waves, *Bull. Seism. Soc. Am.*, **43**, 229–237.

Ben-Menehem, A., and S. J. Singh (1981). *Seismic Waves and Sources*, New York, Springer-Verlag, pp. 840–915.

Birkoff, G., and S. MacLane (1953). *A Survey of Modern Algebra*, New York, Macmillian Company, p. 112.

Bland, D. R. (1960). *The Theory of Linear Viscoelasticity*, New York, Pergamon Press.

Boltzmann, L. (1874). Zur Theorie der elastischen Nachwirkung, *Sitzungsber. Math. Naturwiss. Kl. Kaiserl. Wiss.* **70**, 275.

Borcherdt, R. D. (1971). *Inhomogeneous Body and Surface Plane Waves in a Generalized Viscoelastic Half Space*, Ph.D. thesis, University of California, Berkeley, CA.

Borcherdt, R. D. (1973a). Energy and plane waves in linear viscoelastic media, *J. Geophys. Res.*, **78**, 2442–2453.

Borcherdt, R. D. (1973b). Rayleigh-type surface wave on a linear viscoelastic half-space, *J. Acoust. Soc. Am.*, **54**, 1651–1653.

Borcherdt, R. D. (1977). Reflection and refraction of Type-II S waves in elastic and anelastic media, *Bull. Seismol. Soc. Am.*, **67**, 43–67.

Borcherdt, R. D. (1982). Reflection–refraction of general P- and Type-I S waves in elastic and anelastic solids, *Geophys. J. R. Astron. Soc.*, **70**, 621–638.

Borcherdt, R. D. (1988). Volumetric strain and particle displacements for body and surface waves in a general viscoelastic half-space, *Geophys. J. R. Astron. Soc.*, **93**, 215–228.

Borcherdt, R. D., and L. Wennerberg (1985). General P, Type-I S, and Type-II S waves in anelastic solids: Inhomogeneous wave fields in low-loss solids, *Bull. Seismol. Soc. Am.*, **75**, 1729–1763.

Borcherdt, R. D., and G. Glassmoyer (1989). An exact anelastic model for the free-surface reflection of P and SI waves, *Bull. Seismol. Soc. Am.*, **79**, 842–859.

Borcherdt, R. D., G. Glassmoyer, and L. Wennerberg (1986). Influence of welded boundaries in anelastic media on energy flow and characteristics of general P, SI and SII body waves: Observational evidence for inhomogeneous body waves in low-loss solids, *J. Geophys. Res.*, **91**, 11,503–11,518.

Borcherdt, R. D., M. J. S. Johnston, and G. Glassmoyer (1989). On the use of volumetric strain meters to infer additional characteristics of short-period seismic radiation, *Bull. Seismol. Soc. Am.*, **79**, 1006–1023.

Brand, L. (1960) *Advanced Calculus*, New York, John Wiley and Sons, Inc., p. 454.

Brekhovskikh, L. M. (1960). *Waves in Layered Media*, Orlando, FL, Academic Press, p. 34.

Bullen, K. E. (1965). *An Introduction to the Theory of Seismology*, 3rd edn, 2nd printing, Cambridge, Cambridge University Press.

Cooper, H. F., Jr. (1967). Reflection and transmission of oblique plane waves at a plane interface between viscoelastic media, *J. Acoust. Soc. Amer.*, **42**, 1064–1069.

Cooper, H. F., Jr., and E. L. Reiss (1966). Reflection of plane viscoelastic waves from plane boundaries, *J. Acoust. Soc. Amer.*, **39**, 1133–1138.

Cowan, E. W. (1968). *Basic Electromagnetism*, New York, Academic.

Ewing, J., and R. Houtz (1979). Acoustic stratigraphy and structure of the oceanic crust, in *Deep Drilling Results in the Atlantic Ocean: Ocean Crust*, ed. M. Talwani, C. G. Harrison, and D. E. Hayes, Washington, D.C., American Geophysical Union, pp. 1–14.

Ewing, W. M., W. S. Jardetsky, and F. Press (1957). *Elastic Waves in Layered Media*, New York, McGraw-Hill.

Flugge, W. (1967). *Viscoelasticity*, Waltham, MA, Blaisdell.

Fumal, T. E. (1978). Correlations between seismic wave velocities of near-surface geologic materials in the San Francisco Bay region, California, *U.S. Geological Survey Open-File Report* 78–107.

Fung, Y. C. (1965). *Foundations of Solid Mechanics*, Englewood Cliffs, NJ, Prentice-Hall.

Futterman, W. I. (1962). Dispersive body waves, *J. Geophys. Res.*, **67**, 5279–5291.

Gross, B. (1953). *Mathematical Structure of the Theories of Viscoelasticity*, Paris, Hermann et Cie.

Gupta, I. N. (1966). Response of a vertical strain seismometer to body waves, *Bull. Seism. Soc. Am.*, **56**, 785–791.

Gurtin, M. E. and E. Sternberg (1962). On the linear theory of viscoelasticity, *Archive of Rat. Mech. Analysis*, **II**, 291–356.

Hamilton, E. L., H. P. Buchen, D. L. Keir, and J. A. Whitney (1970). Velocities of compressional and shear waves in marine sediments determined *in situ* and from a research submersible, *J. Geophys. Res.*, **75**, 4039–4049.

Haskell, N. A. (1953). The dispersion of surface waves on multilayered media, *Bull. Seismol. Soc. Am.*, **43**, 17–34.

Haskell, N. A. (1960). Crustal reflection of plane SH waves, *J. Geophys. Res.*, **65**, 4147–4150.

Hudson, J. A. (1980). *Propagation and Excitation of Elastic Waves*, Cambridge, Cambridge University Press.

Hunter, S. C. (1960). Viscoelastic waves, *Progress in Solid Mech.*, **1**, 1–57.

Kanai, K. (1950). The effect of solid viscosity of surface layer on earthquake movements, III, *Bull. Earthquake Res. Inst., Tokyo Univ.*, **28**, 31–35.

Kramer, S. L. (1996). *Geotechnical Earthquake Engineering*, Upper Saddle River, NJ, Prentice-Hall.

Kreysig, E. (1967). *Advanced Engineering Mathematics*, New York, John Wiley.

Lindsay, R. B. (1960). *Mechanical Radiation*, New York, McGraw-Hill.

Lockett, F. J. (1962). The reflection and refraction of waves at an interface between viscoelastic media, *J. Mech. Phys. Solids*, **10**, 53–64.

Love, A. E. H. (1944). *The Mathematical Theory of Elasticity*, Cambridge, Cambridge University Press, 1st edn., 1892, 1893; 4th edn., 1927. Reprinted New York, Dover Publications.

McDonal, F. J., F. A. Angona, R. L. Mills, R. L. Sengush, R. G. van Nostrand, and J. E. White (1958). Attenuation of shear and compressional waves in Pierre Shale, *Geophysics*, **23**, 421–439.

Minster, J. B. and D. L. Anderson (1981). A model of dislocation-controlled rheology for the mantle, *Philos. Trans. R. Soc. London*, **299**, 319–356.

Morse, P. M. and H. Feshbach (1953). *Methods of Theoretical Physics*, New York, McGraw-Hill.

Newman, P. J., and M. H. Worthington (1982). *In situ* investigation of seismic body wave attenuation in heterogeneous media, *Geophys. Prospect.*, **30**, 377–400, 1982.

Rayleigh, Lord, J. W. Strutt (1885). On waves propagated along the plane surface of an elastic solid, *Proc. Roy. Math. Soc.*, **17**, 4–11.

Romney, C. (1964). Combinations of strain and pendulum seismographs for increasing the detectability of P, *Bull. Seism. Soc. Am.*, **54**, 2165–2174.

Stoffa, P. L., and P. L. Buhl (1979). Two ship multichannel seismic experiments for deep crustal studies; expanded spread and constant offset profiles, *J. Geophys. Res.*, **84**, 7645–7660.

Thompson, W. T. (1950). Transmission of elastic waves through a stratified solid, *J. Appl. Phys.*, **21**, 89–93.

Volterra, V. (2005). *Theory of Functionals and of Integral and Integro-differential Equations*, New York, Dover Publications.

Wolfram, S. (1999). *The Mathematica Book*, 4th edn., Cambridge, Wolfram Media/ Cambridge University Press.

Zener, C. (1948). *Elasticity and Anelasticity of Metals*, Chicago, IL, University of Chicago Press, p. 60.

Additional Reading

Becker, F. L., and R. L. Richardson (1972). Influence of material properties on Rayleigh critical-angle reflectivity, *J. Acoust. Soc. Am.*, **51**, 1609–1617.
Buchen, P. W. (1971a). Plane waves in linear viscoelastic media, *Geophys. J. R. Astron. Soc.*, **23**, 531–542.
Buchen, P. W. (1971b). Reflection, transmission, and diffraction of SH-waves in linear viscoelastic solids, *Geophys. J. R. Astron. Soc.*, **25**, 97–113.
Fitch C. E., Jr., and R. L. Richardson (1967). Ultrasonic wave models for non-destructive testing interfaces with attenuation, in *Progress in Applied Materials Research*, vol. 8, ed. E. G. Stanford, J. H. Fearon, and W. J. McGounagle, London, Iliffe Books, pp. 79–120.
Kanai, K. (1953). Relation between the nature of surface layer and the amplitude of earthquake motions, III, *Bull. Earthquake Res. Inst., Tokyo Univ.*, **31**, 219–226.
Kolsky, H. (1958). *Stress Waves in Solids*, Oxford, Clarendon Press.
Krebes, E. S. (1980). Seismic body waves in anelastic media, Ph.D. thesis, University of Alberta, Edmonton.
Press, F. and J. H. Healy (1957). Absorption of Rayleigh waves in low-loss media, *J. Appl. Phys.*, **28**, 1323–1325.
Richards, P. G. (1984). On wave fronts and interfaces in anelastic media, *Bull. Seismol. Soc. Am.*, **74**, 2157–2165.
Rollins, F. R., Jr., (1966). Critical ultrasonic reflectivity: A tool for material evaluation, *Mater. Eval.*, **24**, 683–689.
Schoenberg, M. (1971). Transmission and reflection of plane waves at an elastic–viscoelastic interface, *Geophys. J.*, **25**, 35–47.
Shaw, R. P., and P. Bugl (1969). Transmission of plane waves through layered linear viscoelastic media, *J. Acoust. Soc. Am*, **46**, 649–654.
Silva, W. (1976). Body waves in a layered anelastic solid, *Bull. Seismol. Soc. Am.*, **66**, 1539–1554.
Sokolnikoff, I. S. (1956). *Mathematical Theory of Elasticity*, 2nd edn., New York, McGraw-Hill.
Wennerberg, L. (1985). Snell's law for viscoelastic materials, *Geophys. J. R. Astron. Soc.*, **81**, 13–18.

Index